PRODUCTION ECONOMICS THEORY WITH APPLICATIONS

SECOND EDITION

JOHN P. DOLL

Professor of Economics
University of Missouri—Columbia

FRANK ORAZEM

Professor of Agricultural Economics
Kansas State University

JOHN WILEY & SONS
New York • Chichester • Brisbane • Toronto • Singapore

Library of Congress Cataloging in Publication Data:

Doll, John P.
Production economics.

Includes bibliographies and index.
1. Production (Economic theory) 2. Agriculture—
Economic aspects. I. Orazem, Frank. II. Title.
HB241.D64 1984 338.5 83-21575
ISBN 0-471-87470-1

Printed in the United States of America

10 9 8 7 6 5 4 3 2 1

PREFACE

We have been most encouraged with the acceptance of our book by our colleagues and their students. Thus, the organization and methods of presentation of this, the second edition, remain essentially unchanged. The first edition was based on lectures and course materials developed over many years of teaching both undergraduate and graduate classes of students possessing varying degrees of preparation. In preparing this edition, we have benefited from reviews and comments, both formal and informal, received from readers of the first edition. We believe that this infusion of new ideas has enhanced our original effort.

It is probably the case that no two of us would teach production economics in exactly the same manner. Thus, the changes incorporated into this edition represent a composite of suggestions deemed most desirable by users and reviewers of the first edition. As we anticipated in the original preface, the text is used by students with a variety of backgrounds. When making changes, we have attempted to make the textbook more useful, without diminishing its accessibility to any one group of users. In this way, we hope each teacher will be able to select the material to be incorporated into his or her course.

The introductory chapter has been enlarged to include a more thorough discussion of the subject matter of production economics and its relationship to other branches of economics and disciplines of agriculture. A section has been added to explain the role of assumptions in economic logic. Chapters 2 through 6 remain basically the same, with some new examples, changes in organization, and clarifying points added as needed. Chapter 7, "The Production Process Through Time" (Chapter 6 in the first edition), has been reorganized and expanded to include additional investment criteria. A section on valuing farmland has been added. Chapter 8, "Introduction to Decision Theory," has been completely revised and contains an expanded discussion of modern decision theory. Chapter 9, "Linear Programming," includes some new sections and problems, and the descriptive analysis in Chapter 10 has been updated. Hopefully, the users will find the presentations to be clear and error-free.

Acknowledgment is due to many students, colleagues, and correspondents who have very kindly sent us lists of errata, explanations of obscurities, and pointed out omissions found in the first edition.

Our implicit intellectual debts are numerous and, where possible, we have tried to make them obvious by citing published materials. Beyond that, we have received valuable comments from Earl Swanson, Robert Collins, Burt Sundquist, Earl Kehrberg, Dale Colyer, Gerald White, and Gary Lynne. Further ideas were obtained from James Whittaker's review in the *American Journal of Agricultural Economics* and John Dillon's review in the *Australian Journal of Agricultural Economics.* Finally, we found the anonymous reviews commissioned by John Wiley to be very useful and would like to thank Richard Esposito, Editor of Business and Economics, and the reviewers for their interest. But the errors and shortcomings that always seem to find their way into the written word are ours; please let us know when you encounter them.

We would like to end this preface by addressing the problem of sexism in our language. Because of the unique relationship between the family and the farm, women have always played a major role in American agriculture. The role of women as owners, operators, and, when married, full-time partners with their husbands in the farming operation is well known. Now, due to long-needed changes in our social structure, an increasing number of women are participating in the profession of agricultural economics. We wanted to recognize both roles of women by avoiding the continual use of the masculine gender. Frankly, we didn't find it an easy task, especially in a second edition. We try to use the terms "farmer," "manager," and "decision maker" as much as possible because these terms apply equally to both sexes. When you encounter these terms, please interpret them as applying to either gender. Nevertheless, in other references, we decided it would be difficult to continually insert terms such as "he and/or she" in the text. Thus, although we have occasionally reverted to the use of the male gender, we ask that when you encounter a phrase such as "... the manager must revise his goals ...," you interpret the term "his" as "her" if such a reading is more appropriate for you. By this compromise, we hope to express our concern with the issue of sexism and yet maintain the flow of the language.

John P. Doll
Frank Orazem

PREFACE TO THE FIRST EDITION

In the past decade, production economics has become a standard subject in most undergraduate programs in agricultural economics. Economics is an important and interesting subject. But production economics is, in our opinion, especially important for the student of agriculture. To paraphrase Alfred Marshall, production economics is the study of decisions farmers must make in the course of their everyday life. To be useful, the concepts of production economics must be relevant. Economic theory is more meaningful when it is applied to the analysis of significant problems, and we believe the application of microtheory to agricultural production problems has been particularly fruitful.

Our text is written with the student in mind. We have tried to bridge the gap between abstract theory and its application to real situations in agriculture. Using examples from agriculture, our hope is to entice the student into the world of economic logic while keeping him standing firmly on familiar ground. Our intent is to encourage the student to develop thought processes so that he can better identify economic problems and, more importantly, gather and analyze information that will lead to valid economic decisions.

This book is intended for undergraduate students who have had a course such as principles of agricultural economics but no other exposure to microeconomics. We realize that introductory principles courses cover a large spectrum of topics and that students' understanding of the basic principles of microeconomics is likely to be minimal when they enroll in their first production economics class. In addition, some students enrolling in production economics classes are not majors in agricultural economics and thus have had limited exposure to economics. For these reasons, the beginning chapters include a review of the basic principles of production theory. Students with a good background in microtheory will find this elementary material repetitious and perhaps even tedious. However, we believe it necessary to bring all students up to this same level of understanding; the elements can always be skimmed or skipped by those who find them repetitious.

We have tried to be equally flexible in our manner of presentation. To supplement our written descriptions, we have added geometric and algebraic interpretations and, for the more advanced student, calculus. Our applications stress geometric as well as algebraic steps needed to arrive at solutions. Simple numerical examples are used wherever possible. Chapters 2 through 5 and 9 include "Problems and Exercises" which should further direct students toward the applications of appropriate principles as well as give the instructor additional flexibility with respect to the rigor of the course. While we have not covered all the topics appropriate for all courses in production economics, we have tried to present the basic principles needed for classroom development of other topics.

Many people have contributed to this book. It would be impossible to acknowledge all of them, but we would like to mention a few. Professor Francis Walker of the Ohio State University reviewed the entire manuscript and made numerous useful contributions. Our colleagues at Missouri and Kansas State have been very helpful; among these we would especially like to mention Jerry West, James Rhodes, and John Sjo. We owe a special debt of gratitude to Professor John Nordin, teacher and colleague, for his lectures in microtheory, always organized carefully and delivered superbly. Over the years, students in our classes have contributed to our understanding and aided our methods of exposition. Errors in logic or exposition are ours entirely—please write us when you encounter such stumbling blocks.

We would like to end on a personal note: We were colleagues in the graduate program at Iowa State University. Although we come from widely differing backgrounds—one from a farm in Yugoslavia and one from the Montana plains—we were both attracted to Ames by the same man: Professor Earl O. Heady. The impact of his teaching and writing can be found on every page of this book. Indeed, he pioneered most of the applications of economic theory to agricultural production problems presented herein—production function analysis and linear programming are only two examples. His achievements are in a large part responsible for the fact that the study of production economics has become so widespread in agriculture. It was our privilege to study with Professor Heady, and we dedicate this book to him.

John P. Doll
Frank Orazem

CONTENTS

INTRODUCTION

CHAPTER 1

We live in a complex and continuously changing world. Over the past hundred years, industrialization and mechanization have wrought tremendous changes in the production methods used in agriculture and the rest of the economy. The manager of today's farm must be as efficient and knowledgeable as his counterpart in industry. Strong new interdependencies have been created among the sectors of our economy. Far different from the early settlers on their subsistence farms, the modern farmer is now accustomed to buying many of his inputs and selling most of his products. These interdependencies do not stop at our borders. In recent years, farm prices have been affected by weather conditions and the demand for food in countries as far away as Russia and India. The oil embargo emphasized again agriculture's reliance on inputs supplied not only from off the farm but from outside the United States.

The farm manager must be able to understand and react to these everchanging forces that are, for the most part, beyond his control. To do this, he must master the technical aspects of production on farms—of land and animal husbandry. He must learn new production techniques and be willing to use them. But, more than ever, he must also possess a basic understanding of the economic principles underlying agricultural production.

Economics is broadly concerned with the analysis of the production and consumption of goods and services in our economy. Economic issues include such broad topics as inflation, unemployment, pollution, energy shortages, taxes, imports, and exports. These are, of course, of interest to all citizens. But one particular area of economics is devoted to the analysis of the use of resources in production processes found within individual firms. This topic, often called *production economics*, is the subject of this text. The farm manager must not only understand husbandry, but he must be concerned with costs and returns. He must know how to allocate resources on the farm to meet his goals and at the same time react to economic forces that have their origin far from the farm.

WHAT IS ECONOMICS?

Economics is the study of how resources are used to satisfy the needs and desires of people. It is concerned with individual consumers and producers, as well as with the aggregate of all consumers and producers. Our study will begin by listing certain tenets common to all areas of economics.

We do not live in a world of plenty. Most goods and services, as well as the inputs used in production of goods, are available in limited supplies. Even renewable inputs are limited within a given time span. This suggests the idea of *scarcity*. There are not enough inputs to produce the goods needed to satisfy the nearly insatiable wants of all the people. If there were no scarcity, individuals would not be concerned about the use of resources. Resources or goods available in unlimited supply do not have price tags attached to them; they are free.

Another tenet is the concept of *allocation*—putting resources or products to some use. Allocation is concerned with choosing the "best" alternative use; the word "best" is set in quotes because the "best" use must be consistent with individual or social objectives. Allocation implies that scarce resources must be distributed among competing uses. Examples are the allocation of land between corn and sorghum or the allocation of limited time between study and play.

A third tenet involves *goals* (ends or objectives). Individual desires or wants often appear to be unlimited. Thus, wants compete for scarce resources. Because of this competition, we must decide upon a priority system to order our competing wants and attempt to obtain the most satisfaction from those chosen.

Economics provides a method of analysis that helps choose among alternatives. The choice must be made with the help of an indicator. This *choice indicator* is the criterion, measuring stick, or index that weighs the alternatives according to their value to the consumer or producer. In the parlance of the economist, it is often called the *objective function*. It is usually expressed in monetary terms but could be any other index that reflects likes and dislikes. For example, when shopping, we might ask ourselves whether "this product is worth twice as much or one-half as much as a competing product."

To summarize, economics is a science of choice making. It deals with allocation of *scarce* resources among competing *alternatives*. Scarcity and the alternatives are important ingredients for a problem to be an economic problem. If there is scarcity but no alternatives, choice making is impossible and the problem is not economic in nature; if there are alternatives but no scarcity (goods or resources are free), economics is not required.

Resource allocation is important regardless of the economic and political system of a country. Products must be produced and resources must be allocated. Economic problems touch the subsistence peasant who consumes all he produces as well as the large commercial farmer who sells all of his produce. Economics is as relevant in barter economies as it is in market economies. Only the medium of exchange differs.

Types of Economics

The study of economics can be viewed from different perspectives. Traditionally the study of economics is subdivided into two broad categories: microeconomics and macroeconomics (Figure 1–1).

Microeconomics is the study of specific economic units that make up an economic sector. It concentrates on a single unit or an aggregation of units, always part of a whole. Microeconomics is concerned with the study of individual firms, such as farms, and their relationships to each other, to an entire economic sector, such as agriculture, and to the economy as a whole. The subject of this book, production economics, is one area of microeconomics.

Macroeconomics addresses itself to broader areas. It is concerned with the economy as a whole. Macroeconomics examines the functioning of an economic system as it copes with problems such as inflation, depression, and unemployment. Thus, problems facing agriculture relative to the rest of the economy would fall under the scope of macroeconomics. Macroeconomics, in short, is concerned with the forest and microeconomics is concerned with the individual trees (or, sometimes, with groves of trees).

The study of economics can further be classified as static or dynamic (Figure 1–1). In *statics*, economic phenomena are studied without reference to time—thus, the expression "timeless" economics. In the most restrictive case, the expectation of outcomes in statics is assumed to be single-valued. This is known as "perfect certainty"; 50 pounds of nitrogen would be said to increase the yield of corn by 10 bushels, no more and no less. While uncertainty can be introduced into statics, time will never enter as a variable. For example, a coin flip requires essentially no time but has an uncertain outcome. A static relationship is like a snapshot as distinguished from a motion picture.

Dynamics, on the other hand, takes time into consideration. Economic phenomena are related to preceding or succeeding events. Dynamics implies the possibility of change in an economic relationship over time. Studies of economic trends, for example, fall into this category.

Dynamics not only implies change through time but often associates uncertainty with time. Changes through time could be known with certainty—for example, yields could be said to increase exactly

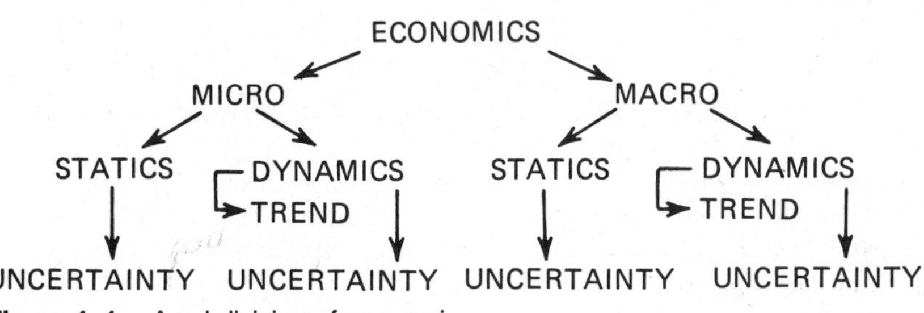

Figure 1–1. A subdivision of economics.

10 percent a year—but more often a change through time also implies uncertainty. When uncertainty is present, the expectations or outcomes are not single-valued but rather are multi-valued. Thus, 50 pounds of nitrogen may increase the yield of corn by 5 bushels, by 10 bushels, by 15 bushels, or somewhere between the extremes, depending on the effect of factors that occur throughout the growing season such as rainfall, wind, sunshine, pests, and weeds. All these factors affect the final outcome—the yield of corn—which will not be known with certainty until the crop is harvested. Static and dynamic models are found in both micro- and macroeconomics.

Positive and Normative Economics

The logic of economics is intended to be used to analyze problems of individual and social significance. Theory has no value if it cannot lead to useful analyses of real problems. Recommendations concerning the *best* way to resolve a particular social or economic problem are termed *policy* decisions. Clearly, the *best* policy solution for a given problem will depend upon the goal of the decision maker(s). In general, many economists believe that they should not specify goals or single out specific policies to meet goals. They believe that these types of decisions should be made by elected public officials or their appointees. This has led to a somewhat different classification of economics.

Positive economics describes the manner in which the economic unit functions. It deals with "what is," apart from value judgments about what "should be." Positive economics describes, for example, the functioning of a household, a firm, a market, or an economic system without attaching statements of "good" or "bad" to those functions. Economic theory can be classified as positive economics (to the extent that producers and consumers actually behave in the ways postulated by economists). Disagreements in the area of positive economics can generally be resolved by logical thought and appealing to the facts.

For example, an economist may study the properties of a particular market—how prices and quantities are determined, the nature of the buyers and sellers, the efficiency of the market, etc. Another economist may disagree with his conclusions. If so, their differences can be resolved by gathering further information about the market. If neither economist makes judgments about such market characteristics as the magnitude of prices (too high or too low) or the distribution of goods among the consumers, they are practicing positive economics. On the other hand, if one of the economists were to make value judgments about one or more properties of the market and appear on a TV talk show to pronounce it good or bad, he would be practicing normative economics.

Normative economics is prescriptive. It deals with what "ought to be" as contrasted with what "is" and necessarily involves statements that are value judgments. Disagreements in normative economics generally cannot be resolved by an appeal to the facts because the disagreeing parties start from different premises. We all have our own likes and dislikes derived from varied philosophical, social, and cultural backgrounds; hence, we all have our own set of goals. Economic policy falls in the area of normative economics.

Positive economics can make important contributions to normative matters because of its descriptive nature. Positive economics identifies and in some instances quantifies the relationships within individual economic units, such as farms, and between different economic sectors, such as agriculture and industry. Positive economics thus serves as a basis for making judgments in normative areas. The economist, using methods of analysis from positive economics, can study alternative policy prescriptions and explain their ramifications to elected officials. Improvements in the description of an economic relationship can result in improvements in the programs designed to affect changes through social policies.

The distinction between positive and normative economics is not always clearly drawn. The economist seeking to describe and analyze must always guard against value judgments creeping into his analysis. As even the casual reader of the daily newspaper knows, economists are often criticized for offering a diverse set of solutions to the same problem.[1] This problem arises because the economists, while perhaps agreeing on the positive economic analysis of a problem, apply different value judgments and thereby arrive at different policy solutions. This situation often occurs when an economist is asked to advise a president, a presidential candidate, or a television reporter. In such cases, the economist provides a set of value judgments, usually presumed to be compatible with those of the policy maker, and then offers a solution. Thus, when it is said that economists disagree, perhaps what should be said instead is that economists disagree much more on social goals and policies to attain them than on matters of economic theory.

To extend this distinction to agriculture, the economist with a farm background may have a nostalgic affinity for enterprises such as chickens and oats. He may value highly the days when he woke to the crow of the rooster and rose to feed Old Gray from the oats bucket. But when it comes time to outline production alternatives to practicing farmers, the economist should put aside his personal value judgments and offer a selection of enterprises designed to aid the farmer attain his profit-maximizing goals in modern agriculture.

In practice, extreme cases are relatively easy to identify. In more subtle instances, however, the observer may find that it is very difficult to distinguish between professional economic advice based on valid

[1]"If there is a problem with the economy and the president calls in 100 economists for advice on what to do, he gets 100 different opinions." Andy Rooney in the *Greenville Piedmont*, October 2, 1981.

research results and prescriptions based on personal value judgments. The student and researcher should always be ready to question the positive–normative distinction. In the study of economics, all value judgments should be clearly identified, lest positive and normative conclusions become confused.

AGRICULTURAL ECONOMICS

The agricultural economist applies the analytical models of economics to agricultural problems. Agriculture, in turn, has traditionally been associated with production on farms; the definition of agriculture as given by a typical dictionary is "the science and art of farming."[2] But to those who participate in agriculture, the word now implies much more. American farmers purchase many inputs and use sophisticated production techniques; their products are marketed worldwide. A major portion of our economy is now involved in the manufacture and sale of inputs to farmers, in farm production, and in the storage, processing, and distribution of agricultural products. The all-encompassing term "agribusiness" is often used to describe this sector, which is estimated to control as much as 60 percent of the total assets of all corporations and farms in the United States.[3]

Agricultural economics has changed with agriculture. Begun initially as the study of costs and returns for farm enterprises, agricultural economics historically emphasized the study of management problems on farms. But the modern field of agricultural economics is composed of specialists who study such topics as agricultural marketing, farm management, agricultural finance and accounting, product transportation, farm cooperatives, and agricultural law. And this list is not exhaustive; many study more than one aspect of the agribusiness industry.

Therefore, as the industry of agriculture has changed, so has the profession of agricultural economics. It is no accident that the disciplines listed above are similar to those found in a college of business. The difference lies in the area of application: Professors of business study the nonagricultural sectors of our modern economy while professors of agricultural economics study the agricultural sector.

By making this comparison between agriculture and business, we can clarify the role of economic logic in agriculture. Just as all disciplines in a business school utilize economics, so do all specialists in agricultural economics.

Both major areas of economics, microeconomics and macroeconomics, have applications in agriculture. The problems of production

[2]*Webster's New World Dictionary*, New York: The World Publishing Company, 1972.

[3]Data presented by Cramer and Jensen, who also present a brief history of agricultural economics. (References listed without further citation in text or footnotes will be found in the Suggested Readings at the end of each chapter.)

on individual farms are important—this text presents principles useful for the management of farm production problems—but these same principles of microeconomics also can be utilized in other areas of agricultural economics. And agriculture is not independent of the rest of the economy. Our economic system is highly interrelated, and budget decisions by the federal government or the monetary policies of the Federal Reserve Board are able to send shock waves back through the market system to every farmer in America. As a result, macroeconomics has become an important area for study by agricultural economists.

The logic of economics is at the core of agricultural economics, but it is not the whole of agricultural economics. To effectively apply economic principles to agriculture, the economist must also understand the biological nature of agricultural production. For example, the economist studying costs for livestock enterprises must understand the nature of costs, as developed in this book, but must also have an understanding of the technical management problems of animal husbandry. From our point of view, one fascinating aspect of agricultural economics is to be found in the unique blend of the abstract logic of economics with the practical management problems of modern agriculture, both on and off the farm. We believe that the best practitioners of the art of agricultural economics, and we use the word "art" advisedly, are those men and women who understand economic theory and know how to use it in a practical setting. In this text, our goal is to help the student master the first half of this dichotomy; but we also suggest that the second half cannot be neglected.

The Goal of Agricultural Economics

The widely accepted goal of study and research in agricultural economics is to increase efficiency in agriculture. Efficiency will be defined in more detail in the chapters to follow, but the basic goal is to produce the needed food and fiber without wasting resources. To meet this goal, the required output must be produced with the smallest amounts of scarce resources, or the maximum possible output must be obtained from a given amount of resources. Thus, the responsibility of teachers and researchers in agricultural economics lies not only in increasing the efficiency, and hence profitability, of farms but also in increasing the efficiency of resource use in all of society. As agricultural efficiency is increased, resources may be freed from production in agriculture and used to produce other goods, or the output of agriculture can be increased to meet the needs of an increasing population.

PRODUCTION ECONOMICS

Production economics is the application of the principles of microeconomics in agriculture. The logic of production economics provides a framework for decision making on the farm. Based on the theory of the firm, the study of the principles of production economics should clarify the concepts of costs, output response to inputs, and the use of resources to maximize profits and/or minimize costs. Therefore, the principles of production economics should be extremely useful to the farm manager seeking profit and efficiency.

But the same principles also have wide application beyond the farm. By combining the study of the technology of production on farms with an analysis of how the profit-seeking manager will respond to changes in such variables as input price, output price, wage rates, or interest rates, the agricultural economist is able to analyze the effects of forces outside of agriculture on a farm or group of farms. For example, agricultural economists are often called on to determine the impact of changing world prices or government policies on input use and production in the farm sector. They do so by using the logic of production economics.

Analyses such as these will be realistic only if the analyst has a clear perception of the problems of technical production on farms. By nature, then, production economics is interdisciplinary. Indeed, the discipline of farm management was initiated by agronomists and horticulturalists who became convinced that they had to look beyond the biology of production. A review of the literature in production economics reveals a wealth of research produced by economists cooperating with agronomists, soil scientists, and livestock specialists.[4] This type of cooperation is not only useful but necessary if the production economist's analysis of farm production is to be valid. The extent and usefulness of the interdisciplinary research resulting from the cooperation of economists and the other scientists in the modern college of agriculture is perhaps unmatched by any other sector of the modern university.

Production economics plays a unique role in farm management. In the organization of business schools, management and economics are separate disciplines. The study of management involves learning and using the logic of economics, but management extends to disciplines other than economics, such as finance, accounting, and personnel management. We believe the same distinction can be usefully made in agriculture. Production economics develops principles important for good management, but good management includes more than sound economics. Similarly, the logical analyses of production economics are applicable to farms and enterprises within farms but also have application to machinery dealers, grain buyers, meat packing companies, and other types of firms in the agricultural sector as well as to the entire agricultural sector.[5]

[4]Jensen presents a short history of production economics and farm management in his review paper, pp. 3–7, and a very complete reference list, pp. 71–89.

[5]Jensen reviews the distinction between production economics and farm management and presents some of the philosophical arguments underlying the distinction.

Studying Production Processes

In this section, we hope to provide the student who is unfamiliar with the art of abstract reasoning, which is what we regard economic theory to be, with an intuitive understanding of the methods used in economics. The assumptions made in the study of production will be presented and discussed at this time. We hope this discussion will provide a perspective that will prove useful throughout the text. Although the assumptions will be introduced here in general, in later chapters they will be tailored as needed for specific instances.

To illustrate the nature and extent of the problems encountered in economics research, we will first present a set of questions to be considered and, we hope, answered in upcoming chapters. These questions all focus on individual farms, but the same questions could be addressed to aggregations of farms.

What is efficient production?

How is the most profitable amount of input determined?

How will farm production respond to a change in the price of an output, say, corn price?

What enterprise combinations will maximize farm profits?

How much can a farmer pay for a durable input, such as a tractor?

What should a manager do when he is uncertain about yield response?

How will technical change affect output?

The complexity of the agricultural production process is well known. Crops grow in seasonal cycles and are affected by many inputs, some of which are controlled by the manager while others, especially weather, occur at random. Time is important. Crops such as winter wheat are not harvested in some parts of the United States until almost a year after planting. Livestock production can take even longer; three years can pass between the time a cow is bred and a feeder is marketed. How does the economist deal with such complexity?

When the soils scientist wishes to determine the effects of fertilizer on crop yields, he selects a plot of ground, lays out an experiment including varying rates of fertilizer, and measures the resulting yields. In a like manner, the chemist or the physicist may establish fundamental relationships through experimentation under carefully controlled laboratory conditions. The agricultural economist is not as fortunate; he must study a system of which he is a part and over which he has no control. When the economist would like to determine the impact of a change in corn prices on the amount of commercial fertilizer a farmer would use, he cannot change the prevailing market price of corn and then stand aside to watch farmers react.

Instead, economists have developed a logical reasoning process that in every sense parallels the experiment of the soil scientist. This process, embodied in the theory of the firm, predicts how the profit-maximizing manager will respond to a change in corn price, given that other variables do not change, just as in the experimental process. One important difference is that the experiment can be observed and measured while the reasoning of the economist is abstract. This abstraction makes the discipline initially appear to be more difficult than it actually is.

When viewing a production process in agriculture, the theory of production begins by classifying inputs into three general categories. Some inputs can be controlled by the manager—these are usually the ones chosen for study. These inputs are analogous to the one the soil scientist chooses to vary in his experiment, the rate of fertilization, and are called *variable inputs*. A second category of inputs is classified as fixed. A *fixed input* may represent a dairy cow, in which case the variable input could be feed, or it may be an acre of land, with a variable input of fertilizer.

To continue the analogy with the soil scientist, once a site is selected, the soil type chosen for the experimental site, say Putnam silt loam, is a fixed input and the experimental results will apply only to Putnam silt loam. We do not intend to suggest that the fixed input is never changed, just that it is not changed for purposes of the analysis. The third general category might be called *random inputs*. These inputs are often associated with nature but also result from economic forces beyond the farm. The soil scientist also must deal with random inputs. For example, the results of the experiment are valid only for one growing season, the one that occurred during the year of the experiment; but growing seasons are unique and, in fact, may never be repeated exactly in nature.

The soil scientist's interpretation of results must be restricted to the conditions of the experiment: a specific crop variety, grown on Putnam silt loam with a three percent slope, in the 1983 growing season, with predetermined planting rates, and using selected cultivation methods. The economist parallels this type of solution by delineating the nature of the production response under study. This is accomplished by stating a carefully structured set of assumptions about the production and then reasoning from these assumptions to determine how the manager should act to attain his goals. In economic logic, these assumptions are analogous to the experimental site and cultural methods selected for use by the soil scientist.

The assumptions are designed to enable the economist to eliminate the effect of some variables while studying the production response to others. We will now present some of the typical assumptions and discuss the purposes of each. They are all important and not presented in any particular order.

The first set of assumptions is designed to make the production process stable during the period of study. Technology is assumed constant; the manager selects the most efficient known technology and doesn't change it during the production period. Institutional factors such as land ownership and government programs do not change.

A second set of assumptions is required to make the analysis of the production process easy to conceptualize for beginning students. Production functions are drawn as smooth, well-behaved curves. Inputs are assumed to be homogeneous, equivalent in quality, easily divisible, and mobile. Such assumptions are quite reasonable for some inputs, such as commercial fertilizer, but unrealistic for others, such as labor. The same assumptions of homogeneity and perfect divisibility are also applied to products.

A third category of assumptions is required to deal with the randomness of some inputs. Initially, the assumption of perfect certainty is made. This assumption is restrictive and implies that the manager has perfect foresight. Because of this, price and yield uncertainty are removed. Clearly, the impact of random variables on agricultural production is important; but our technique is to study the complexities of the production process without uncertainty in the early chapters of the book and then reintroduce it in Chapter 8.

A fourth category of assumptions is required to deal with time. Perfect certainty removes the random elements associated with time, but another problem remains. A dollar at planting time has a different value than a dollar at harvest. This is called *time discounting* and is not initially considered. In the presence of perfect certainty and in the absence of time discounting, production can be abstracted from time and, for all practical purposes, considered instantaneous. Economic analyses through time are introduced in Chapter 7.

Finally, the goals of the manager must be considered. Our initial assumption is that the manager is motivated by profits and is rationally seeking to maximize profits. This assumption is relaxed in Chapter 8, but more general goals such as the quality of life or leisure are not explicitly considered.

Starting from these assumptions, the logic of economics can be developed and, as a result of these assumptions, our analysis is static. After the general static theoretical framework is developed, each of the assumptions can be relaxed in turn. The purpose of the assumptions is not to deny the existence of real-world forces on agricultural production but rather to simplify the analysis to a point where a reasonable starting point can be identified. After the elementary theory is developed, each additional source of complexity can be introduced and analyzed in turn. This is similar to the soil scientist who, after the initial experiment, may decide to conduct a similar set of experiments over a wide range of soil types, varieties, cultural practices, and for a span of several years. The results from such a series of experiments will have a wider application but will also be much more difficult to interpret than the simple experiment. That is why, faced with so many variables, economists have developed a consistent body of theory.

Role of Economic Theory

Economic logic represents an abstraction from the total economic milieu of the real world. Theory does not attempt to describe the real world in minute detail. Indeed, little advantage would be gained by developing a theoretical system as complicated as the real world, for the purpose of theory is to provide propositions of general usefulness about the real world. The world is full of facts. Some facts are useful to a problem; some are not. The role of theory is to identify useful facts and unite them so that meaningful conclusions may be derived. Analyzing economic problems without a theoretical framework would be like shooting in the dark.

The principles presented represent a body of knowledge that economists must know before they can undertake economic analysis. Thus, just as a football player must know the rules of the game and the techniques of blocking and tackling before he can play on a professional football team, so must the aspiring agricultural economist master the economic principles and their implications before he can undertake the job of an agricultural economist. It is important to remember that the economic principles to be studied are essentially a planning framework. We are like a field general who is trying to determine beforehand all the possible strategies he could use to win a big battle. In the case of the farmer, the battle is for survival and profits, and he is armed with production functions, input and output prices, and opportunity costs principles. Economics is not a study of what has been but rather a study of how to proceed in the future.

Economists present their principles in different ways. Some use geometry and graphs, some use mathematical symbols, and some prefer verbal discussions. Regardless of the method of presentation, they all seek the same thing: a set of principles useful in coordinating economic facts and making predictions and evaluations beneficial to society.

The Goals of Production Economics

The economics of agricultural production presents an analysis of the manner and mechanisms through which individual farms adjust to changing economic forces. Although some farms are large relative to others, agriculture is composed of many farms that are small relative to the industry. In this type of market, one of pure competition, the adjustments of individual farms can be analyzed. How will individual farms react to change? How can the agricultural economist charged with advising farmers determine how they should react?

Changing consumer demands affect the prices offered for farm products as well as the prices charged for agricultural inputs. How is farm output affected by these changing prices? When will an input price change cause change in the method of production? Will the output of small farms and large farms react similarly to a price change?

Agriculture is typified by technological change. What is the impact of new inputs and new methods of production upon various segments of agriculture? How will production respond? What farms are most likely to be hurt or helped by a technological advance? When will specialization result? Individual farmers must be prepared to cope with the changes caused by the weather, credit availability, leases, government programs, etc.

Commercial farming can now be categorized as big business. High land prices, mechanization, improved farming practices, and high incomes from job opportunities off the farm have contributed to this growth. Farm families have developed a new dependence on the market system by increasing both their sale of output and their purchases of production supplies. This dependence on the market contributes to the complexity of farming today and demands greater skills from the farm operators. Few professionals make as many different kinds of executive decisions as do today's farmers. They are their own buyer, accountant, price analyst, salesperson, machinst, and labor relations director.

As our knowledge of agriculture and the production alternatives available to farmers has increased, decision making has become more complex. The days when farms and other so-called small businesses could be developed successfully in the same fashion as tasty home-cooked dishes—a pinch of this and a scoop of that—seem to have passed. This increased complexity has also put competitive pressure on farmers who are reluctant to adjust their farming operation and who fail to keep in step with more efficient farmers. Mistakes made in farming prove more costly every day. As a result, new procedures and research tools are being developed and applied by agricultural economists to replace intuitive judgments and rules of thumb, and to develop recommendations for farm management decisions.

For example, one technique widely used in agricultural economics research, linear programming, has been used to analyze many aspects of the farm business. The capital available to a farmer plays an important and critical role in the organization of his business. This, of course, surprises neither farmers nor bankers. But linear programming applications have provided new insights into the use of capital in agriculture.

These methods and machines, of course, will never substitute for good judgment; skilled managers will continue to be scarce. And these techniques may not always be used by farmers in day-to-day operations, but they enable researchers to find answers to pressing problems.

Efficiency goals were defined above for agricultural economics, but to end this section, we present a listing of the general goals of production economics. In this text, the theory of the firm is to be applied to the management problems found on farms. But similar efficiency goals could also be applied to farm supply, farm marketing, and other firms found in the agricultural sector. The overall goals of production economics are to:[6]

1. Assist farm managers in determining the best use of resources, given the changing needs, values, and goals of society.
2. Assist policy makers in determing the consequences of alternative public policies on output, profits, and resource use on farms.
3. Evaluate the uses of the theory of the firm for improving farm management and understanding the behavior of the farm as a profit-maximizing entity.
4. Evaluate the effects of technical and institutional changes on agricultural production and resource use.
5. Determine individual farm and aggregated regional farm adjustments in output supply and resource use to changes in economic variables in the economy.

THE ECONOMIC SETTING

This section will digress from the mainstream of thought to review the type of market in which farmers buy and sell. A more complete discussion of market structure can be found in any elementary principles text. Readers familiar with the theory of markets in pure competition may proceed directly to Chapter 2.

To begin with, the analysis of a farm business can be divided into three parts:

1. The characteristics of the markets in which the firm purchases its inputs.
2. The technical and economic characteristics of the production process within the firm—that is, the nature of costs, profits, and so on.
3. The type of market in which the firm sells its products.

These three facets of the firm's economic existence are not independent. Even though our study is primarily concerned with production within the individual farm, the markets in which farmers buy and sell cannot be neglected, for they will influence their buying and selling policies.

Types of markets are classified according to the number and size of the firms which sell the product, the number of buyers who purchase the product and the amounts they purchase, the similarity or lack of similarity among products sold by the firms, and the ease with which firms may enter into or cease production of the product. Consider the farmer as a seller. Farmers all over the nation produce and sell wheat, corn, beef, pork, etc. For any one product—corn, for example—so

[6]These are paraphrased from Jensen's review article, pp. 3–4. Jensen presented these as goals for both farm management and production economics.

many sellers exist that the quantity sold by any one seller is infinitely small compared to the quantity sold by all other sellers. The quantity of corn produced by one farmer is minute compared to the total amount of corn produced annually in the United States.

There are many sellers of agricultural products—but that is not all. Agriculture products of a given type are similar in appearance and quality. Farmers sell a homogeneous product. The corn or hogs raised by one farmer may be of higher quality than a neighbor's but will be similar to the corn or hogs produced by a large number of other efficient farmers. The important thing, in any event, is that buyers of the agricultural products are either unaware of minor differences or regard them as unimportant. To the buyer, hard red winter wheat with a given protein content is the same regardless of who grew it, and the number two corn is number two corn regardless of where it originated. In general, buyers do not prefer the products of one farmer to the products of a second. The result is that individual farmers are unable to create a unique demand for their own particular products by special marketing methods such as advertising.

These two conditions—many sellers and homogeneous products—mean that an individual farmer cannot influence the market price of his product. If he holds his product off the market, prices will not increase due to reduced supply, because his production is an insignificant portion of the total production. On the other hand, he does not need to offer to sell his product at less than market price because he can sell all he has at the market price. The farmer is often described as a price taker, meaning that he must take the market price as given.

Consider now the farmer as a buyer. In general, he is one of many buyers, all of whom need the same type of input. Thus, if an individual farmer decides not to buy a tractor or a bag of fertilizer, the market price of tractors or fertilizer is not affected because that farmer's purchases would represent an infinitesimal quantity relative to the total number of tractors or fertilizer sold. Therefore, the farmer must also regard the price of inputs as given.

There are exceptions to the general rule that farmers are price takers. Farmers who supply a particular type of product or purchase a given type of input may be able to influence the market prices of inputs or products if they are good merchandisers or hagglers. Local markets can sometimes be developed for products such as eggs. Farmers who buy labor from their teenage sons may be able to determine both price and quantity, the latter being more doubtful than the former. Usually, however, these farmers are able only to influence market prices for themselves in a local market and not the general price level facing all farmers.

The type of market described above, with many buyers or sellers and a homogeneous product, is called a *purely competitive market.* Many of the markets in which farmers buy and sell approximate pure competition. The term "pure competition" could have been better chosen. There is nothing pure or impure about the type of market just described; the name "pure competition" simply describes a type of market. Further, there is no competition among individual businesses which are pure competitors. In fact, the reverse is true. The amount of production by one farmer has no effect on either the amount another farmer may sell or the price he receives for it. The only competition among firms in pure competition is the indirect competition that occurs through increases in efficiency and reduction of costs. An alternative term, "atomistic competition," is often used interchangeably with "pure competition" to avoid the connotations of the word "pure."

Two more conditions must be mentioned before pure competition is completely defined. First, businesses must be free to enter into or cease production as they please. There must be no patents or other legal barriers placed before prospective producers. If the price of corn promises to be high relative to the price of soybeans, farmers who grew soybeans last year may decide to switch to corn this year. If they do decide to make the shift, there should be no restrictions in their path.

Second, for pure competition in the purest sense, there should be no artificial restrictions placed on the supply of or demand for products. Thus, prices are free to vary, and equilibrium levels will be reached in the marketplace. In this way, increased production by the industry results in a lower market price and an increase in the quantity demanded and vice versa.[7]

Pure competition, as just described, is rarely found in the real world. It is considered worthy of study, nonetheless, for the following reasons. First, it comes closer to approximating market conditions in agricultural production than any other type of market. Second, one of the characteristics of an industry selling in a purely competitive market is that profits are the result of natural rather than contrived scarcity. Thus, ignoring farm programs for the moment, the profits earned by an Illinois corn farmer are due him because he owns corn land that is naturally scarce. Profits from contrived scarcity would result if a farmer were able to make corn supplies artificially scarce, perhaps by purchasing all corn land in the nation and removing most of it from production, thereby increasing corn prices. In an industry selling in a purely competitive market, however, no individual producer owns enough of any productive resource to be able to create artificial shortages.

Price determination in pure competition can be depicted on a graph. In Figure 1–2, the graphs in the left-hand column illustrate typical supply and demand curves for the total amount of a particular agricultural commodity produced in the United States for a given time span—number one red winter wheat for a year's time. The aggregate

[7]By requiring also perfect mobility of labor and other resources and perfect knowledge by consumers and producers, pure competition could be improved to a state called "perfect competition."

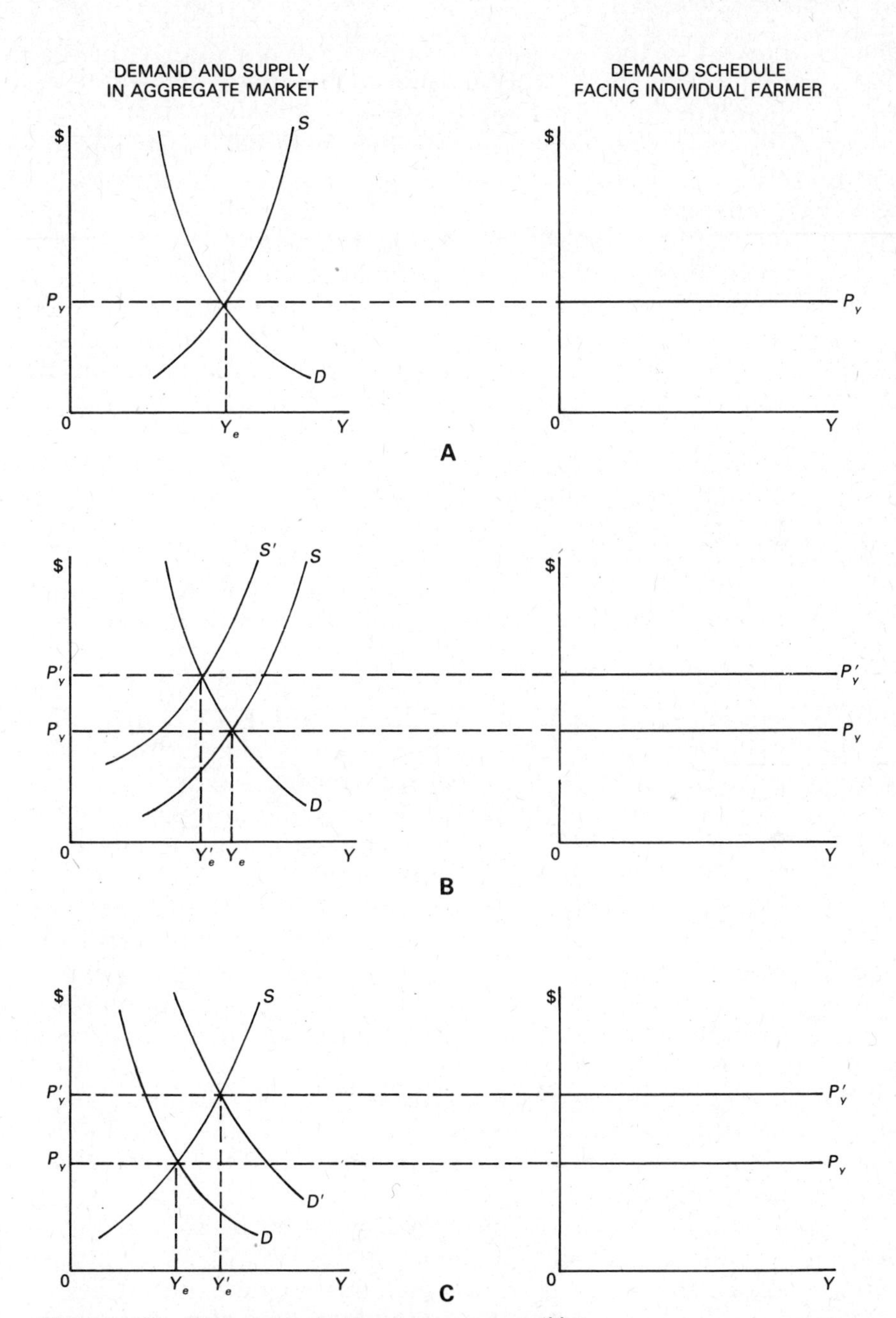

Figure 1–2. Price determination in pure competition.

demand curve lists the quantity consumers will buy at each price; as the price falls, consumers will buy more. The demand curve slopes downward and to the right. Aggregate demand is determined by population size, income, prices of other products in the economy, the tastes and preferences of consumers, and exports to foreign countries. The aggregate demand curve is a function of all the individual demands in society—albeit not necessarily a simple function.

Aggregate supply is the sum of the amounts supplied by individual farmers. As the price of a commodity rises, farmers already producing the commodity will find ways to produce more and new producers will enter the market. Hence, the aggregate supply curve has a positive slope.

The equilibrium or market-clearing price and quantity occur where aggregate demand equals aggregate supply. The equilibrium quantity is labeled Y_E and the equilibrium price, P_Y. This is the price facing the individual farmer. Because he cannot produce enough output to affect aggregate supply and thus equilibrium price, his demand curve for output can be depicted as a horizontal line, set at the level of equilibrium price. This is depicted in the graph in the right-hand column of Figure 1–2A.

The individual farmer cannot change the prices in the market; only changes in aggregate forces can affect prices. For example, if the weather is generally unfavorable across the United States for the crop in question, supplies will be reduced. Farm output will be less at every price; the supply curve will shift to the left. Equilibrium price will rise, and equilibrium quantity will fall. The price at the farm level will also rise. Thus, in Figure 1–2B, price rises from P_Y to P'_Y.

Shifts in demand can also change the farm price. Figure 1–2C depicts a situation in which the aggregate demand curve shifts to the right, perhaps due to a new source of demand such as increased exports to foreign countries to meet an unexpected famine. Equilibrium price again rises at the farm level, although in this situation aggregate equilibrium quantity increases.

We have emphasized price increases, but changes in aggregate supply and demand can also reduce prices. Thus, in Figure 1–2B, a favorable weather year can increase supplies from S' to S, thus causing price to drop from P'_Y to P_Y.

Supply curves for inputs purchased by farmers also take the appearance of the horizontal price lines in the right-hand column of Figure 1–2. In the input markets, the relevant demand curves are composed of the aggregate of the demands of individual farmers for inputs. As will be shown later in this book, these individual demand curves will have a negative slope. The supply curve for these inputs is less certain. Many agricultural inputs are produced by one or a few large corporations, and the manner in which these companies might respond to changes in price is not generally known—at least, the response cannot be adequately depicted on a graph as a supply curve. Indeed, many such firms can set prices. Thus, we must content ourselves to say that input supply functions facing individual farmers can be represented on a graph by horizontal lines, suggesting that the farmer can buy all he wishes at a price set by forces elsewhere in the economy.

Suggested Readings

Bilas, Richard A. *Microeconomic Theory*, Second Edition. New York: McGraw-Hill Book Company, 1971, Chapter 1.

Cohen, K. J. and Cyert, R. M. *Theory of the Firm*. Englewood Cliffs, N.J.: Prentice-Hall, Inc., 1965, Chapters 1 and 2.

Cramer, Gail L. and Jensen, Clarence W. *Agricultural Economics and Agribusiness*, Second Edition. New York: John Wiley and Sons, Inc., 1982, Chapters 1, 2, and 7.

Ferguson, C. E. and Gould, J. P. *Microeconomic Theory*, Fourth Edition. Homewood, Ill: Richard D. Irwin, Inc., 1975, Introduction.

Jensen, Harald R. "Farm Management and Production Economies, 1946–70." In Lee R. Martin, ed., *A Survey of Agricultural Economics Literature*, Part I, Volume I. Minneapolis: University of Minnesota Press, 1977.

Johnson, Glenn, L. "Stress on Production Economics." In Colyer, Dale K., Finley, Robert M., and Headley, J. C., eds., *Readings in Production Economics*. New York: MSS Educational Publishing Co., 1969.

Leftwich, Richard H. *The Price System and Resource Allocation*, Third Edition. Hinsdale, Ill: Holt, Rinehart and Winston, 1966, Chapters 1 and 2.

Linder, Staffan B. *The Harried Leisure Class*. New York: Columbia University Press, 1970.

Mundell, Robert. *A Man and Economics*. New York: McGraw-Hill, Inc., 1968, Chapter 1.

PRODUCTION AND COST FUNCTIONS

CHAPTER 2

This chapter presents some basic input-output and cost relationships that are important to understanding resource allocation problems in agriculture. The agricultural production process is complex and continually changing as new technologies appear. Agricultural research not only contributes to development of new varieties, breeds, and quality of inputs but also affects the use and combinations of these inputs. Thus the input-output relationships are in a continuous state of flux, changing the rates at which inputs or resources are transformed into products or outputs.

No product is produced with a single input. However, the effect a single input has on output can be determined when that input is varied while all other inputs are held constant. In this chapter we study the impact of one variable input on the output of one product. It is the purpose of this chapter to provide a theoretical foundation that will be expanded in subsequent chapters.

The concept of a production function is discussed first. Next, a production function is presented and its characteristics discussed; this is followed by presenting the nature of costs and their derivation from production functions. Finally, different types of production functions will be presented, along with a derivation of associated cost functions.

Many textbooks treat production functions and cost functions in separate chapters; this often obscures the fact that both are derived from the same technical input-output relationship. By introducing both concepts in one chapter, we hope to establish an understanding of the dichotomy underlying production theory that extends through to Chapter 9, *Linear Programming*.

THE CONCEPT OF A PRODUCTION FUNCTION

The production function portrays an input-output relationship. It describes the rate at which resources are transformed into products. There are numerous input-output relationships in agriculture because the rates at which inputs are transformed into outputs will vary among soil types, animals, technologies, rainfall amounts, and so forth. Any given input-output relationship specifies the quantities and qualities of resources needed to produce a particular product.

A production function can be expressed in different ways: in written form, enumerating and describing the inputs that have a bearing on the output; by listing inputs and the resulting outputs numerically in a table; in the form of a graph or a diagram; and as an algebraic equation.

Symbolically, a production function can be written as

$$Y = f(X_1, X_2, X_3, ..., X_N)$$

where Y is output and $X_1 \ldots X_N$ are different inputs that take part in the production of Y. The functional symbol "f" signifies the form of the relationshhip that transforms inputs into output. For each combination of inputs, there will be a unique amount of output. For example, Y may represent corn yield; X_1, fertilizer; X_2, soil moisture at seeding time; X_3, plant population, X_4, rainfall during the growing season; and so forth. The above symbolic relationship simply lists the inputs. In its present abstract form, it does not specify their importance or their contribution to the production process.

The above notation for a production function does not specify which inputs are variable and which are fixed. For example, feed or fertilizer often represent variable inputs that are applied to a fixed input such as an acre of land or a dairy cow. The fixed inputs clearly play an important role in agricultural production; they are often called *technical units*. Technical units have varying capacities to absorb and transform variable inputs into outputs. A sandy soil, for example, can absorb less water than a clay soil; the stomach capacity of an average Holstein cow is greater than that of a Jersey cow. Symbolically, fixed inputs can be included in the notation for a production function by inserting a vertical line between the fixed and variable inputs. For example,

$$Y = f(X_1, X_2, X_3, ..., X_{N-1}|X_N)$$

states that X_N is the fixed input (technical unit) while all other inputs are variable.

The subsequent paragraphs will present an empirical example of a production function that depicts output response to one variable input. In terms of the notation introduced above, this production function would be written

$$Y = f(X_1|X_2, X_3, ..., X_N)$$

This notation signifies that only one input, X_1, is to be varied by the manager, but that a large number of other inputs, X_2 through X_N, while assumed fixed or constant in amount for this example, also determine the level of output.

Perhaps the best way to introduce the concept of a production function is with an example. Following the empirical example, certain assumptions to be used in this chapter will be reviewed: perfect certainty, the level of technology, and the length of time period. These assumptions were first presented in the Introduction; in the discussion below, we will develop them in more detail.

An Empirical Example

Agricultural scientists are vitally concerned with production functions; much of the research in agriculture in the recent past has attempted to find the relationships between the amount of feed a dairy cow consumes and her milk output, the amount and composition of hog feed and the rate of gain on the hog, the amount of fertilizer applied to a field and the resulting crop yield, and so forth.

The actual estimation of a production function is an empirical task. The input-output relationships are generally derived using data from physical and biological experiments, or from farm records. Data of this type provide knowledge of the production function that enables producers to improve decisions concerning resource allocation.

Fertilizer response curves, for example, are determined from experimental data. Ordinarily, the more rates of fertilization and the more replicates that are utilized in one experiment, the more accurate the resulting response curve will be. Proper care in planning and locating experiments enables researchers to determine fertilizer practices that should be profitable for the crop in question.

Agricultural production is a complex biological process. But production functions relating output to inputs are real. The existence of a functional relationship between crops and different soil nutients is recognized and well established. However, biological laws of growth are not uniform. Consequently, an algebraic function that best explains a causal relationship among different fertilizer treatments and a crop in one situation may be a poor device for explaining a causal relationship in another situation. In addition to nutrient supplies, numerous other factors affect crop yields; weather, plant population, soil type, and management practices represent only a few. A study which considers all the factors would be prohibitive. For this reason studies are limited to a given soil type, climatic conditions, and farming practices. Causal relationships of different factors on yields of crops can be investigated only by the use of experiments that isolate the effects of a limited number of variables at one time.

The point to be emphasized is that production functions are not obtained without considerable effort. Because of the complexity of agricultural production processes, the true mathematical forms of production functions are not known. In this text, we will use tables, graphs, or algebraic equations to represent production functions. These are, in fact, rather simplified approximations of the true response functions.

An example of a production function for corn has been presented by Heady, Pesek, Brown, and Doll.[1] These scientists wished to determine the response of corn to applications of nitrogen and phosphoric acid (P_2O_5) fertilizer. To do this, a controlled experiment was conducted on Ida silt loam soil in western Iowa; nine rates of nitrogen and P_2O_5 ranging from zero to 320 pounds per acre were applied to experimental plots. Plot yields were measured at the end of the growing season. The experimental data were then used to estimate the response of corn to applications of nitrogen and P_2O_5 fertilizer on a per acre basis. The data in Table 2–1 show the response obtained from fertilizer, where a unit of fertilizer is 40 pounds of nitrogen and 40 pounds of P_2O_5. All other inputs were held constant. Thus, the technical unit would be an acre of farmland.

The data in Table 2–1 represent a production function in tabular form. They relate the response of corn yields to applied fertilizer. The unit of nitrogen and P_2O_5 represents the variable input, while all the other inputs needed to produce corn—seed, labor, fuel, land—are the fixed inputs. From examination of Table 2–1 it can be seen tht large increases in corn yields result from the initial fertilizer applications, but the yield increases become very small at high levels of application. A maximum yield of 139.9 bushels occurred when six units of fertilizer, 240 pounds each of nitrogen and P_2O_5, were applied. Yields decrease when larger amounts were applied. The curve in Figure 2–1, drawn by plotting the points in Table 2–1 and drawing a smooth curve through them, depicts the same information.

Table 2–1 Response of Corn Yields to Fertilizer Applications, Ida Silt Loam, Iowa

Units of Fertilizer Applied Per Acre[a]	*Bushels of Corn Per Acre*
0	0
1	44.9
2	83.6
3	110.1
4	127.3
5	136.9
6	139.9
7	137.1
8	129.2

Source: Heady, Earl O. and Dillon, John L. *Agricultural Production Functions*. Ames: Iowa State University Press, 1961, Chapter 14.

[a]A Unit of fertilizer is 40 pounds of nitrogen and 40 pounds of P_2O_5.

[1]Heady, Earl O., Pesek, John T., Brown, William G., and Doll, John P., "Crop Response Surfaces and Economic Optima in Fertilizer Use," Chapter 14 in Heady, E. O. and Dillon, John L., *Agricultural Production Functions*, Ames: Iowa State University Press, 1961.

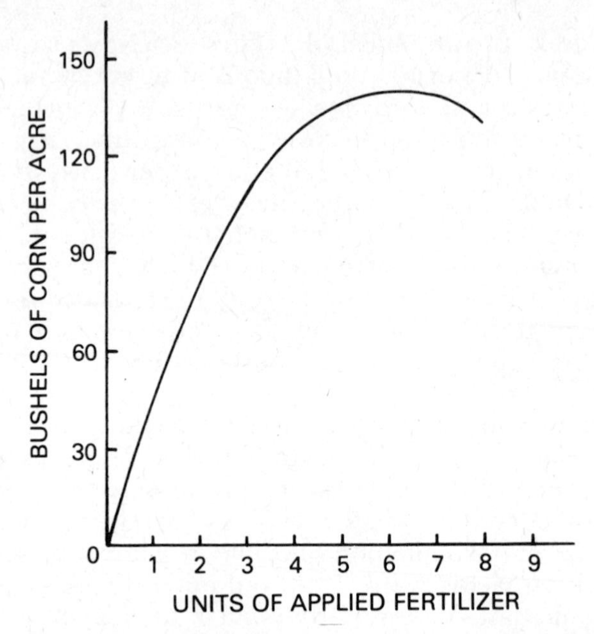

Figure 2–1. Response of corn yields to fertilizer applications, Ida Silt Loam, Iowa.

One aspect of the production function in Table 2–1 is that no corn yield is forthcoming when fertilizer applications are zero. This is unusual and occurred in part because of a lack of fertility in the soil used for the experiment. In general, if a variable input is essential for plant growth and is not present in the soil, corn yields would be zero when the amount of the variable input is zero. Most soils do contain some residual amounts of nitrogen and P_2O_5. If, however, the variable input had been plant population, then yield would have been zero when the variable input amount was zero, that is, when the crop is not planted. Therefore, yield may or may not be zero when the amount of the variable input is zero, depending upon the particular production process being studied.

Production functions in agriculture may be determined experimentally, as in this case, or through the study of records from farm businesses. Many farmers probably have mental pictures of production functions for their farms, although they do not, of course, give them the name "production function." The farmer's knowledge of his production functions is the result of years of practical experience.

Production functions in agriculture are highly specialized. The production function in Table 2–1 applies only to a certain soil type, soil fertility level, seed variety, technology, growing season, level of fixed inputs, etc. A change in any of these "givens" may cause a change in the production function. Thus, some Iowa farmers might find that the production function in Table 2–1 applied quite well to conditions on their farm, at least for some years. On other Iowa farms, soil and growing conditions may differ considerably from those of the experimental site, and the production function in Table 2–1 would be of little value. To be useful, the production function must be appropriate for the production process and growing conditions being studied.

ASSUMPTIONS

Perfect Certainty

Production economics is a study of the economic principles to be used when making management decisions. Decisions must guide future rather than past production. When determining the amount of fertilizer use for next year's corn crop, the farmer needs to know the production function for the coming season. The response of corn yield to fertilizer during the past year will be useful for next year's planning only if it represents or at least approximates next year's production function. Thus, to be useful in planning future action, the production function must be looked upon as a future or expected relationship. It is a planning device.

For businesses employing automatic man-made machinery operating under carefully controlled environmental conditions, the production function for the last production period may provide an excellent estimate of the production function most appropriate for coming time periods. Thus, the manager of a wheat milling plant or a cotton gin may be reasonably sure of the performance of his machinery year in and year out. He has control of all the inputs and knows exactly how the machinery is built and operates. Agricultural production, however, involves many inputs that cannot be controlled, such as rainfall and other weather variables, and many processes that are not completely understood, such as animal and plant nutrition or photosynthesis. Thus, for agriculture, last year's production response may be a poor estimate of this year's production response. The Iowa farmer who bases his production decisions on the corn yield response in Table 2–1 may be badly disappointed if the growing season for the coming year, for which he is planning, was not like that for the year in which the production function was estimated. The problem in this case, of course, is that the weather enters the production function in the form of a variable input not under the control of the farmer. A change in the weather may shift the function up or down.

Management decisions based on production functions can be only as accurate as the production functions themselves. This does not negate the production function as a useful concept; it does mean that care must be taken to select the appropriate production function for any given situation. Part of the manager's job is to select the appropriate production function or accept the consequences.

Problems arising because the future is unknown, called problems of "risk and uncertainty," will be introduced in Chapter 8. Study of such problems comprise an ever increasing portion of economic theory, but even an introduction to the subject is complex. To avoid these complexities, the following discussion will assume that the farmer knows the eventual outcome of the production process at the beginning of the production period. Under this assumption, the farmer knows the corn yields that will result from the use of varying amounts of fertilizer, he knows the number of pigs that will farrow and the number of hogs that will be marketed, and so on. Along with this, he knows the cost of fertilizer, the cost of feed, and the prices that corn and hogs will bring at the end of the production period. This assumption is known as the *perfect certainty* assumption. It simplifies our beginning analysis and permits a clear development of the basic principles of production economics. Perfect certainty, when combined with the assumption of no time discounting, amounts to an assumption of static, instantaneous production.

Level of Technology

The production process implies a method or manner of production. Usualy, a product can be produced many different ways. In the general discussion of the production process, the common assumption is that the farm manager uses the most efficient process available to him, that is, the one that results in the most product from a given amount of input. This assumption is not necessary, but it is reasonable. The method of production used is often called the *level of technology* or *the state of the arts*.

Length of Time Period

The production function depicts the output resulting from the production process during a given unit of time. Before discussing time periods in the production process, the general classification of resources used in production will be reviewed. A resource is called a *fixed resource* if its quantity is not varied during the production period. A resource is called a *variable resource* if its quantity is to be varied at the start or during the production period. All resources used are either fixed or variable. The group of all fixed resources at a particular time is often referred to as the *plant*, an analogy drawn directly from manufacturing firms.

Resources may be fixed for several reasons. First, the manager may be using exactly the right amount of a resource, meaning that an increase or decrease in the quantity used would lower his profits. In this case, he would be foolish to change the amount of the resource he is using. Second, the time period involved in the production process may be so short that the farmer is unable to change the amount of resource he has. Land is a good example. A farmer may realize he needs more land but is unable to purchase it immediately because it is located too far away, because he cannot arrange for suitable financing, or because it is unavailable. Third, the farmer may not want to vary the amount of input. For example, a dairy farmer may change the rations fed to a milk cow to determine the effect on the quantity of milk produced by the cow. For this evaluation, a cow is a fixed resource and the feed is a variable resource. He could change the size of his herd, but such a change is not relevant to the question he is investigating. In the long run, a manager has the *opportunity* to change the level of usage of all inputs. In the short run, he may not.

Fixed and variable resources are used to classify the length of the production period as follows:

Very Short Run—time period so short that all resources are fixed.

Short Run—time period of such length that at least one resource can be varied while other resources are fixed.

Long Run—time period of such length that all resources can be varied.

The above classification of time is not completely satisfactory. The long-run situation would seem most likely to occur before any resources are purchased. Even in this situation, however, the management ability of the farmer could be limiting. Once resources are committed to a production process, the farmer must move expeditiously from one short-run situation to another; that is, he will always be faced with some fixed inputs. The distinction between short-run and long-run periods is often nebulous.

Short-run or long-run situations defined above serve to classify production situations and, in some cases, to simplify economic analyses. They should never be regarded as shackles placed on a manager. That is, the farmer doesn't say to himself, "Gee, I could sure make money by buying another tractor, but this is the short run and tractors are assumed fixed." Instead he sizes up his farm business, decides which inputs should be increased or decreased, and makes the appropriate changes. Even when he decides to purchase a tractor, however, some time is required to locate, purchase, and bring home a new tractor; during that short period, tractors available on the farm are fixed in number.

THE CLASSICAL PRODUCTION FUNCTION

A production function for a specific production process, one involving corn production and fertilizer on an Iowa soil, has been presented. A discussion of all the production functions that now exist in agriculture would involve more space than any book can provide. Agricultural researchers can never hope to measure and record all production func-

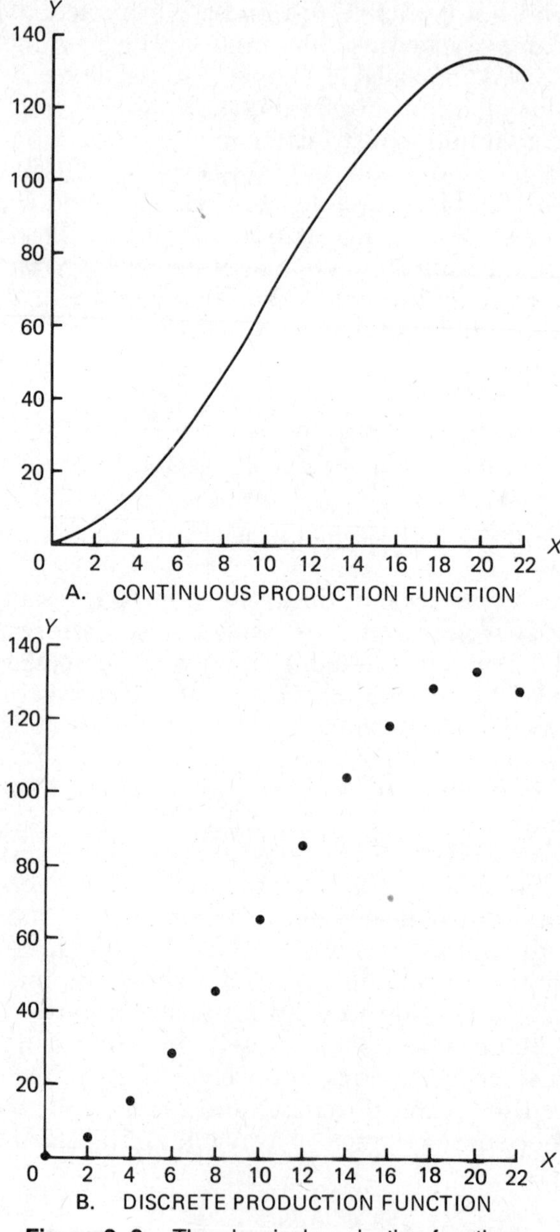

Figure 2–2. The classical production function.
A. Continuous production function.
B. Discrete production function.

tions in existence. The purpose of research on agricultural production functions is to provide a better understanding of input-output relationships and thus provide general guides and indications useful to farm managers; each manager must ultimately determine the appropriate production response for his locale and his farm.

Because all possible production functions cannot be discussed here, it becomes necessary to seek, in the manner of the agricultural research worker, some general principles that are applicable in all situations, regardless of the specific form of the production function. To begin with, two measures derived from the production function—the average physical product and the marginal physical product—will be defined and computed. In the following section, costs of production, obviously of great importance to a manager making production decisions, will be defined and derived from a production function.

The production function to be studied in detail is presented in Figure 2–2 and Table 2–2. This production function will be used to demonstrate the general principles important in the economic analysis of production. The reasons underlying the shape of this production function will be discussed later, but the form presented in Figure 2–2 is thought to be quite general. It is termed the *classical* production function because it displays all the characteristics necessary for a study of production functions.

The shape of the production function describes the change in output, Y, as increasing amounts of variable input, X, are added to a bundle of fixed factors (the technical unit and associated technologies). In Figure 2–2A, output is zero when input is zero. Output increases at an increasing rate as the first few units of input are added; it continues to increase, but at a decreasing rate at higher input levels. The maximum yield is 133.3 units of Y, resulting from application of 20 units of X. For higher input levels, output decreases.

Table 2–2 The Classical Production Function

(1)	*(2)*	*(3)*	*(4)*	*(5)*	*(6)*
			Marginal Physical Product MPP		
Input X	*Output Y*	*Average Physical Product APP*	*Exact*	*Average*	*Elasticity of Production MPP/APP*
0	0	—	0		—
2	3.7	1.9	3.6	1.9	1.9
4	13.9	3.5	6.4	5.1	1.8
6	28.8	4.8	8.4	7.5	1.8
8	46.9	5.9	9.6	9.1	1.6
10	66.7	6.7	10.0	9.9	1.5
12	86.4	7.2	9.6	9.9	1.3
14	104.5	7.5	8.4	9.1	1.1
16	119.5	7.5	6.4	7.5	0.8
18	129.6	7.2	3.6	5.1	0.5
20	133.3	6.7	0.0	1.9	0.0
22	129.1	5.9	−4.4	−2.1	−0.7

Total physical product function: $Y = X^2 - (1/30)X^3$
Average physical product function: $APP = X - (1/30)X^2$
Marginal physical product function: $MPP = 2X - (1/10)X^2$

The production function as graphed in Figure 2–2A implicitly demonstrates two more of the general assumptions first presented in Chapter 1. First, the production function is drawn as a continuous curve (in mathematics, a curve is continuous if it can be drawn without lifting pencil from paper). The function can have this property only if the inputs and outputs are perfectly divisible. In practice, inputs such as fertilizer, gasoline,and water are perfectly divisible while inputs such as tractors and cows are not. Inputs not divisible are termed *discrete*, and often called "lumpy."

When the input is perfectly divisible, the resulting output is also assumed perfectly divisible, such as corn yield response to fertilizer, and the smooth curve in Figure 2–2A results. Unlike some other assumptions that are vital to the economic analysis to be presented, the assumption of perfect divisibility is mostly for convenience. It permits the use of smooth curves. Suppose the inputs and outputs listed in Table 2–2 were lumpy; output would only result from using units of 2, 4, 6, 8, and so on. Then the production function would appear as in Figure 2–2B. The discrete function is not as tractable mathematically as the continuous function; therefore, the continuous case will be used for most discussions below. Nonetheless, the economic logic to be developed in this text does apply to all discrete cases without qualification.

The second assumption embodied in the production functions illustrated in Figure 2–2 is the assumption of homogeneity of inputs and outputs. All inputs represented on the X axis are equally productive. Yield does not increase rapidly between 8 and 10 units of input, and drops between 20 and 22 units of input because these inputs are of different quality. All inputs are homogeneous, and the units from 8 to 10 could be switched with those from 20 to 22 with no change in the production function. To state it another way, the order in which homogeneous inputs are applied is not important. The same assumption is made about resulting output. All units of output on the Y axis are homogeneous and of equal marketable value. If Y is corn, then all bushels are number two; if it is winter wheat, then all bushels are number one with the same protein content, etc. This assumption is crucial to the following analysis and is difficult to relax.

The continuous production function in Table 2–2 and Figure 2–2 can be expressed algebraically. Its equation is

$$Y = X^2 - \frac{1}{30}X^3 \tag{2.1}$$

where Y is the units of output resulting from the use of some number of units of X. Using the X values listed in Table 2–2, the corresponding Y values can be "predicted" by substituting into this equation. In addition, it will also predict the value of output for any other value of X, such as 2.5, 7.91, 14.357, or 15.333. There are also many other uses for this equation.

Output, Y, is often called *total physical product* to distinguish it from the *average physical product* and *marginal physical product* defined below. In the following discussion, the terms "total physical product," "yield," "output," or the symbol Y may all be used to denote the output of the production process. Similarly, the terms "input," "resource," "factor," or the symbol X will be used to denote the resources used in the production process.

Average Physical Product

APP is obtained by dividing the total amount of the output, Y, by the total amount of the variable input, X. From Table 2–2, when $X = 2$ and $Y = 3.7$, $APP = 3.7/2 = 1.9$; when $X = 10$ and $Y = 66.7$, $APP = 66.7/10 = 6.7$, and so on. The term "physical" means that average product is measured in physical units such as pounds, bales, kilograms, tons, bushels, or a number of eggs, rather than in value units, such as dollars.

Average physical product, Y/X, is defined geometrically in terms of the slope of a particular straight line. That slope represents the average rate at which the input, X, is transformed into product, Y. The straight line (*ray*) must always pass through the origin and intersect the production function. For example, in Figure 2–3 the straight line crosses the *TPP* curve at points A and C, at input levels $X = 8$ and $X = 22$, respectively. Since points A and C on the *TPP* curve lie on the same line through the origin and thus have a common slope, the two average physical products are identical. Geometrically, the slope can be calculated and expressed as a ratio of distances:

$$\frac{A - B}{B - O} = \frac{C - D}{D - O} \quad \text{or} \quad \frac{46.9}{8} = \frac{129.1}{22} = 5.9$$

Because the slope of the ray (straight line passing through the origin) is the numerical value of *APP*, then *APP* must increase as the rays move counterclockwise. Any ray drawn below the ray *OC* in Figure 2–3 will intersect the production function at points where the *APP* is less than 5.9. Rays above *OC* will determine larger *APP*s. Ray *OE* determines the maximum value of *AP*, which is 7.5 when $X = 15$. This is because a ray steeper than *OE* would not touch the production function.

The equation for *APP* can be derived from the equation for the production function. In particular, it would be

$$APP = \frac{Y}{X} = \frac{1}{X}\left(X^2 - \frac{1}{30}X^3\right) = X - \frac{1}{30}X^2$$

Again, by substituting the appropriate X values into this equation, the values of *APP* in Table 2–2 can be derived. Because dividing by zero

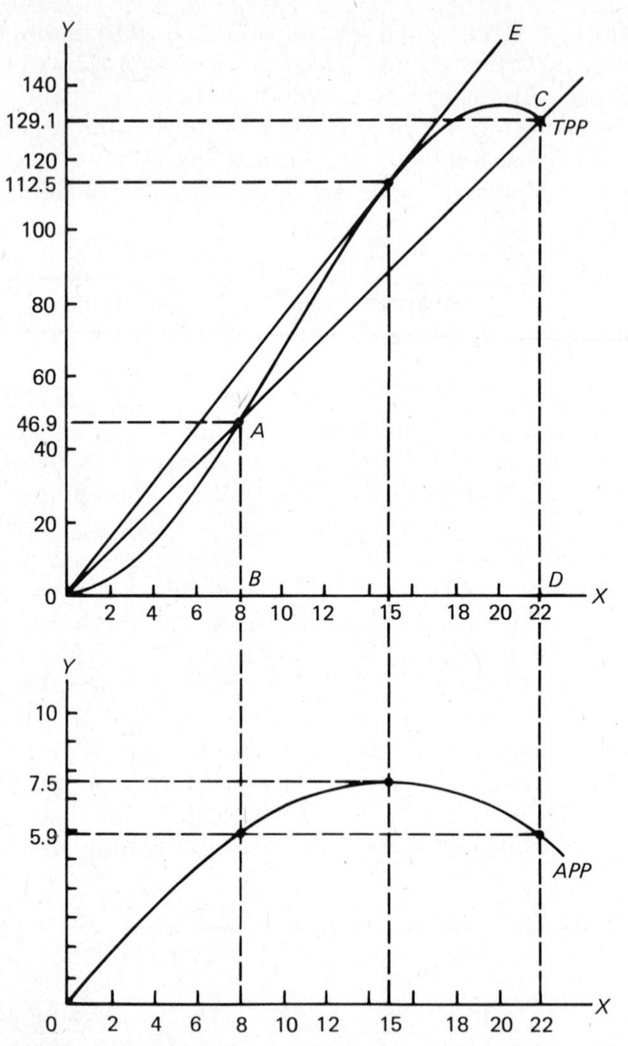

Figure 2–3. Geometric relationship between the total and average physical products.

is not possible, note that *APP* is not defined when *X* is zero. Thus, the value of zero predicted by the *APP* equation when *X* is zero must be regarded as a limit as *X* approaches zero.

The average physical product measures the average rate at which an input is transformed into a product. In Figure 2–3, it can be seen that the *APP* increases until $X = 15$, where $APP = 7.5$, and total product equals 112.5, and from that point on it decreases. The shape of the *APP* curve depends on the shape of the total physical product (*TPP*) curve or production function. Thus, *APP* could have a different shape, depending on how output varies with input.

One of the concerns of agricultural economics is the efficient use of resources. Further, efficiency is measured as output divided by input. Therefore, *APP* measures the efficiency of the variable input used in the production process. Output is the result of combining the variable input with the fixed factors. For the production function in Figure 2–2, *APP* is undefined when $X = 0$. While efficiency of a factor not being used cannot be measured, it is clear that if production is to be undertaken at all, then using no variable input with the fixed inputs is not an efficient situation. As more and more of the variable input is combined with the fixed inputs, efficiency of the variable input increases and eventually decreases.[2]

Marginal Physical Product

MPP is the change in output resulting from a unit increment or unit change in variable input. It measures the amount that total output increases or decreases as input increases. Geometrically, *MPP* represents the slope of the production function.

There are two ways to compute *MPP*: average and exact. The average method is used when working with tabular data and does not require calculus. The exact method uses calculus and therefore requires that the production function be expressed as an equation. Average *MPP* will be discussed first.

Average *MPP* is computed by dividing the change in output by the causal amount of input, that is, by the increment or change in input that caused the change in output. Algebraically, this can be expressed as

$$MPP = \frac{\Delta Y}{\Delta X}$$

where ΔY is read as "change in the amount of output" and ΔX is read as "change in the amount of input." In Table 2–2, column 5, *MPP* between the input amounts $X = 10$ and $X = 12$ is equal to

$$MPP = \frac{86.4 - 66.7}{12 - 10} = \frac{19.7}{2} = 9.9$$

Between the input amounts of 10 and 12, an added unit of input increases total output by 9.9 units.

MPP can be negative. Between the input amounts 20 and 22

$$MPP = \frac{129.1 - 133.3}{22 - 20} = \frac{-4.2}{2} = -2.1$$

Therefore, the addition of one additional unit of input when 20 units are already being used will cause total output to decrease by 2.1 units.

[2]*APP* measures the physical or technical efficiency of the variable input, which is distinct from economic efficiency.

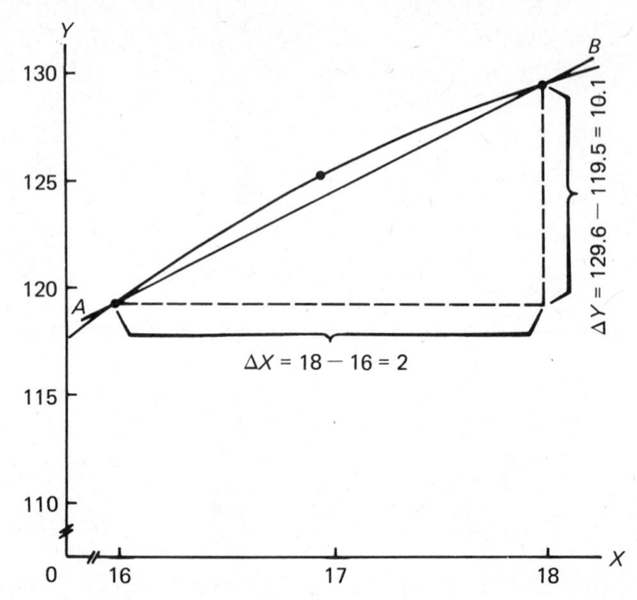

Figure 2–4. Geometric meaning of the marginal physical product.

The marginal physical product represents the slope of the total product curve. To demonstrate this, in Figure 2–4 the portion of the production function between $X = 16$ and $X = 18$ is shown greatly enlarged. *MPP* between $X = 16$ and $X = 18$ has been shown to be $\Delta Y/\Delta X = 10.1/2 = 5.1$. Examination of Figure 2–4 shows that 5.1 is also the slope of the straight line *AB*, drawn between the points ($X = 16$, $Y = 119.5$) and ($X = 18$, $Y = 129.6$). ΔY measures the vertical distance, or *rise*, in the production, and ΔX measures the horizontal distance or the *run*. The slope is equal to *rise/run*.

The quantity 5.1 is not the marginal product at 16 units of input or at 18 units of input; rather, it is the marginal product between these amounts. The marginal physical product is defined as a slope, but what slope? Between the input amounts $X = 16$ and $X = 18$ numerous slopes exist along the total physical product curve. The *MPP* of 5.1 represents the average of all the slopes on the *TPP* curve between the inputs $X = 16$ and $X = 18$, the input is transformed into product at a rate of 5.1 per unit increase. Because the *MPP* as computed represents the average of all the slopes between the two levels of input (an arc on the *TPP* curve), 5.1 is referred to as the average *MPP* between 16 and 18.

An equation for the exact *MPP* can be derived from the production function. The production function is

$$Y = X^2 - \frac{1}{30}X^3$$

and the *MPP* equation is the first derivative of the production function taken with respect to the variable input. The equation for the exact *MPP* is

$$\frac{dY}{dX} = 2X - \frac{1}{10}X^2 \qquad (2.2)$$

This equation defines the slope of the *TPP* curve or the exact *MPP* at any level of X. For example, when $X = 12$, the exact *MPP* is $2(12) - 0.1(144) = 9.6$; when $x = 14$, the exact $MPP = 8.4$. The average of the two exact *MPP*'s approximates the "average" *MPP* of 9.1 between the inputs $X = 12$ and $X = 14$.[3]

Table 2–2 shows both sets of marginal physical products—the average *MPP* and the exact *MPP*. The differences between the two are not large because the intervals between the inputs or arcs on the *TPP* curve are not large. In this case, the average *MPP* approximates the exact *MPP* better than it would when intervals between the amounts of X are large.

An examination of Table 2–2 and Figure 2–2 also shows that *MPP*, like *APP*, is not constant for the classical production function but varies with the amount of input use. The shape of the *MPP* curve depends on the shape of the production function. For the production function under consideration here, *MPP* increases to a maximim at 10 units of input when Y is 66.7 and $X = 10$ (at the point of inflection of the production function), and decreases as input is increased further. *MPP* is equal to zero at 20 units of input, where output is a maximum 133.3, and is negative for larger input amounts. When *MPP* is increasing, total output is increasing at an increasing rate. When *MPP* is decreasing but positive, total output is increasing at a decreasing rate. When *MPP* is zero, the total output curve attains a maximum. When *MPP* is negative, total output decreases. This relationship is shown in Figure 2–5.

LAW OF DIMINISHING RETURNS AND THE THREE STAGES OF PRODUCTION

The *law of diminishing returns* was developed by early economists to describe the relationship between output and a variable input when other inputs are constant in amount. It is believed to have widespread application. The law can be stated as follows.

If increasing amounts of one input are added to a production process while all other inputs are held constant, the amount of output added per unit of variable input will eventually decrease.

[3] The exact marginal product at a point is the slope of the tangent to the production function at that point. Appendix I contains a more detailed explanation.

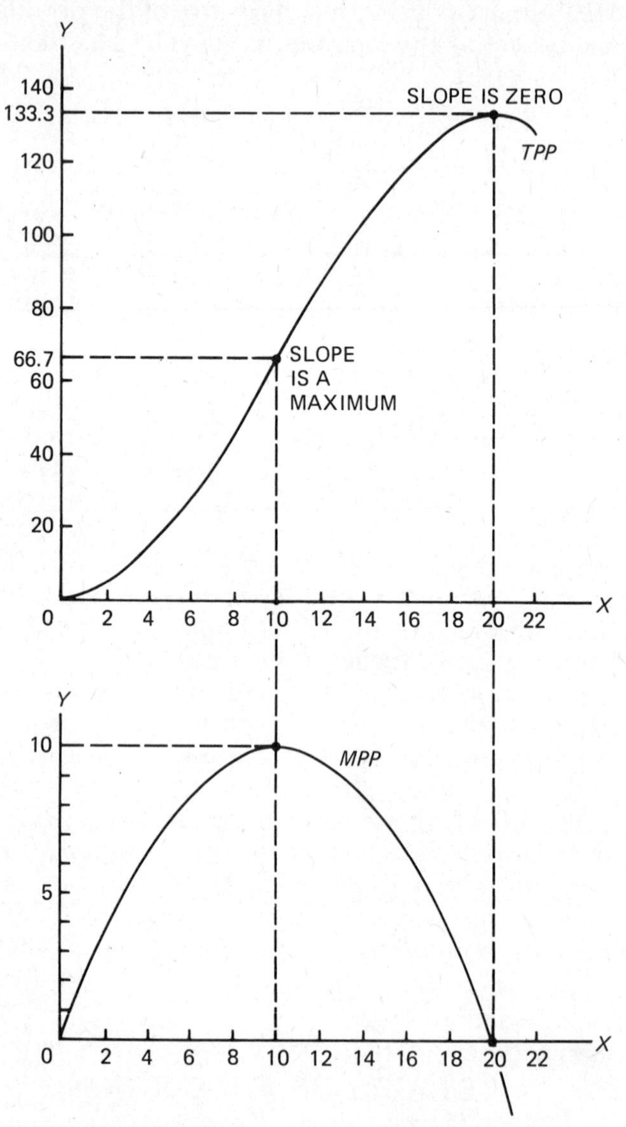

Figure 2–5. Geometric relationship between the total and marginal physical products.

The law suggests that there is some "right" amount of variable input to use in combination with the fixed input. The manager should neither use too much nor too little of the variable resource. Methods for determining the right amount of input, from the economic point of view, will be developed in the next chapter.

The law of diminishing returns requires that the method of production does not change as changes are made in the amount of variable input. It refers to the changing proportions between the variable and fixed inputs and does not apply when all inputs are varied. The law of diminishing returns is often referred to as the *law of diminishing productivity* or the *law of variable proportions*.

Application of the law of diminishing returns to the production function concept can result in a production function of the classical type (Figure 2–2). This production function displays increasing marginal returns first and then decreasing marginal returns. The law of diminishing returns says nothing about increasing marginal returns when the variable input is used in small quantities, but it does specify that marginal returns eventually decrease. Thus, marginal returns could decrease beginning with the first applications of the variable input; the production function is determined by biological and physical conditions that limit animal and plant growth. Here we seek general principles useful in all production situations—no generality is lost by assuming the classical production function.

Three Stages of Production

The classical production function can be divided into three regions or stages, each important from the standpoint of efficient resource use. The three stages are shown in Figure 2–6. Stage I occurs when *MPP* is greater than *APP*. *APP* is increasing throughout Stage I, indicating that the average rate at which variable input, X, is transformed into product, Y, increases until *APP* reaches its maximum at the end of Stage I.

Stage II occurs when *MPP* is decreasing and is less than *APP* but greater than zero. In Figure 2–6, Stage II falls between, and includes, the input quantities of 15 and 20. The physical efficiency of the variable input reaches a peak at the beginning of Stage II, for an input amount of 15. The dashed line on the left shows this boundary. (On the other hand, the efficiency of the fixed input is greatest at the end of Stage II, when the variable input equals 20. This is because the number of units of the fixed inputs is constant—usually at one. Therefore, the output per unit of fixed input must be the largest when total output from the production process is a maximum.)[4]

Stage III occurs where *MPP* is negative. Stage III occurs when excessive quantities of the variable input are combined with the fixed input, so much, in fact, that total output begins to decrease. The dashed line on the right in Figure 2–6 shows the boundary between Stages II and III. (The student should show that the production function in Figure 2–1 displays only Stages II and III.)

[4]Ferguson, C.E., *Microeconomic Theory,* Third Edition. Homewood, Ill: Richard Irwin, Inc., 1972, pp. 150–158. This delineation of the three stages is based on the assumption that the long-run production function is homogeneous to degree one. This will be discussed further in Chapter 6.

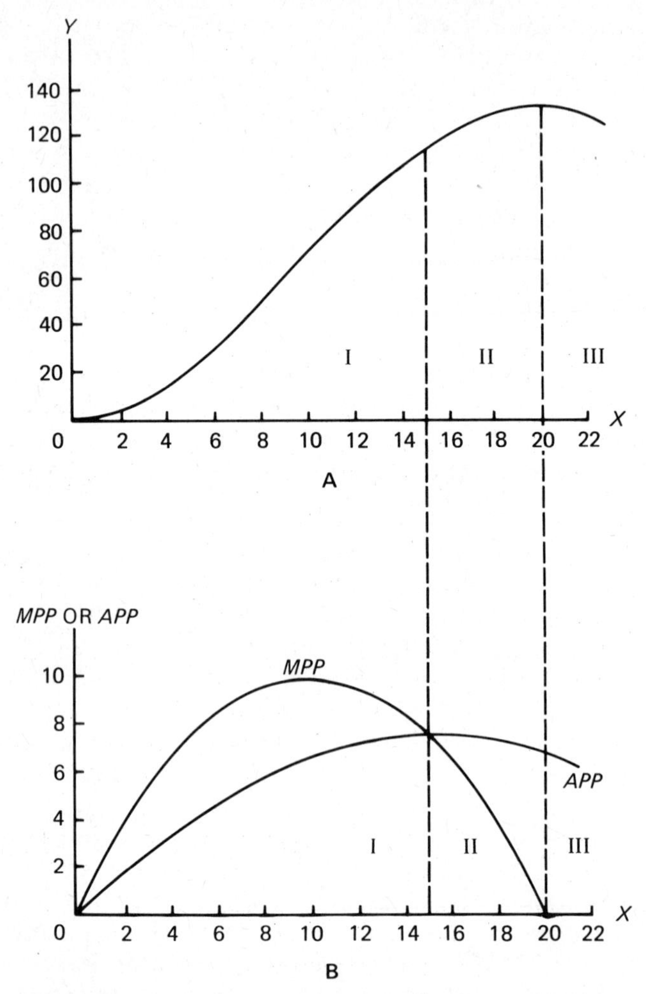

Figure 2–6. The classical production function and the three stages of production.

Stages of Production and Economic Recommendations

In the next chapter, production functions will be used to determine the most profitable amount of variable input and, simultaneously, the most profitable amount of output. Prices of the inputs and outputs must be known to make a complete economic analysis. However, when the technical relationship between input and output—the production function—is known, some recommendations about input use can be made even though prices are not specified.

First, if the product has any value at all, input use, once begun, should be continued until Stage II is reached. That is because the physical efficiency of the variable resource, measured by *APP*, increases throughout Stage I; it is not reasonable to cease using an input when its efficiency in use is increasing. For the production function in Figure 2–6, at least 15 units of input should be used.

Second, even if the input is free, it will not be used in Stage III. The maximum total output occurs on the upper boundary of Stage II; further input increments decrease output. It is not reasonable to increase input use when total product is decreasing. Thus, in Figure 2–6 the largest amount of variable input that would be used is 20 units.

Stage II and its boundaries define the area of economic relevance. Variable input use must be somewhere in Stage II, but the exact amount of input can be determined only when choice indicators, such as input and output prices, are known. If production is undertaken at all, then the objective will be achieved somewhere in Stage II for firms that buy and sell in pure competition.

Stages of Production—Algebraic Interpretation

The interpretation of the three stages of production and their delineation on the basis of the relationships between the *APP* and *MPP* can be deduced from the data of Table 2–2 and Figure 2–6. The same conclusions can be reached using algebraic calculations.

The slope of *TPP* is zero when *TPP* reaches its maximum. Since the *MPP* equation defines the slope at any level of input, X, the amount of X at which *TPP* reaches its maximum can be obtained by equating the *MPP* equation to zero:

$$\begin{aligned} MPP &= 2X - 0.10X^2 = 0 \\ &= X(2 - 0.10X) = 0 \end{aligned}$$

which has as solutions $X = 0$ and $X = 20$. But when $X = 0$, Y is also zero; thus, when $X = 20$, *TPP* reaches its maximum. This gives the boundary between Stages II and III. It locates the point where the tangent to the production function has a zero slope.

Similarly, the first derivative of the *APP* equation defines the slope at any level of input, X, on the *APP* curve. When *APP* is at its maximum, the slope of the *APP* equals zero. From equation (2.1), the classical production function,

$$APP = X - \frac{1}{30}X^2$$

$$\frac{dAPP}{dX} = 1 - \frac{1}{15}X = 0, \text{ from which } X = 15$$

APP reaches its maximum when $X = 15$. At that point *APP* also equals *MPP*. By substituting $X = 15$ into the *MPP* and *APP* equation, it can be shown that *MPP* and $APP = 7.5$. At this point, the average and the marginal rate at which input, X, is transformed into product,

Y, is equal, and a line through the origin is tangent to the production function. This result can be derived by differentiating the general expression for APP, that is, by differentiation of $f(X)/X$ with respect to X.[5]

Elasticity of Production and Point of Diminishing Returns

A discussion of the law of diminishing returns and the classical production function inevitably leads into the determination of the "point" of diminishing returns, meaning the input and yield amount at which returns begin to diminish. But what is that point? The law itself is ambiguous. Study of Figure 2–6 shows that marginal physical product begins to decrease at an input level of 10, the point of inflection on the production function, where MPP is at the maximum. Average physical product begins to decrease at 15 units of input, and total physical product begins to decrease at 20 units. Clearly, the point of diminishing returns depends on which of these three measures is being discussed.

To avoid this, some writers apply the law of diminishing returns directly to the marginal product. That is, they call it the law of diminishing marginal returns and specify in the definition that, as successive units of the variable input are added, marginal returns will eventually decrease. It is appropriate to define the law of diminishing returns in terms of the marginal product. Some ambiguity is caused, however, because the point of diminishing marginal returns, occurring at an input of 10 in Figure 2–6, differs from the boundary of Stage II, occurring where X is 15. A solution using the elasticity of production has been suggested by Cassels.

The elasticity of production is a concept that measures the degree of responsiveness between output and input. The elasticity of production, like any other elasticity, is independent of units of measure. Elasticity of production (ϵ_P) is defined as

$$\epsilon_P = \frac{\text{Percent change in output}}{\text{Percent change in input}}$$

From this, the elasticity of production is determined to be

$$\epsilon_P = \Delta Y/Y \div \Delta X/X = \frac{X}{Y} \cdot \frac{\Delta Y}{\Delta X} = \frac{MPP}{APP}$$

[5]As this discussion implies, the amount of X at which APP is a maximum can also be found by equating APP to MPP and solving for X.

In Stage I, *MPP* is greater than *APP*. Therefore, ϵ_P is greater than one. In Stage II, *MPP* is less than *APP* and ϵ_P is less than one but greater than zero. In Stage III, *MPP* is negative and ϵ_P is negative (see right hand column in Table 2–2).

The elasticity of production figures in Table 2–2 represent exact elasticities as opposed to average, or "arc," elasticities that can be calculated by dividing average *MPP* by *APP*. The exact ϵ_P is derived by dividing the exact marginal physical product by the average physical product at any level of *X*.

$$\epsilon_P = \frac{dY}{dX} \cdot \frac{X}{Y}$$

From the classical production function (2.1),

$$MPP = \frac{dY}{dX} = 2X - \frac{1}{10}X^2$$

and

$$\frac{1}{APP} = \frac{X}{Y} = \frac{1}{x - \left(\frac{1}{30}\right)X^2}$$

and

$$\epsilon_P = \frac{60X - 3X^2}{30X - X^2}$$

When $X = 10$, $\epsilon_P = 1.5$; when $X = 15$, $\epsilon_P = 1.0$, as shown in Table 2–2. If the elasticity of production is equal to one, a 1 percent change in input will produce a 1 percent change in output. If ϵ_P is greater than (less than) one, then a 1 percent change in input will bring about a greater than (less than) 1 percent change in output.

The "point" of diminishing returns can be defined to occur where $MPP = APP$ and ϵ_P is one—the lower boundary of Stage II. This is the minimum amount of variable input that would be used, and it occurs where the efficiency of the variable input is at a maximum. Using this definition, it can be argued without knowing input or output prices that input use will always be extended to the point of diminishing returns. At the other boundary of Stage II, *MPP* equals zero and hence ϵ_P also equals zero. Thus the relevant production interval for a variable input is that interval wherein $0 \leqq \epsilon_P \leqq 1$.

Table 2–3 Cost Curves Derived from the Classical Production Function (P_X = \$100)

(1) Input X	*(2)* Output Y	*(3)* Total Fixed Costs TFC	*(4)* Total Variable Costs TVC	*(5)* Total Costs TC
0	0.0	$1000	0	$1000
2	3.7	1000	$ 200	1200
4	13.9	1000	400	1400
6	28.8	1000	600	1600
8	46.9	1000	800	1800
10	66.7	1000	1000	2000
12	86.4	1000	1200	2200
14	104.5	1000	1400	2400
16	119.5	1000	1600	2600
18	129.6	1000	1800	2800
20	133.3	1000	2000	3000

COSTS OF PRODUCTION

Costs are the expenses incurred in organizing and carrying out the production process. They include outlays of funds for inputs and services used in production. In the short run, total costs include fixed and variable costs. In the long run, all costs are considered variable costs because all inputs are variable.

Fixed and Variable Costs

A resource or input is called a fixed resource if its quantity is not varied during the production period. A resource is a variable resource if its quantity is varied at the start of or during the production period. Most inputs have costs associated with them. Costs of fixed inputs are called fixed costs, while costs of variable inputs are called variable costs.

Fixed costs do not change in magnitude as the amount of output of the production process changes and are incurred even when production is not undertaken. Thus, fixed costs are independent of output. In farming, cash fixed costs include land taxes, principal and interest on land payments, insurance premiums, and similar costs. Noncash fixed costs include building depreciation, machinery and equipment depreciation caused by the passing of time, interest on capital investment, charges for family labor, and charges for management.

Fixed costs are usually associated with fixed inputs (technical units) because when the amount of a resource is fixed, the costs associated with it are also fixed. Some care must be taken with this definition because it may not always be true. A fixed cost requires only that the cost to the farmer be constant over the production period, not that

(6) Average Fixed Cost AFC	(7) Average Variable Cost AVC	(8) Average Total Cost ATC	(9) Marginal Cost MC Average	(10) Marginal Cost MC Exact
—	—	—	—	—
$270.3	$54.1	$324.4	$54.1	$27.8
71.9	28.8	100.7	19.6	15.6
34.7	20.8	55.5	13.4	11.9
21.3	17.1	38.4	11.0	10.4
15.0	15.0	30.0	10.0	10.0
11.6	13.9	25.5	10.1	10.4
9.6	13.4	23.0	11.0	11.9
8.4	13.3	21.7	13.3	15.6
7.7	13.9	21.6	19.8	27.8
7.5	15.0	22.5	54.1	—

the amount of input used be constant. For example, consider the farmer who subscribes to a rural electric utility and agrees to pay a monthly charge regardless of the amount of electric power consumed. In this case, the cost of electricity is fixed, but the amount used can be varied.

Computation of Total Costs of Output

Table 2–3 includes the computation of costs for the classical production function presented in Figure 2–2 and Table 2–2. The production function is presented in columns 1 and 2 of Table 2–3. The cost per unit of variable input is assumed to be $100. Fixed costs are assumed to be $1000. For simplicity, the $1000 is assumed to represent exactly the costs associated with the fixed inputs used in the production process, and the only variable cost is that incurred in purchasing the variable input.

Total fixed costs, *TFC*, are shown in the third column of Table 2–3. These costs are the same for all output levels. Thus, once computed from the farm or enterprise budget, *TFC* are known and unchanging. *TFC* are shown graphed in Figure 2–7A. *TFC* are $1000 for all output levels; this is represented by a straight line parallel to the horizontal or *Y* axis and located 1000 units up the vertical scale.

Total variable cost, *TVC*, is computed by multiplying the amount of variable input used by the price per unit of input.[6] In symbolic

[6]The farmer is assumed to be operating in pure competition. The input price is, therefore, constant regardless of the amount purchased. A brief discussion of pure competition is included in Chapter 1.

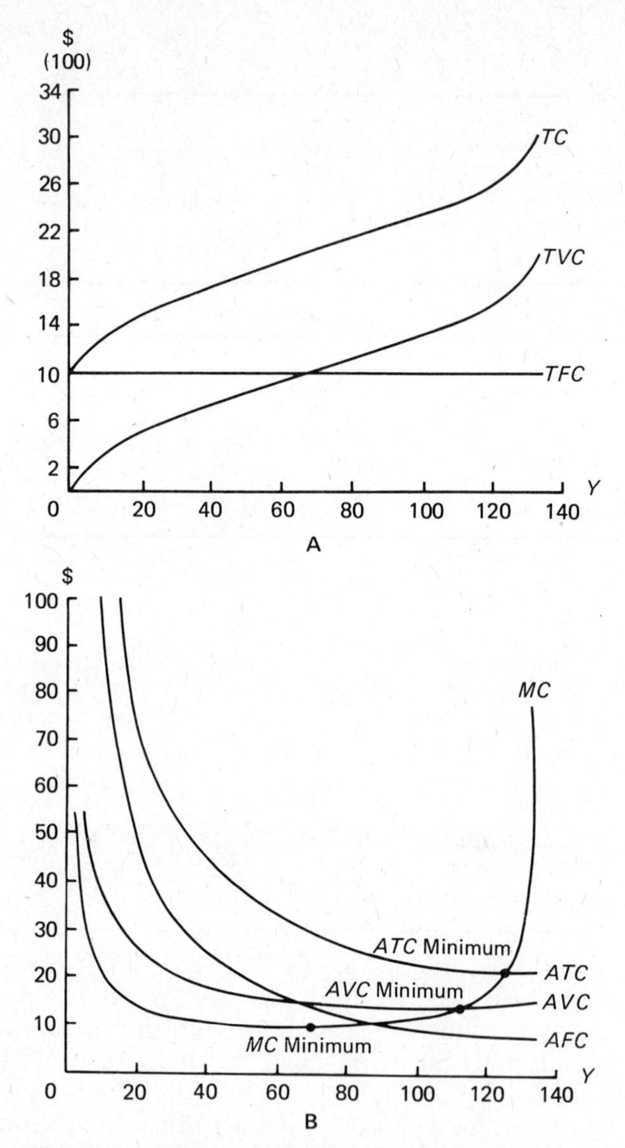

Figure 2–7. Cost curves for the classical production function.

notation, if X is the amount of variable input used and P_X is the price or cost per unit of input, then

$$TVC = P_X X$$

From Table 2–3, when 4 units of input are used, $TVC = \$100 \cdot 4 = \$400$. When 12 units of input are used, $TVC = \$100 \cdot 12 = \$1200$, etc. A graph of TVC, computed in column 4 of

Table 2–3, is shown in Figure 2–7A. *TVC* is zero when output, and consequently the input, is zero. It increases as output increases. The shape of the *TVC* curve depends on the shape of the production function; for the classical production function, *TVC* is always shaped as in Figure 2–7A.

Total costs, *TC*, are the sum of total variable costs and total fixed costs. They are presented in column 5 of Table 2–3 and obtained by adding *TVC* and *TFC* for any output level. For an output of 86.4 units, *TC* are \$1000 plus \$1200, or \$2200. *TC* for other output levels are obtained similarly. When no variable input is used, $TC = TFC$. Total costs are graphed in Figure 2–7A. The *TC* curve is equal to the vertical addition of *TFC* and *TVC*. It is shaped exactly like the *TVC* curve, but in this example it is always 1000 units higher on the vertical axis. The shape of the *TC* curve, like that of the *TVC* curve, depends on the production function. In symbolic notation, *TC* can be written

$$TC = TFC + TVC = TFC + P_X X$$

Average Fixed Costs, Average Variable Cost, Average Total Costs

Average fixed costs, *AFC*, are computed by dividing total fixed costs by the amount of output. *AFC* varies for each level of output; as output increases, *AFC* decreases. Thus, when economists refer to increasing output as a method of "spreading fixed costs," they mean increasing production to divide total costs among an increased number of units of output, thereby reducing costs per unit. *AFC* are presented in column 6 of Table 2–3 and graphed in Figure 2–7B. *AFC* for output amounts of 28.8 and 119.5 are:

$$Y = 28.8\text{: } AFC = \frac{TFC}{Y} = \frac{\$1000}{28.8} = \$34.72$$

$$Y = 119.5\text{: } AFC = \frac{TFC}{Y} = \frac{\$1000}{119.5} = \$8.36.$$

When output is zero, *AFC* cannot be computed because division by zero is not permissible. *AFC* always has the same shape regardless of the production function. Its general location on the graph depends on the magnitude of total fixed costs.

Average variable costs, *AVC*, is computed by dividing total variable costs by the amount of output. *AVC* varies depending on the amount of production; the shape of the *AVC* curve depends on the shape of the production function. The height of the *AVC* curve depends on the unit cost of the variable input. *AVC* is computed in column 7 of Table

2–3 and graphed in Figure 2–7B. *AVC* for two different output amounts is

$$Y = 28.8: AVC = \frac{TVC}{Y} = \frac{P_X X}{Y} = \frac{\$100(6)}{28.8} = \$20.83$$

$$Y = 119.5: AVC = \frac{\$100(16)}{119.5} = \frac{\$1600}{119.5} = \$13.39$$

As *AFC*, *AVC* cannot be computed when output is zero.

Average variable cost is inversely related to average physical product. When *APP* is increasing, *AVC* is decreasing; when *APP* is at a maximum, *AVC* attains a minimum, when *APP* is decreasing, *AVC* is increasing. Compare the cost curves in Figure 2–7B to the production function in Figure 2–6 from which the cost curves were derived. *APP* attains a maximum at an input level of 15, when output is 112.5; in Figure 2–7B *AVC* is a minimum at 112.5 units of output. For output amounts between 0 and 112.5, *APP* is increasing and *AVC* is decreasing. For output levels larger than 112.5, *APP* is decreasing and *AVC* is increasing. Thus, for a production function, *APP* measures the efficiency of the variable input; for cost curves, *AVC* provides the same measure. When *AVC* is decreasing, the efficiency of the variable input is increasing; efficiency is at a maximum when *AVC* is a minimum and is decreasing when *AVC* is increasing. The relationship between *APP* and *AVC* can be demonstrated algebraically as follows:

$$AVC = \frac{TVC}{Y} = \frac{P_X X}{Y} = P_X \cdot \frac{X}{Y} = \frac{P_X}{APP}$$

because $(X/Y) = (1/APP)$.

Average total costs, *ATC*, can be computed in two ways. Total costs can be divided by output or *AFC* and *AVC* can be added. *ATC* is presented in column 8 of Table 2–3 and in Figure 2–7B. The shape of the *ATC* curve depends on the shape of the production function. In Figure 2–7B, *ATC* decreases as output increases from zero, attains a minimum, and increases thereafter. *ATC* is often referred to as the unit cost of production—the cost of producing one unit of output. The intial decrease in ATC is caused by the spreading of fixed costs among an increasing number of units of output and the increasing efficiency with which the variable input is used (as indicated by the decreasing *AVC* curve). As output increases further, *AVC* attains a minimum and begins to increase; when these increases in *AVC* can no longer be offset by decreases in *AFC*, *ATC* begins to rise. For the output amounts of 28.8 and 119.5, *ATC* are computed as follows:

$$Y = 28.8: ATC = \frac{TC}{Y} = \frac{\$1600}{28.8} = \$55.55$$

or

$$ATC = AFC + AVC = \$34.72 + \$20.83 = \$55.55$$

and

$$Y = 119.5{:}\ ATC = \frac{\$2600}{119.5} = \$21.75$$

or

$$ATC = \$8.36 + \$13.39 = \$21.75$$

Marginal Costs

Marginal cost, *MC,* is defined as the change in total cost per unit increase in output. It is the cost of producing an additional unit of output. *MC* is computed by dividing the change in total costs, ΔTC, by the corresponding change in output, ΔY. Examples of marginal cost are presented in Table 2–3 and Figure 2–7B. Between the output amounts of 3.7 and 13.9, *MC* is computed as follows:

$$MC = \frac{\Delta TC}{\Delta Y} = \frac{\$1400 - \$1200}{13.9 - 3.7} = \frac{\$200}{10.2} = \$19.61$$

Between the output amounts of 129.6 and 133.3,

$$MC = \frac{\$2000 - \$1800}{133.3 - 129.6} = \frac{\$200}{3.7} = \$54.05$$

By definition, the only change possible in total costs is the change in variable cost, because fixed cost does not vary as output varies. Thus, $\Delta TC = \Delta TVC$. Therefore, *MC* could also be computed by dividing the change in total variable cost by the change in output.

Geometrically, *MC* is the slope of the *TC* curve and the *TVC* curve. The marginal cost of \$19.61 is the average of all the slopes on *TC* between the points $Y = 3.7$, $TC = \$1200$, and $Y = 13.9$, $TC = \$1400$. Within these output limits each unit of output added to total output will cost \$19.61 to produce. The "average" *MC* between the outputs of 129.6 and 133.3 is equal to \$54.05. The *MC* calculated in Table 2–3, column 9, should be viewed as the slope between the corresponding levels of output and not the exact slope at the indicated levels of output. The data in Table 2–3 column 10, on the other hand, show the exact *MC* for each corresponding level of output. The average *MC* and the

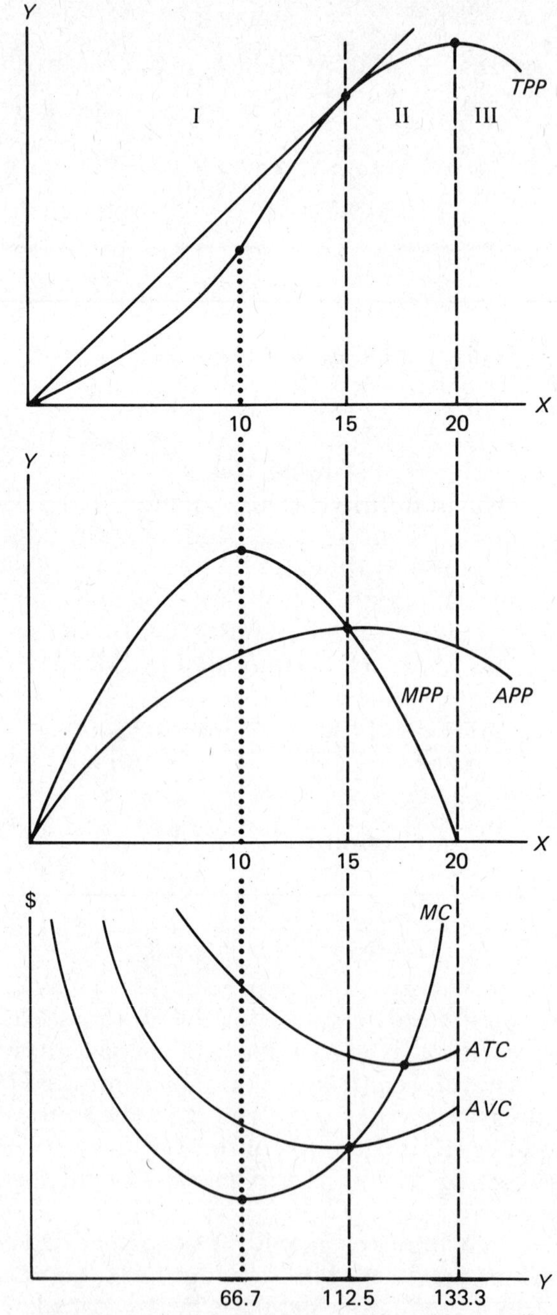

Figure 2–8. Physical and cost relationships in the three stages.

exact MC are corollaries to the average MPP and exact MPP in Table 2–2.

The shape of the MC curve is in an inverse relationship to that of MPP. Compare the MC curve in Figure 2–7B to the MPP curve in Figure 2–6. MPP is a maximum at 10 units of input; output at this point is 66.7. In Figure 2–7B, MC is a minimum at 66.7 units of output. For lower levels of output, MC is decreasing while MPP is increasing. For output levels above 66.7, MPP is decreasing while MC is increasing. Algebraically, the relationship between MPP and MC can be shown as

$$MC = \frac{\Delta TC}{\Delta Y} = \frac{\Delta TVC}{\Delta Y} = \frac{P_X(\Delta X)}{\Delta Y} = P_X \frac{(\Delta X)}{(\Delta Y)} = \frac{P_X}{MPP}$$

where the change in variable costs between two output amounts, TVC, is equal to the change in the variable input used, ΔX, multiplied by the price of the input. The term $(\Delta X/\Delta Y)$ is, of course, the inverse of the MPP. Thus, the exact marginal cost figures in Table 2–3 were computed by dividing the input price, P_X, by the exact marginal product at each output level.

MC and AVC are equal when 15 units of input are used and output is 112.5; this is the same point at which MPP is equal to APP. For output amounts lower than 112.5, MC is less than AVC; for higher outputs, MC is greater than AVC. As long as there is some fixed cost, MC crosses ATC at an output greater than the output at which AVC is at the minimum and MC is equal to ATC at the latter's minimum point.

To summarize the relationships described above, the total, average, and marginal product curves are presented in Figure 2–8 along with the average variable, average total, and marginal cost curves (for additional clarity, Figure 2–8 is not drawn to scale). Study of Figure 2–8 shows more clearly the nature of the cost curves in Stage II of the production function. Students should note that the graphic analysis used to derive average product from total product, Figure 2–3, can also be used to derive AFC from TFC, AVC from TVC, and ATC from TC. And, just as marginal product is the slope of total product, marginal cost is the slope of total cost (or total variable cost) and can be derived in an analogous manner (see Figure 2–4). Because it would be repetitive, these graphic analyses will not be presented here, but students should work them through as an exercise.

Comments on Costs

Costs are usually computed as a function of output, that is, a manager is usually interested in the total cost of producing an output or in the unit cost at a level of output. Thus, the cost curves in Figures 2–7 and 2–8 are graphed with dollars on the vertical axis and units of output on the horizontal axis. Costs are graphed as a function of input (with units of input on the horizontal axis) only in special situations.

Costs need be computed and graphed for input and output amounts only in Stages I and II of the production function; Stage III is an area in which no rational manager would produce. Stage II begins at the point where $MC = AVC$ and continues to the point where output is a maximum. In Figure 2–8, these limits are 112.5 and 133.3 units of output, inclusively. On the boundary between Stage II and III, *MPP*, as it was shown earlier, is zero. Therefore, on the same boundary, $MC = P_X/MPP = P_X/0$, which is an undefined quantity. Intuitively, the *MC* curve at this point would be vertical and *MC* would cease to have meaning. For this reason the *MC* data in Table 2–3 is omitted when *TPP* is decreasing, or when *X* exceeds 20 units.[7]

Marginal cost is a widely used concept in agricultural economics. Strictly defined, it is the increase in total cost resulting from a one unit increase in output. Any other definition of marginal cost is not valid. That is, one could define a cost concept that measures the change in total cost caused by a one unit increase in input. Such a cost concept is useful but is not marginal cost; it is often called the marginal factor cost or the marginal expense of the variable input. In pure competition, the marginal factor cost, P_X is a constant.

MORE ON COST AND PRODUCTION FUNCTIONS

In practice, there are two ways researchers estimate cost functions for a farm enterprise. One way is to estimate the cost functions directly. For example, by observing costs and output data for a large sample of similar farms, the "typical" relationship between costs and output for, say, the corn enterprise or the dairy enterprise can be developed. Once the total cost functions are obtained, the other cost functions can be determined.

The second way is to estimate the cost functions from the production function. The production function, when known, can be used along with fixed costs and input prices to derive all cost functions.

Cost Function Known

When the total cost function can be estimated directly, then other cost functions and their properties can be derived. In this section we will present an algebraic illustration. A total cost function with the shape or form that corresponds to a general classical production function may also be expressed as a cubic equation. For example, with dollar signs omitted, such a function for total cost could be

$$TC = 100 + 6Y - 0.4Y^2 + 0.02Y^3 \qquad (2.3)$$

[7] The *MC* equation approaches a vertical asymptote located at the maximum output, 133.3. As the *MPP* approaches zero, *MC* increases without limit. For outputs above 133.3, *MC* is negative. The *MC* equation has an "infinite discontinuity" where *MPP* is zero.

When graphed, this equation would have the same general shape as the TC function in Figure 2–7A (it is not the equation of the TC curve in the graph, however). In this cost function,

$$TFC = 100 \text{ and } TVC = 6Y - 0.4Y^2 + 0.02Y^3$$

The 100 is constant and does not change as output, Y, changes. The remaining terms in the function, $6Y - 0.4Y^2 + 0.02Y^3$, change with every change in output, Y.[8]

From the total cost function, other costs can be derived. Thus, average variable cost would be

$$AVC = \frac{TVC}{Y} = \frac{6Y - 0.4Y^2 + 0.02Y^3}{Y}$$
$$= 6 - 0.4Y + 0.02Y^2 \quad (2.4)$$

Average fixed cost would be

$$AFC = \frac{100}{Y}$$

Marginal cost would be

$$MC = \frac{dTVC}{dY} = 6 - 0.8Y + 0.06Y^2 \quad (2.5)$$

Because the total cost function corresponds to the classical production function, it has average variable cost and marginal cost curves that decrease, reach a minimum, and then increase. The level of output, Y, at which the average variable cost and marginal cost reach their minima can be calculated.

At the point where AVC is at the minimum, its slope equals zero. Thus, from (2.4),

$$\frac{dAVC}{dY} = -0.4 + 0.04Y = 0 \quad \text{or} \quad Y = 10$$

The average variable cost, in the above example, reaches its minimum when output, Y, is 10 units. At this level of output the average variable cost equals marginal cost. By substituting value of 10 for Y in

[8]When $TC = ay^3 + by^2 + cy + d$ is used to represent total cost, then the following restrictions must hold: $a > 0$, $b < 0$, $c > 0$, $d > 0$, and $b^2 < 3ac$. Chiang, A. C. *Fundamental Methods of Mathematical Economics,* Second Edition, New York: McGraw-Hill, Inc. 1974, p. 264.

the average variable cost equation, (2.4) and marginal cost equation, (2.5) the corresponding costs, \$4, are equal.

Similarly, the value of Y at which marginal cost is a minimum can be calculated by equating the MC slope to zero. From (2.5),

$$\frac{dMC}{dY} = -0.8 + 0.12Y = 0 \quad \text{or} \quad Y = 6.67 \tag{2.6}$$

Thus, the minimum of MC occurs where output equals 6.67 units, and this is, as would be expected, at a lower output than the point at which AVC is a minimum.

The average total cost, ATC, reaches its minimum at an output larger than the minimum of the average variable cost because of the effect of the declining average fixed cost on average total cost. From the above example,

$$ATC = \frac{TC}{Y} = 100Y^{-1} + 6 - 0.4Y + 0.02Y^2$$

and

$$\frac{dATC}{dY} = -100Y^{-2} + 0.04Y - 0.4$$

or

$$\frac{-100}{Y^2} + 0.04Y - 0.4 = 0$$

Solving this equation by approximation, the value of output, Y, when the slope of the ATC curve equals zero, is 17.85 units. At this level of output ATC is at the minimum and equals the value of MC. The value for both MC and ATC is 10.83.

Finally, the slope of the MC curve at any point can be determined by substituting different values of Y into (2.6). When $Y = 10$, AVC is a minimum. Substituting $Y = 10$ in (2.6) shows that MC has a positive slope at that point. The same is true where ATC is a minimum. Thus, the MC curve is increasing through those points. The properties of cost curves displayed in Figure 2–7 can be derived using calculus.

Deriving Cost Functions from Production Functions

Cost functions incorporate both the production function and the fixed and variable costs of production. Production functions are purely technical and depict only what happens to output as variable inputs are applied to a technical unit. But economic analysis of the production process requires the consideration of costs and returns. Among the many production alternatives that exist on the farm, the farmer must consider those that will be most profitable. Market forces far from the farm will help determine that profitability. In our development of the economic principles applicable to the analysis of farm enterprises, cost functions are the first step in incorporating the impact of the marketplace on the farm enterprise. We will now develop the unique manner in which production functions and inputs prices can be combined to derive cost curves.

Cost curves usually are expressed so that output, Y, is the so-called independent variable. The costs themselves represent the costs of inputs, either fixed or variable. Combining these two statements, cost curves express the cost of fixed and variable inputs as functions of the amount of output.

Production functions, on the other hand, express output as a function of input. But input costs are input quantities multiplied by input prices. Therefore, cost functions and production functions are by nature inversely related to each other. Knowledge of one implies knowledge of the other—when input prices are known.

In symbolic notation, total variable cost has the equation

$$TVC = P_X X$$

while the production function is expressed as

$$Y = f(X)$$

But the standard concept of the cost relationship demands that variable cost be expressed as a function of output, not input.

This dilemma is solved by expressing input, X, as a function of output, Y, for input and output amounts restricted to Stages I and II. When this is done, the resulting function is called the inverse production function and is expressed as

$$X = f^{-1}(Y)$$

where f^{-1} denotes the inverse function of the original production function, f. This inverse function is not a cause-effect relationship; it does not imply output causes input. It tells the minimum amount of input needed to produce a given level of output.

When the equation for a production function is available, it may be possible to solve for the inverse production function. In any event, the shape of the inverse function can be determined on a graph. Some algebraic examples will be presented first.

A simple but very useful algebraic form for a production function, based on the parabola, is called the quadratic production function. An example is

$$Y = 8X - \tfrac{1}{2}X^2 \tag{2.7}$$

where Y and X are as previously defined. For this production function, APP and MPP are linear functions found as follows:

$$APP = \frac{Y}{X} = 8 - \tfrac{1}{2}X$$

and

$$MPP = \frac{dY}{dX} = 8 - X$$

Total product takes a maximum value of 32 when $X = 8$, the same value of X for which marginal product is zero. The zone of economic relevance for this function is $0 \leqq X \leqq 8$ with Y values that have the limits $0 \leqq Y \leqq 32$. This function along with the APP and MPP curves are shown in Figure 2–9. (Note: Figure 2–9 is not drawn to scale.)

The inverse production function for (2.7) is found using the quadratic formula. The production function can be written in standard form as

$$0 = -\tfrac{1}{2}X^2 + 8X - Y$$

which, when substituted into the quadratic formula, reduces to[9]

$$X = 8 - \sqrt{64 - 2Y} \quad \text{for} \quad 0 \leqq Y \leqq 32$$

Variable cost for this production function can be expressed as

$$TVC = P_X X = P_X(8 - \sqrt{64 - 2Y}),\ 0 \leqq Y \leqq 32$$

and marginal cost would be

$$MC = \frac{dTVC}{dY} = \frac{P_X}{\sqrt{64 - 2Y}},\ 0 \leqq Y < 32$$

[9] The formula for solving quadratic equations is

$$X = \frac{-b \pm (b^2 - 4ac)^{1/2}}{2a}$$

where $a = -1/2$, $b = 8$, and $c = -Y$. The solutions resulting when the radical is positive are too large and are ignored.

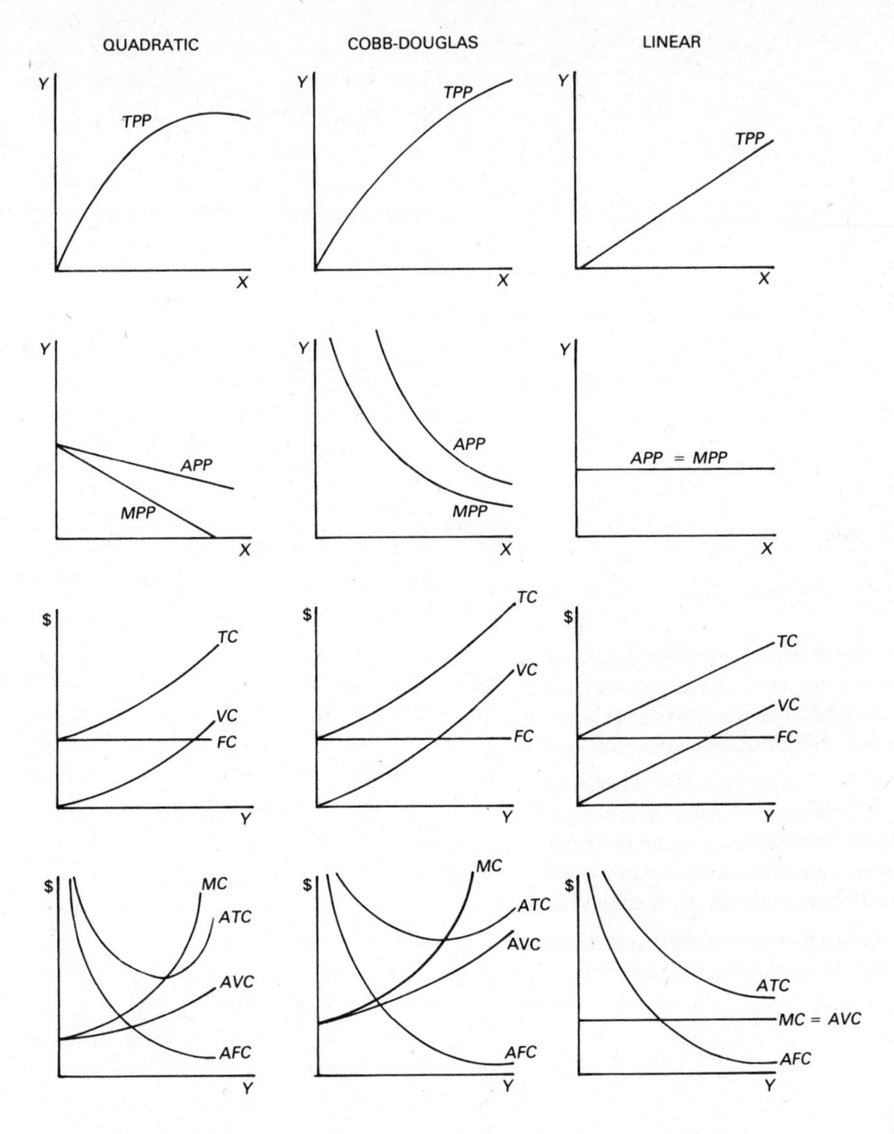

Note: Not drawn to scale.

Figure 2–9. Cost curves derived from three types of production functions.

which is undefined when Y equals 32. Notice that MC becomes very large and will increase without limit as Y approaches 32; thus, the line $Y = 32$ becomes a vertical asymptote for MC. These curves are also presented in Figure 2–9.

An example of an algebraic production function that is easier to solve is the famous Cobb-Douglas equation.

$$Y = 6X^{1/2} \tag{2.8}$$

where the exponent of X is the elasticity of production and usually restricted to an interval between zero and one. The marginal and average product equations for this function would be

$$APP = \frac{6}{\sqrt{X}}$$

$$MPP = \frac{dY}{dX} = \frac{3}{\sqrt{X}}$$

The inverse production function in this case is

$$X = \frac{1}{36} Y^2$$

so that

$$TVC = P_X X = \frac{P_X}{36} Y^2$$

$$AVC = \frac{P_X}{36} Y$$

and

$$MC = \frac{dTVC}{dY} = \frac{P_X}{18} Y$$

This function is also illustrated in Figure 2–9.

A particularly simple example is given by the linear production function. An example would be

$$Y = 2X$$

For this function,

$$APP = MPP = 2$$

$$X = \tfrac{1}{2}Y$$

$$TVC = \tfrac{1}{2}P_X Y$$

and

$$AVC = MC = \tfrac{1}{2}P_X$$

This function is also depicted in Figure 2–9.

Much of this chapter has been devoted to analysis of the classical production function and associated cost curves. As yet, the inverse function for this function has not been derived. The reason for this is that we chose a cubic equation to represent the classical function in our example, and a simple algebraic expression for the inverse of the cubic equation cannot be found. This situation sometimes happens in mathematics. The inverse function *exists* in Stages I and II of the classical function expressed as a cubic; it just cannot be expressed as a simple algebraic formula. In such cases the shape of the inverse production function and hence the variable and total cost curves can be determined visually. The shape is often all that is needed for expository purposes.

To determine the shape of the variable cost curve (or total cost curve) for any conceivable production function, hold a graph of the production function up to a mirror with output on the horizontal axis. If a mirror isn't handy, this same "upsidedown and backwards" view can be obtained from a graph. Imagine that a graph of the production function is transparent and centered on an axis that passes through the origin at a 45 ° angle. Visualizing the 45 ° axis as an axle, imagine rotating the graph 180°. The X axis and the Y axis exchange places. Further, each point on the production function that was represented by (X_o, Y_o) becomes the point (Y_o, X_o). Points falling on the 45° line stay in place because it is the axis of rotation. An example of this for the classical function is depicted in Figure 2–10.

In Figure 2–10, the following points shift places as the X and Y axes are reversed.

$$(5, 3) \rightarrow (3, 5)$$
$$(10, 8) \rightarrow (8, 10)$$
$$(14, 14) \rightarrow (14, 14)$$
$$(16, 20) \rightarrow (20, 16)$$
$$(18, 25) \rightarrow (25, 18)$$
$$(22, 32) \rightarrow (32, 22)$$
$$(27, 34) \rightarrow (34, 27)$$

The inverse production function is defined to end at the point (34, 27) where the production function achieves a maximum. To go further would move into Stage III and also would cause the relationship to relate two X values to each Y value. This violates the definition of a function and is infeasible from the viewpoint of efficiency. The least amount of input would always be used to produce a given output.

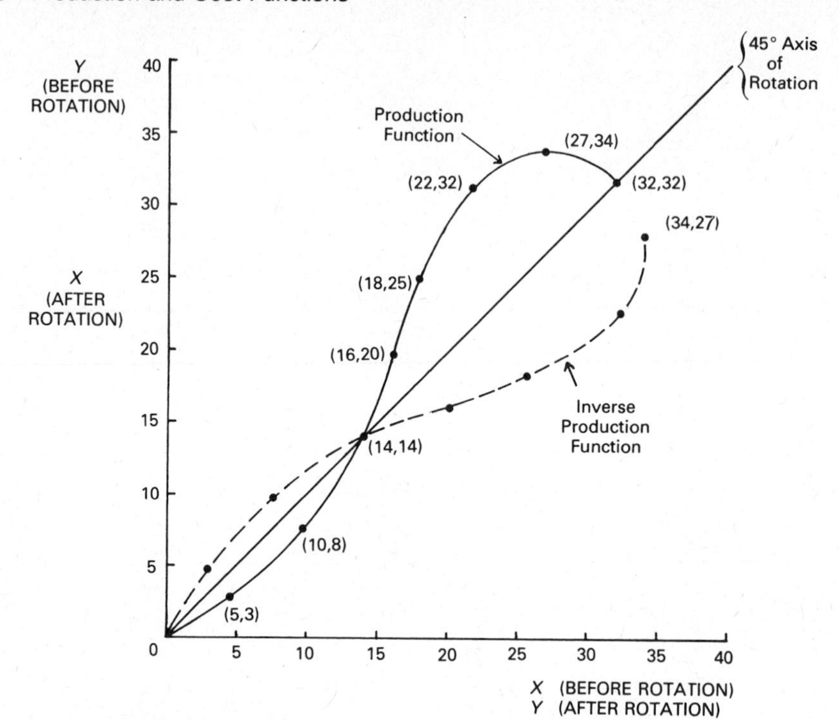

Figure 2–10. Derivation of the inverse production function.

Problems and Exercises

2–1. Consider the production function $Y = X + 4X^2 - 0.2X^3$. (a) Derive the exact marginal physical product and average physical product equations. (b) At what levels of X do the *MPP*, *APP*, and *TPP* reach their maximums? (c) At what levels of X does Stage II begin and end? (d) Plot the *TPP*, *APP*, and *MPP* curves.

2–2. The production function is $Y = 0.5X^\beta$. (a) Prove the β is the elasticity of production. (b) Has this function a point of inflection? (c) If $\beta = 1/2$, what are *MPP* and *APP* when $X = 4, 9, 16$, and 25?

2–3. Average total cost $ATC = 100/Y - 3Y + 4Y^2$. Calculate: (a) Total fixed cost. (b) Average variable cost when $Y = 2$. (c) Total variable cost when $Y = 2$. (d) Marginal cost when $Y = 2$. (e) Level of Y at which *AVC* is at the minimum.

2–4. The total cost function is $TC = 2Y - 2Y^2 + Y^3$. (a) Find the average variable cost function. (b) Find the level of Y where the *AVC* is at the minimum. (c) Find the marginal cost function. (d) Show that, at the minimum of average cost, average cost is equal to marginal cost. Plot the average cost and marginal cost curves.

2–5. The production function fitted to agronomic experimental data showing the corn response, Y, in bushels, to nitrogen, X, in pounds, on a per acre basis is $Y = 37 + 0.8X - 0.001X^2$. (a) Estimate the per acre yield of corn when $X = 0, 40, 80, 120, \ldots, 160, \ldots, 480$ pounds. (b) What is the *APP*, average *MPP*, and exact *MPP* at the above nitrogen application rates? (c) At what level of nitrogen is the total yield, *TPP*, maximum? (d) Calculate the arc (average) and exact elasticities of production at 40 pounds of nitrogen intervals. (e) Plot the *TPP*, *APP*, and *MPP* curves.

2–6. Calculate the appropriate figures for the blank colums.

X	*TFC*	*TPP*	*APP*	*MPP*	*TVC*	*AVC*	*MC*	*AFC*	*ATC*
0	$40.00	0	—	—	$ 0	—	—	—	—
1	40.00	4	—	—	5.00	—	—	—	—
2	40.00	10	—	—	10.00	—	—	—	—
3	40.00	15	—	—	15.00	—	—	—	—
4	40.00	18	—	—	20.00	—	—	—	—
5	40.00	20	—	—	25.00	—	—	—	—

2–7. Consider a U-shaped *AVC* curve. Draw the kind of *MPP*, *APP*, and *TPP* curves you expect to find. Explain your procedure carefully.

2–8. $Y = 10X - X^2$. Find the exact or point elasticities of production when $X = 2$, 4, and 5.

2–9. Only one resource, *X*, is used in producing an output, *Y*. As *X* is increased, total physical product, *Y*, increases at a decreasing rate, reaches a maximum level and then decreases. Show graphically the relation between the total physical product curve, the marginal physical product curve, the average physical product curve and their respective cost curves.

2–10. $AVC = Y^2 - 2Y + 2$. (a) Derive the *TVC* and *MC* equations. (b) At how many units of *Y* is *AVC* at the minimum? (c) At how many units of *Y* is *MC* at the minimum? (d) At how many units of *Y* does the *MC* curve become the supply function for the product *Y*?

2–11. Draw the classical production function, *APP* and *MPP*, and carefully delineate the three stages of production.

2–12. There is some fixed cost. Does the minimum average total unit cost occur at an output less than, equal to, or greater than the output at which the minimum average variable cost occurs? Explain your answer.

2–13. What can be said about the elasticity of production, assuming the classical production function, at the output at which:

(a) *MPP* is at maximum
(b) *APP* is at maximum
(c) *MPP* is zero
(d) $MPP = APP$
(e) *MPP* is negative
(f) *MPP* is less than *APP*

2–14. For the production function $Y = f(X)$, show, in general, using calculus that $APP = MPP$ where *APP* is a maximum. (Hint: Differentiate $f(X)/X$.)

2–15. The *AFC* curve is called a rectangular hyperbola, which means that any rectangle with two sides on the axes, a corner on the curve, and sides parallel to the axes will have a constant area. What is the implication of that for our study of cost curves? (Hint: Would a similar interpretation hold for any rectangle drawn under the *AVC* curve? What do the areas of these rectangles represent?)

2–16. Use calculus to show that the *AFC* curve always has a negative slope. Will the slope of *ATC* curve always equal the sum of the slopes of the *AFC* and *AVC* curves?

2–17. A firm has a short-run total variable cost curve given by $TVC = ay^2 + by$ where $a > 0$ and $b > 0$. Show that the output level, Y, at which $SRATC$ is a minimum is a direct and positive function of fixed costs, TFC. Graph the solution for $a = 1$, $b = 4$.

2–18. The graphs in Figure 2–7 are not drawn to scale. Choose numerical amounts for P_X and TFC, derive all cost curves, and plot the graphs to scale using the production functions given in the text.

2–19. Plot the production functions used in Figure 2–7 accurately on graph paper. On the same graph, plot the inverse production function in each case, using the technique described in the text. Be sure to use the same scale on both axes.

2–20. Is the result of problem 2–17 true in general? Show that the same solution can be found by equating $MC = ATC$ and solving for Y.

Suggested Readings

Allen, C. L. *The Framework of Price Theory*. Belmont, Cal: Wadsworth Publishing Co., Inc., 1967, Chapters 8 and 13.

Baumol, W. J. *Economic Theory and Operations Analysis*. Englewood Cliffs, N.J.: Prentice-Hall, Inc., 1961, Chapter 11.

Cassels, John M. "On the Law of Variable Proportions," *Readings in the Theory of Income Distribution*, Philadelphia: The Blakiston Co., 1951, Chapter 5.

Cohen, K. J. and Cyert, R. M. *Theory of the Firm*. Englewood Cliffs, N.J., Prentice-Hall, Inc., 1965, Chapters 3 and 7.

Cramer, Gail L. and Jensen, Clarence W. *Agricultural Economics and Agribusiness*, Second Edition, New York: John Wiley and Sons, Inc., 1982, chapters 4 and 6.

Ferguson, C. E. and Gould, J. P. *Microeconomic Theory*, Fourth Edition, Homewood, Ill: Richard D. Irwin, Inc., 1975, Chapters 5 and 7.

Heady, Earl O. *Economics of Agricultural Production and Resource Use*. New York: Prentice-Hall, Inc., 1952, Chapters 2 and 3.

Heady, Earl O. and Dillon, John L. *Agricultural Production Functions*. Ames: Iowa State University Press, 1961, Chapters 1, 3, and 14.

Leftwich, Richard H. *The Price System and Resource Allocation*, Third Edition. Hinsdale, Ill: Holt, Rinehart and Winston, 1965, Chapter 7.

Mansfield, E. *Microeconomics*, Third Edition. New York: W. W. Norton and Company, Inc., 1979, Chapters 6 and 7.

ALLOCATION OF ONE VARIABLE INPUT

CHAPTER 3

Problems associated with the allocation of one variable input are often referred to as the factor-product relationship. The objective of the factor-product relationship is to determine the quantity of the variable input that will be used in combination with the fixed inputs. Questions such as: How much fertilizer to apply per acre? How much irrigation water to use? How many steers on a given pasture? How many hens in a hen house of a given size? All are within the scope of the factor-product relationship.

The answers to these types of allocation problems will depend on the manager's objectives. The manager has limited amounts of resources to use on the farm. His problem is to use these resources to achieve his goal. But, as mentioned in Chapter 1, resolution of an economic problem must depend on the specification of a goal or set of goals—often referred to in more abstract terminology as the *objective function*. In the study of production economics, the commonly assumed goal of the farm manager is economic efficiency, which subsumes the narrower goal of profit maximization.

ECONOMIC EFFICIENCY

Economic efficiency refers to the combinations of inputs that maximize individual or social objectives. Economic efficiency is defined in terms of two conditions: necessary and sufficient.

Necessary Condition

This condition is met in a production process when there is (a) no possibility of producing the same amount of product with fewer inputs and (b) no possibility of producing more product with the same amount of inputs. In production function analysis, this condition is met in Stage II, that is, when the elasticity of production is equal to or greater than zero and equal to or less than one ($0 \leq \epsilon_P \leq 1$).

The necessary condition refers only to the physical relationship. It is universal because it is applicable in any economic system. No one would knowingly produce in Stage III because the same or larger output could be obtained by moving to Stage II and using less input. In a given input-output relationship, many input-output combinations will satisfy the necessary condition. For this reason, an additional condition is needed to single out one alternative from the many that meet the necessary condition.

Sufficient Condition

Unlike the necessary condition, which is objective, the sufficient condition for efficiency encompasses individual or social goals and values. It is subjective in nature and, thus, may vary among individuals as their objectives or their likes and dislikes vary.

In abstract theory the sufficient condition is often called a *choice indicator*. This choice indicator helps the manager determine input use compatible with his objectives. Thus, in subsistence agriculture, for example, a family that prefers potatoes to turnips will place greater emphasis on potato production.

The sufficient condition for an individual striving for high yields per acre will be different from that of an individual whose objective is maximization of profits per acre. In either of these cases, while the choice indicators satisfying the sufficient condition vary, economic efficiency is met because the manager is achieving his goals.

In the United States, commercial farming is a big business. Farmers have become increasingly dependent on the market system to obtain inputs and sell outputs. Farming is also a highly competitive business. A farmer, like any businessman who wants his business to survive, has to be mindful of costs of inputs and prices of products that those inputs produce. The manager of a highly competitive commerical enterprise has to include input-output price ratios in his decision-making process. A choice indicator based on the assumption of profit maximization from an enterprise seems to be a valid criterion for a businesslike farm operation.

Physical input-output relationships by themselves do not specify the level of resource use for a producing unit. Economic efficiency can be defined only when the relevant end or objective is at a maximum as specified by a choice indicator. While economic ends and choice indicators may differ, the principles of production economics are always the same. The necessary and sufficient conditions become the goals (maximum net returns, maximum family satisfactions, or other ends of individuals and groups) toward which analyses in agricultural production economics are directed.[1]

PROFIT MAXIMIZATION FOR AN ENTERPRISE

Throughout this chapter, unless otherwise indicated, the assumption is made that farmers buy inputs and sell products in purely competitive markets. It is further assumed that farmers want to maximize net returns or profits from variable inputs. Prices and input-output rela-

[1]The necessary and sufficient conditions for economic efficiency were set forth by Heady, pp. 103–104, for firms in pure competition. They should not be confused with the terms "necessary and sufficient conditions" as used in mathematics.

tionships are assumed to be known with certainty. This chapter is an extension of the preceding chapter and, thus, consideration is given only to one variable input applied to fixed inputs.

Methods of Determining the Optimum

The problem is to determine the most profitable point of operation for an enterprise in the short run. This can be done by determining either the most profitable amount of input or the most profitable level of output. Because the production function relates input to output in a unique manner in Stage II, either method results ultimately in the same answer. In economic terminology, the "most profitable" amount can also be called the "optimum" amount; both terms will be used here.

The optimum amount of a variable input is that amount that maximizes short-run profits from the production process. Once the optimum amount of variable input is being used in the short run, the only possible way to increase profits would be to change technologies or quantities of fixed inputs.

Determining the Optimum Using Total Value Product and Total Costs. Total costs, *TC*, were defined in the last chapter. Another measure of importance in the production process is the total value product, *TVP*, the total dollar value of the production of an enterprise. In symbols, $TVP = P_Y \cdot Y$ where P_Y is the price per unit of output, and Y is the amount of output at any level of input, X. For the production function in column 1 of Table 3–1, the same classical production function discussed in Chapter 2, *TVP* is shown computed for all output levels; the

Table 3–1 Determining the Optimum Point of Production Using Total Costs and Total Value Product (P_Y = \$30, P_X = \$100, and *TFC* = \$1,000)

(1)		(2)		(3)
Input X	Output Y	Total Costs $TC = TFC + P_X(X)$	Total Value Product $TVP = P_Y(Y)$	Profit $TVP - TC$
0	0.0	\$1000	\$30(0.0) = \$ 0	−\$1000
2	3.7	1200	30(3.7) = 111	− 1089
4	13.9	1400	30(13.9) = 417	− 983
6	28.8	1600	30(28.8) = 864	− 736
8	46.9	1800	30(46.9) = 1407	− 393
10	66.7	2000	30(66.7) = 2001	1
12	86.4	2200	30(86.4) = 2592	392
14	104.5	2400	30(104.5) = 3135	735
16	119.5	2600	30(119.5) = 3585	985
18	129.6	2800	30(129.6) = 3888	1088
20	133.3	3000	30(133.3) = 3999	999
22	129.1	3200	30(129.1) = 3873	673

price of Y is \$30 per unit. When the amount of input is 10, Y is 66.7 and TVP is $\$30 \cdot 66.7$ or \$2001. When X is 14, Y is 104.5 and TVP is $\$30 \cdot 104.5 = \$3135$. In pure competition, the farmer can sell all his product at the prevailing market price; thus P_Y is \$30 per unit for all levels of output. Because the farmer cannot sell enough product to influence the market price, he can regard the output price as a constant.

The production function is vital to the computation of profit. It relates total value product to the amount of input and total costs to the amount of output. That is, total value product is easily related to output; it is price times quantity. In the same way, total costs may be easily computed for various input amounts. But only the production function can relate *inputs to revenues* and *outputs to costs*.

Total value product minus total costs gives profit, which also is often called net returns or net revenue. Using an input cost of \$100 per unit and a fixed cost of \$1000, profit is presented in column 3 of Table 3-1. When input is zero, output is zero and profit is −\$1000, the amount of fixed costs. As output increases, profits increase and reach a maximum of \$1088 at 18 units of input and 129.6 units of output.

The simple profit model can be expressed algebraically as

$$\begin{aligned} Profit &= TVP - TC = TVP - TVC - TFC \\ &= P_Y Y - P_X X - TFC \end{aligned}$$

This equation describes the process used to compute the numbers in column 3 of Table 3–1. When an algebraic expression is available for the production function, $Y = f(X)$, then profit is expressed as a function of the input, X, as follows:

$$Profit = P_Y \cdot f(X) - P_X X - TFC$$

Thus, the search for the optimum amount of variable input leads directly to the problem of maximizing profit as a function of X. This procedure will now be studied in detail.

Determining the Optimum Amount of Input. Many output amounts result in a profit; in fact, Table 3–1 shows that any input amount between 10 and 22 earns a profit. The optimum amount, however, is that amount resulting in the largest profit.

All methods of determining the optimum amount of input can be derived from the study of total value product and total costs. The optimum amount of input for the production function in Table 3–1, computed by subtracting total costs from total revenue, is 18 units; the resulting maximum profit is \$1088. The same solution is presented in Figure 3–1A by graphing total value product and total cost as functions of X. Profits are a maximum in the graph when total value product

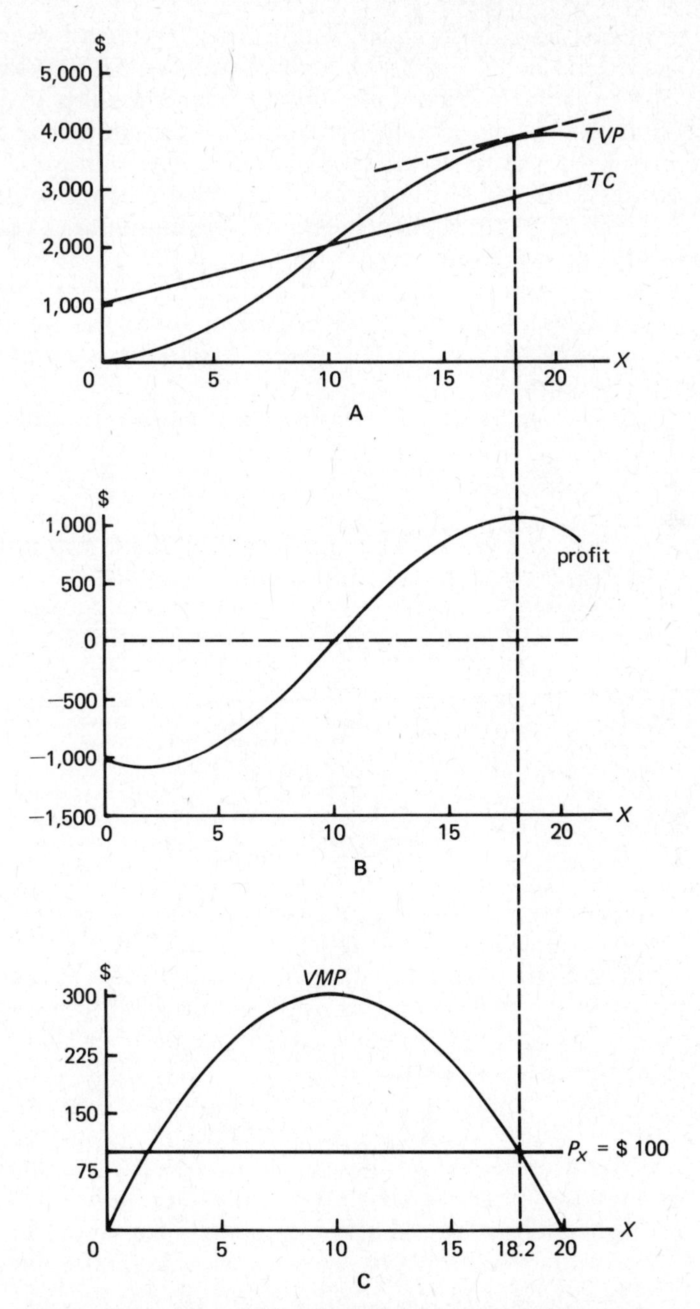

Figure 3–1. Determining the optimum amount of input using total curves, the profit curve, or marginal curves.

exceeds total cost and the vertical distance between the two is a maximum. This occurs in Figure 3–1A at 18.2 units of output—a slightly more accurate result than can be obtained from the tabular data.

A second way to determine the optimum amount of input is to consider profit directly as a function of input. The optimum amount of input occurs where profit is a maximum. Profits from column 3 of Table 3–1, are graphed in Figure 3–1B, and the maximum profit again occurs at 18.2 units of input.[2]

The marginal criterion for determining the optimum amount of input is derived, as might be expected, from the slopes of the total value product and total cost curves, when those curves are plotted as functions of the input, X.

First, consider the profit equation as a function of input:

$$Profit = P_Y f(X) - P_X X - TFC \tag{3.1}$$

To maximize this function with respect to the variable input, the first derivative would be set to zero as follows:

$$\begin{aligned} \frac{d\,Profit}{dX} &= P_Y \frac{dY}{dX} - P_X = 0 \\ &= P_Y MPP - P_X = 0 \end{aligned}$$

or

$$P_Y \cdot MPP = P_X \tag{3.2}$$

But the term $P_Y \cdot MPP$ is the slope of the TVP curve and is called the value of the marginal product (VMP). The term P_X is the slope of the total cost function. In pure competition, P_X will always be a constant. Profits will be maximized when the slope of the TVP function equals the slope of the TC curve.[3] In this example, remember that both are expressed as functions of input.

Figure 3–1A determines the optimum amount of input, which is again 18.2 units, where the slopes of TVP and TC are equal. Figure 3–1B depicts the profit equation; Figure 3–1C shows the solution where $VMP = P_X$. Figure 3–1C is based on data calculated in Table 3–2. The data in Table 3–2 show that VMP exceeds P_X for 18 units of X—suggesting again that 18.2 units of X would be more profitable. Again, the graph is more accurate.

[2]The fact that the break-even point occurs approximately at $X = 10$ (where MPP is a maximum) is coincidental; a change in fixed costs would shift the break-even point.

[3]It is possible, of course, that the solution could be a minimum rather than a maximum. To ensure that a maximum is obtained, the second derivative should be examined for the appropriate (negative) sign.

Table 3–2 Determining the Most Profitable Amount of Variable Input Using the Production Function (P_Y = $30 and P_X = $100)

(1)			(2)		(3)	
X	Y	MPP	$VMP_X = P_Y(MPP)$	P_X	$P_X(\Delta X)$	$P_Y(\Delta Y)$
0	0.0					
2	3.7	1.9	$ 57	$100	$200	$111
4	13.9	5.1	153	100	200	306
6	28.8	7.5	225	100	200	447
8	46.9	9.1	273	100	200	543
10	66.7	9.9	297	100	200	594
12	86.4	9.9	297	100	200	591
14	104.5	9.1	273	100	200	543
16	119.5	7.5	225	100	200	450
18	129.6	5.1	153	100	200	303
20	133.3	1.9	57	100	200	111
22	129.1	−2.1	−63	100	200	−126

Algebraically, the optimum amount of X can be calculated when the production function is known. The classical function used in Chapter 2 was

$$Y = X^2 - \frac{1}{30}X^3$$

Substituting its derivative into (3.2) gives

$$\$30(2X - \frac{1}{10}X^2) = \$100$$

where P_X = $100 and P_Y = $30. Solving $(0.1X^2 - 2X + 10/3 = 0)$ using the general quadratic formula gives[4]

$$X = \frac{2 + \sqrt{4 - (4/3)}}{0.2} = 18.2$$

[4]The second solution for X equals 1.8. While it, too, equates the slopes of *TVP* and the price ratio line (see Figure 3–1C), it results in negative profit. When allowance is made for the $1000 total fixed costs, the profits are −$1088.6 when X is 1.8 and $1088.7 when X is 18.2 (see Table 3–1). For profits to be maximum, the second derivative of the equation must be negative. Thus, $d^2Profit/dX^2$ is negative when $X = 18.2$ and positive when $X = 1.8$. When $X = 2$, Table 3–1 shows a loss of $1089. The difference is due to rounding; output when $X = 2$ is actually 3.7333 or 3 11/15 and profit is −$1088.

Equation (3.2) represents the main principle of the factor-product relationship; that is, the slope of the *TVP* curve has to equal the slope of the *TC* curve if maximum profit from the variable input is the objective. Dividing both sides of (3.2) by P_Y gives

$$MPP = \frac{P_X}{P_Y}$$

Therefore, another method of stating the marginal criterion is to say that the marginal product of the variable input (measured in physical units of output) must equal the inverse ratio of the number of units of input that can be purchased from the sale of one unit of output (as represented by the input-output price ratio). For example, if the unit price of an input is three times the unit price of output, then a unit of input must produce three units of output at the "margin."

The intuitive meaning of the precise calculus solution can be demonstrated by returning to the concept of finite increments. First, consider the slope of the total value product function

$$\frac{\Delta TVP}{\Delta X} = \frac{\Delta(P_Y Y)}{\Delta X} = P_Y \frac{\Delta Y}{\Delta X} = P_Y \cdot MPP = VMP$$

because P_Y will be a constant. Thus, the slope of the *TVP* curve represents the value of the increment in yield resulting from the use of one additional unit of output. (Remember that the average *MPP* is defined for one unit increments in input.)

Similarly, the slope of total cost as a function of input is given by

$$\frac{\Delta TC}{\Delta X} = \frac{\Delta TVC}{\Delta X} = \frac{\Delta(P_X X)}{\Delta X} = P_X \frac{\Delta X}{\Delta X} = P_X$$

because P_X is a constant. Thus, the slope of the total cost curve is P_X—the increment to total cost caused by using an additional unit of input.

When short-run profits are maximized, the manager will continue to increase the amount of the variable input used in the production process as long as the addition to revenue exceeds the addition to cost. When added returns equal added cost, the optimum is attained.

The value of the marginal product curve in Stage II is considered to be the producer's input demand curve. By equating *VMP* to the input price, the manager determines the amount of input to be purchased. As the input price changes, the producer's purchases move along the value of the marginal product curve. Changes in output price will cause this "demand" curve to shift.

Determining the Optimum Amount of Output. Using the data in Table 3–1, the optimum output was determined to be 18 units of input and the resulting output was 129.6. Using Figure 3–1, the optimum amount of input was found to be 18.2, the 0.2 difference being caused by the increased accuracy of the algebraic calculations and of the graphs.

The optimum output can also be determined directly by comparing total revenue and total cost at each output amount. *TR* and *TC* computed in Table 3–1 are graphed in Figure 3–2 as functions of output.[5] Because the farmer sells in a purely competitive market, *TR* is a straight line when plotted against output. Each additional unit sold adds exactly the same amount, P_Y, to total revenue. The shape of the *TC* curve is determined by the production function. A profit occurs whenever *TR* is greater than *TC*; a break-even point occurs at an output of 65 (compare with Figure 3–1B where profit is zero at an input of just under 10 units of *X*). Profit is the greatest where the vertical distance between *TR* and *TC* is the largest ($TR - TC$ equals a maximum). This point occurs at an output of 129.6 in Table 3–1 and is also determined to be 129.6 in Figure 3–2A.

Profit can also be considered directly as a function of output. Profits, from column 3 of Table 3–1, are graphed in Figure 3–2B: Again the maximum occurs at an output of 129.6. The solution is analogous to that presented above for inputs.

The marginal conditions for the maximization of profit as a function of output can be derived from the profit function. Now the variables in the equation must be regarded as functions of output. Profit would be

$$\begin{aligned} Profit &= TR - TC \\ &= P_Y Y - P_X X - TFC \\ &= P_Y Y - P_X f^{-1}(Y) - TFC \end{aligned}$$

where the concept of the inverse production function must be used to express X as a function of Y. That is, $X = f^{-1}(Y)$ in Stages I and II. Taking the derivative of profit with respect to Y results in

$$\frac{d\,Profit}{dY} = P_Y - P_X \frac{dX}{dY} = 0$$

[5]The revenue or value of output is called *total value product* (*TVP*) when expressed as a function of input and is called *total revenue* (*TR*) when expressed as a function of output. That is, although $P_Y \cdot Y = P_Y \cdot f(X)$, the first is called *TR* and the second *TVP*. In a similar fashion, total expenses are sometimes called total factor costs when expressed as a function of input and total costs (*TC*) when expressed as a function of output. But, because we have previously used the acronym *TFC* to represent *total fixed costs*, we have opted to use *TC* to represent total expenses in both contexts.

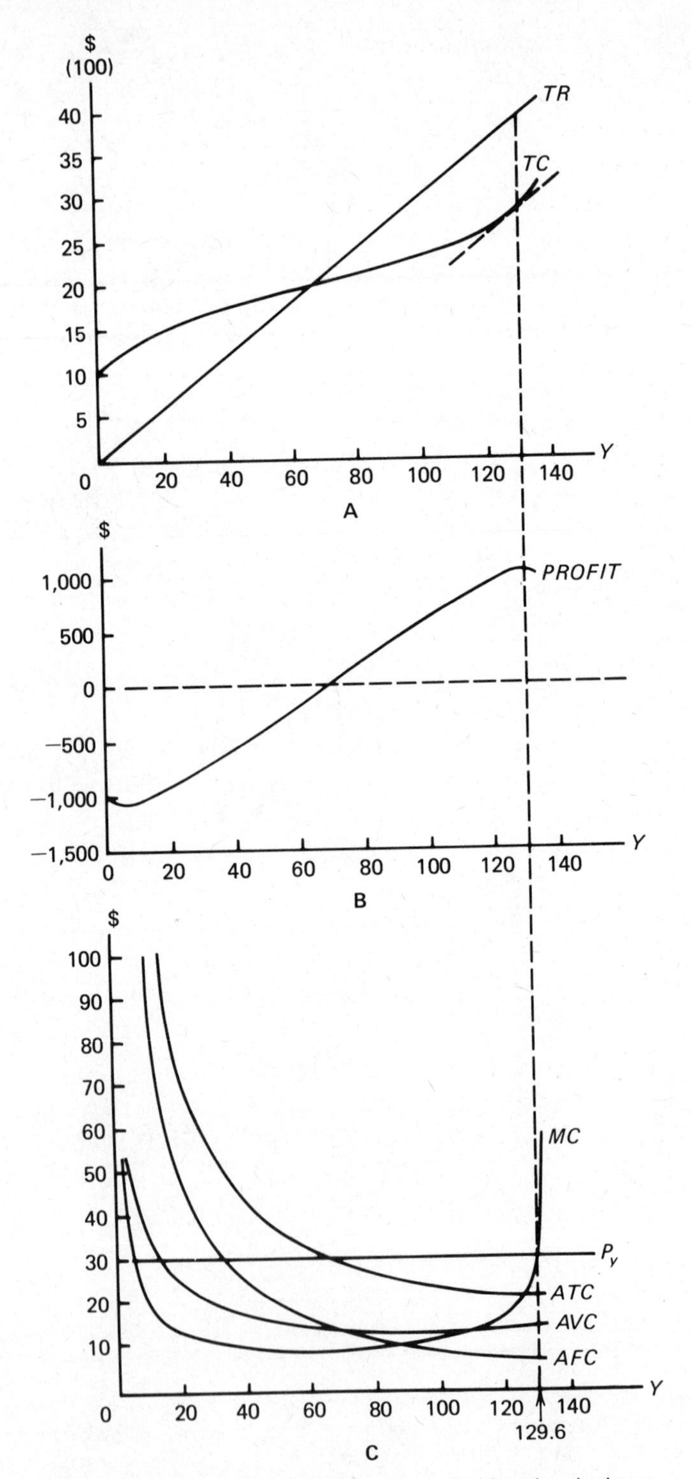

Figure 3–2. Cost and revenue curves for the classical production function.

where dX/dY is the derivative of the inverse function with respect to Y. But in Stages I and II

$$\frac{dX}{dY} = \frac{1}{\frac{dY}{dX}} = \frac{1}{MPP}$$

so the marginal condition can be written

$$\frac{d\ Profit}{dY} = P_Y - \frac{P_X}{MPP} = 0$$
$$= P_Y - MC = 0$$

Therefore, $P_Y = MC$ at the optimum. More generally, differentiation of the profit equation with respect to Y would give

$$\frac{d\ Profit}{dY} = \frac{dTR}{dY} - \frac{dTC}{dY} = 0$$

or

$$\frac{dTR}{dY} = \frac{dTC}{dY}$$
$$MR = MC$$

where the change in TR with respect to Y is defined as marginal revenue, MR, while the change in TC with respect to Y is, of course, the definition of marginal cost, MC. In pure competition, $MR = P_Y$.

Marginal costs and P_Y are graphed in Figure 3–2C. The price of output, P_Y, is represented by a straight line parallel to the horizontal axis and 30 units up the vertical ($) scale. Because the farmer sells in a purely competitive market, the output price is constant. MC and P_Y intersect at an output of 129.6; the vertical dashed line shows that optimum output is the same for both the total, the profit, and the marginal curves in Figure 3–2ABC.

The producer will use inputs as long as the selling price of the output is equal to or greater than average variable cost. When the price line, P_Y, falls below AVC, the manager is losing all fixed costs and a part of his variable costs. The optimum output always occurs where the price line, P_Y, intersects MC at a point on or above the AVC curve. This is the portion of MC that is in Stage II of the production function.

The cost curves in Figure 3–2C are relevant only for the specified cost of the variable input, $P_X = \$100$. For this cost, the lowest output price at which the farmer would produce is $13.3, the magnitude of

MC at the minimum AVC. This minimum will change when P_X changes. Under no circumstances would more than 133.3 units of output, the maximum, be produced.

In the short run, the farmer will produce as long as he is able to cover variable costs. For this reason, the marginal cost curve above average variable cost is considered to be the supply function for the producer. As price of the product, P_Y, increases or decreases, the optimum levels of output increase or decrease along the MC curve.

Comparison of Input and Output Criterion. All methods of determining the most profitable level of output or input lead to comparable answers.

Notice that the input criterion, $VMP = P_X$, can be written as follows:

$$P_Y \cdot MPP = P_X$$

$$P_Y \cdot \frac{\Delta Y}{\Delta X} = P_X$$

or

$$P_Y \cdot \Delta Y = P_X \cdot \Delta X$$

The expression $(P_Y \cdot \Delta Y)$ measures returns added by an increase in output while $(P_X \cdot \Delta X)$ measures the cost added by the increases in input. But, from the output criterion, $P_X = MC$, the same added cost—added return expression—can be derived:

$$P_Y = MC = \frac{P_X}{MPP} = P_X \frac{\Delta X}{\Delta Y}$$

or

$$P_Y \cdot \Delta Y = P_X \cdot \Delta X$$

Thus, the two methods of determining the optimum are comparable.

Marginal revenue and marginal cost refer to units of output. An added unit of output increases TR by the amount of MR; it increases TC by the amount of MC. MR and MC should never be used when referring to added returns or costs per unit of input. In a like manner, VMP and marginal factor cost (marginal expense of a unit of input) are never used when referring to added units of output. Another point to remember is that cost and total revenue curves are conventionally plotted with output on the horizontal axis. Again, this is done because managers are usually interested in costs per unit of output.

Short-Run Equilibrium. The firm operating where marginal cost equals marginal revenue is said to be in equilibrium. When fixed factors are present and the firm is operating in a short-run situation, this equilibrium is called a short-run equilibrium. Over a long period of time, when the firm is able to make changes in the fixed as well as the variable inputs, it is said to be in a long-run situation and attempting to attain a long-run equilibrium. All that is required in the short run is that variable inputs be used such that $MR = MC$; in the long run, all inputs must be used according to that criterion.

When pure competition exists in the markets, the individual farmer, either as a buyer of inputs or a seller of products, is unable to influence the market prices. Once the price is determined by the market, the only hope of the manager is to organize his productive resources in such a way that satisfactory profits can be made. If he is unable to do this, he will eventually go out of business. In this way the free enterprise system promotes productive efficiency.

In the short run, the manager's primary concern is staying in business. Thus, short-run equilibrium is determined by the output where $P_Y = MC$, as long as MC is equal to or greater than AVC. In the short run, the manager will operate even if he cannot cover fixed costs because these costs must be paid regardless of the amount of production. In the long run, a business that does not repay all costs cannot be operated; therefore, all costs must be paid in the long run.

In the short run, the firm may sometimes operate where $P_Y = MC$, but P_Y is greater than ATC. When the manager operates where $P_Y = MC = ATC$, all costs are being paid and the manager's return is exactly that needed to keep him in business in the long run. When P_Y is greater than ATC, the manager is making a profit in excess of that needed to keep him in business. This excess profit, sometimes called rent, accrues only in the short run. In the long run, additional producers begin supplying the product, existing producers expand output, the supply of the product is more plentiful, and its price falls. As a result, market prices facing the pure competitor are forced down to the point where $P_Y = MC = ATC$; total costs are covered, the manager is earning exactly that amount needed to induce him to stay in business, and because there is no excess profit, new producers are not tempted to enter into production.

Choice Indicators: Maximum Profit Versus Maximum Yield. The solutions presented above are based on the assumption that the manager's goal—his choice indicator—is to maximize profit. Once profit maximization is selected as a goal, the choice indicator automatically selects the input amount. Given the data, a computer could make the appropriate choice every time. The same is true of other choice indicators, such as yield maximization.

Maximum profits from an enterprise do not occur where physical output is a maximum. For the example in Table 3–1, output reaches a maximum of 133.3 at 20 units of input; profit at this point is only $999. *Therefore, the goal of maximum yield from a fixed input, such as bushels of corn per acre or pounds of milk per cow, is not compatible with the goal of maximum profit per unit of fixed input as long as there is a price tag attached to the variable input.* The reason, of course, is that the efficiency of production decreases in Stage II of the production function and, beyond the maximum profit point, the added inputs cost more than they are able to earn. Thus, the quest for maximum yields, while perhaps an admirable goal, is not consistent with income maximization unless variable inputs are free.

The difference between the concepts of maximizing profit and maximizing yield can best be shown on a graph of the production function. The objective of the factor-product relationship is to find the amount of X where the slope of TPP is equal to the price ratio, P_X/P_Y, when the goal is profit maximization. Figure 3–3 shows the production function and the price ratio line with a slope of 3.33, for $P_Y = \$30$ and $P_X = \$100$.

The price ratio line can be interpreted as an equal-profit line (Problem 3–13). It can be shifted up or down the Y axis, so long as the slope is maintained; an infinite number of profit lines will exist and all are parallel to each other, but only one will be tangent to the production function at the point where profits are maximized.

Profit or net return from the use of the variable input increases as the price ratio lines move higher. The price ratio line that is tangent to the total physical product curve is the highest attainable price ratio line. A higher line would not touch the production function. The line that is tangent intersects the Y axis at the highest attainable point and thus selects the amount of X consistent with the profit-maximizing objective.

The production function and the price ratio line are tangent at $X = 18.2$ in Figure 3–3. The tangency maximizes the distance OK on the Y axis, which represents the relative magnitude of profit before fixed costs from the variable input. (Distance KB represents the relative cost of the variable input.) Any deviation from the optimum amount of 18.2 units would decrease the distance represented by OK on the Y axis and thus lower profits.

If the objective of the manager were to maximize yield, the price ratio line would have to pass through point Z, where $X = 20$ and $MPP = 0$. If the objective were to maximize APP, then the price ratio line would shift downward to pass through point W where $X = 15$. Both of these downward shifts would reduce profit. Thus, maximum yield from the fixed input or maximum efficiency of the variable input are not compatible with maximizing short-run profits.

As an aside, one virtue of this particular exposition of the optimum is that it shows clearly the importance of relative prices. If P_X and P_Y

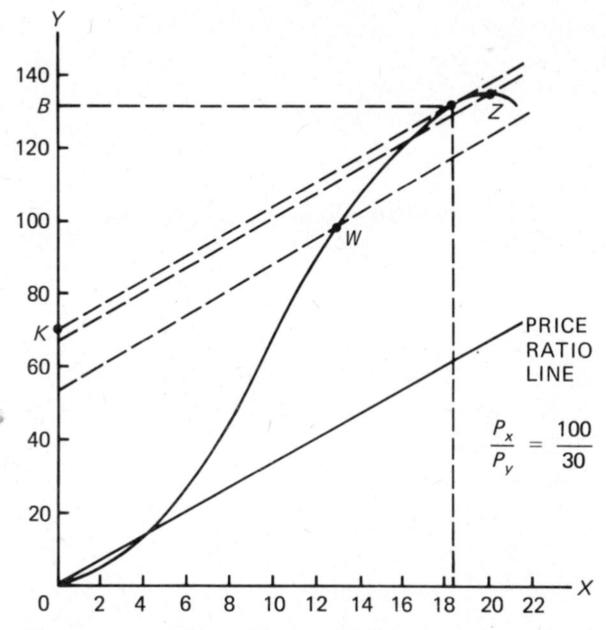

Figure 3–3. Determining the optimum amount of input using the total physical product curve and choice indicators.

both double, the slope of the price ratio line will not change. If P_X declines (increases) relative to P_Y, then the price line becomes flatter (steeper) with a numerically smaller (larger) slope. If the variable input is free, the price ratio line will be horizontal and maximum yield will become compatible with maximum profit.

An Empirical Example

Since the production function depicting corn yield response to fertilizer applications on an Iowa soil was presented in Chapter 2, the discussion has centered around theoretical concepts and hypothetical examples, apparently far removed from agriculture. On the contrary, however, these concepts represent the types of relationships inherent in agriculture and, as developed later in this book, provide the basis for both the problems and the prosperity of agriculture. A production function for corn, similar to that presented in Table 2–1 of Chapter 2, will be used to demonstrate the application of some of the principles developed thus far to an agricultural enterprise.

The estimated production function to be used is the quadratic form

$$Y = 65.54 + 1.084X - 0.003X^2 \tag{3.3}$$

where Y is corn in bushels and X is nitrogen in pounds.

This function was estimated using ten years of data obtained from fertilizer experiments on irrigated land at the Tribune Branch of the Kansas Agricultural Experiment station. Nitrogen rates were 0, 40, 80, 120, 160, and 200 pounds per acre. Soil moisture was maintained by irrigation. Consistent production practices were followed each year so that seasonal variations in yields resulting from environmental conditions were small.

Proper use of fertilizer increases the yield of the crop to the extent that it not only returns its own cost but other costs of production that might not be met without the fertilizer applications. Table 3–3, as an example, shows returns at various levels of nitrogen application assuming the price of corn at $2.75 per bushel and nitrogen at 20 cents a pound. Column 1 shows the level of nitrogen application, and columns 2 through 5 show, respectively, the estimated yields, the increase in yield, the value of additional increases in yield, and the cumulative net returns (profits) from various levels of nitrogen applied to corn. (Columns 4 and 5 can be modified for other nitrogen-crop price relationships.)

Table 3–3 illustrates the importance of nitrogen to corn on irrigated land in extreme western Kansas. Nitrogen helps to more than double corn yields under irrigation. The estimated yield of irrigated corn with no nitrogen is 65.54 bushels per acre; the first 20 pounds of nitrogen increase the yield of corn by 19.81 bushels. Successive 20-pound units of nitrogen keep increasing yields but by smaller and smaller increments. The 20 pounds of nitrogen in the 160–180 pound range boost the yield of corn only 0.90 of a bushel.

Table 3–3 Estimated Returns At Various Rates of Nitrogen Applications to Irrigated Corn per Acre (Corn Price $2.75 per Bushel and Nitrogen Price $0.20 per Pound)

(1) Nitrogen Application X (Pounds)	(2) Estimated Total Yields TPP (Bushels)	(3) Increase in Yields MPP (Bushels)	(4) Value of Additional Increase in Yield VMP ($)	(5) Cumulative Net Returns to Nitrogen (Profit)[a] ($)
0	65.54			
20	85.35	19.81	54.48	50.48
40	103.05	17.70	48.67	95.15
60	118.36	15.31	42.10	133.25
80	131.26	12.90	35.47	164.72
100	141.76	10.50	28.87	189.59
120	149.86	8.10	22.27	207.86
140	155.56	5.70	15.67	219.53
160	158.87	3.31	9.10	224.63
180	159.77	0.90	2.47	223.10

[a]Value of additional yield from nitrogen, column 4, minus the cost of nitrogen.

Column 5 in Table 3–3 shows cumulative net returns to nitrogen per acre. The returns keep increasing up to 160 pounds of nitrogen. At those nitrogen levels with the assumed prices, nitrogen returns $224.63 per acre, about $7 for each dollar spent for nitrogen on the average.

Variations in relative corn and fertilizer prices affect the returns to fertilizer as well as the suggested optimum fertilizer application. This can be shown by recalculating data in Table 3–3 using different nitrogen-corn price relationships. Table 3–4 shows the optimum amounts of nitrogen under different prices for corn and nitrogen.

For example, the data in Table 3–4 suggest that 168 pounds of nitrogen would be used when corn is $2 a bushel and nitrogen 15 cents a pound. For other corn-nitrogen price relationships, other amounts of nitrogen are recommended. If corn were $2.50 a bushel and nitrogen 30 cents a pound, for example, 161 pounds of nitrogen is suggested. The data, pounds of nitrogen per acre, in Table 3–4 were estimated from production function equation (3.3) using the $VMP = P_X$ criterion. For example, the *MPP* equation from (3.3) is,

$$\frac{dY}{dX} = 1.084 - 0.006X \tag{3.4}$$

When price of corn is $2.50 per bushel and price of nitrogen 25 cents per pound then

$$1.084 - 0.006X = \frac{0.25}{2.50} \tag{3.5}$$

or

$$X = \frac{2.71 - 0.25}{0.015} = 164$$

Table 3–4 Optimal Amounts of Nitrogen per Acre for Different Corn and Nitrogen Prices

Nitrogen Price per Pound of N ($)	*Corn Price per Bushel ($)*				
	1.50	*2.00*	*2.50*	*3.00*	*3.50*
0.05	175	176	177	178	178
0.10	170	172	174	175	176
0.15	164	168	171	172	174
0.20	158	164	167	170	171
0.25	153	160	164	167	169
0.30	147	156	161	164	166
0.35	142	151	157	161	164
0.40	136	147	154	158	162

The 164 pounds of nitrogen per acre meet the economic efficiency when the criterion or choice indicator is $VMP = P_X$ or when the P_X/P_Y ratio is 0.10. This is true not only when P_X = \$0.25 and P_Y = \$2.50 but also when P_X = \$0.15 and P_Y = \$1.50; P_X = \$0.20 and P_Y = \$2.00; P_X = \$0.30 and P_Y = \$3.00; and when P_X = \$0.35 and P_Y = \$3.50, indicating that in resource allocative decision making the relative prices are as important as absolute prices (see Table 3–4). Lower P_X/P_Y price ratios call for larger amounts of nitrogen and higher P_X/P_Y ratios call for smaller amounts if net returns or profits from nitrogen are to be maximized.

If the objective were to maximize the total physical product or the yield per acre, the criterion or choice indicator would be $MPP = 0$. The amount of fertilizer meeting this objective is 180.7 pounds per acre.

Derived Demand for Inputs

Production functions can be used to derive the farmer's demand curves for inputs. Just as the marginal cost curve above the average variable cost curve serves as a supply function for the product, the value of the marginal product curve can be interpreted as the demand schedule for the variable resource. In general, the "derived" demand curve for the variable input is the *VMP* curve in Stage II.

Whenever returns to variable input are decreasing, the value of marginal physical product curve is sloping downwards to the right because of the diminishing effectiveness of nitrogen as more of it is used. The negative slope denotes that larger quantities of input will be purchased only at lower input prices.

Figure 3–4 represents a more general derived demand curve for

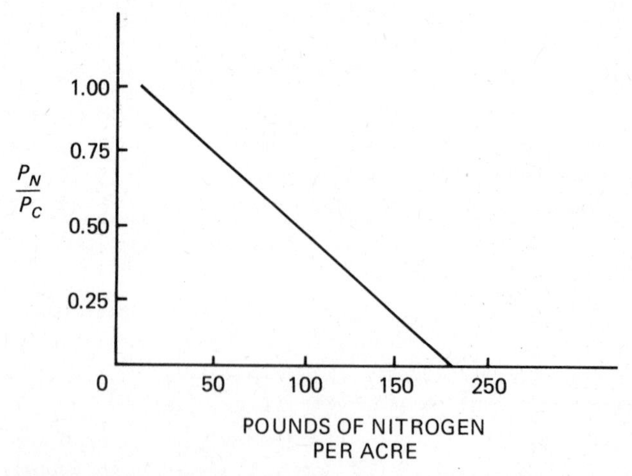

Figure 3–4. Derived demand curve for nitrogen on irrigated corn in western Kansas.

nitrogen on irrigated corn land. It is based on data presented in Table 3–3 and equation (3.5); it expresses quantity demanded as a function of the nitrogen-corn price ratio. The nitrogen-corn price ratios are plotted on the vertical axis and the rates of nitrogen on the horizontal axis. Thus, when P_X (nitrogen) is 15 cents per pound and P_Y (corn) is 1.50 dollars per bushel, the P_X/P_Y ratio is 0.10 and the optimum rate of nitrogen is 164 pounds per acre. This rate does not change as long as the P_X/P_Y ratio remains at 0.10. If both prices double, the price ratio stays the same, $0.30/3.00 = 0.10$, and the optimum rate of nitrogen remains at 164 pounds per acre. If the price of nitrogen increases relative to that of corn, the price ratio becomes larger and consequently the quantity of nitrogen demanded decreases (see Figure 3–4 and Table 3–4).

FARM INCOME AND COSTS

In this and the previous chapter we have discussed production functions, cost curves, and the choice of the optimal amount of a variable input. In this section, these concepts will be integrated with traditional farm budgeting techniques. A further discussion of costs will be included.

Table 3–5 presents a complete budget for a typical farm enterprise—irrigated corn in western Kansas. This budget is for one acre, but the same concepts could be used to develop a budget for a farm with several enterprises.

Net Farm Income

Total revenue, $412.50 per acre, is computed in line one of Table 3–5. This represents an average of the revenue to be expected over a period of time; annual revenues will vary about this expected amount. Cash earnings are the total incoming cash flow for the year. Total revenue and cash earnings are identical for this example but need not be in general. For example, the farmer who sells 99 hogs and keeps one to eat has the same total revenue as the farmer who sells 100 hogs. The total cash receipts of the two farmers differ, however, by the value of one hog. The hog kept for home consumption represents a noncash portion of total revenue. Other noncash items, such as value of firewood and garden products, are relevant when computing total revenue for a farm. Noncash returns are often difficult to estimate accurately, but should be included if possible when computing the total revenue for a farm.

Costs in Table 3–5 are categorized as fixed and variable. These have been defined in Chapter 2. Variable costs are listed in lines 2 to 12 and fixed costs in lines 13 to 17. In addition, costs, like revenue, can be either cash or noncash. Cash costs are incurred when resources are purchased and used immediately in the production process. For the most part, cash costs result from purchases of nondurable inputs

Table 3–5 An Example of Total Cost and Returns Calculated for Irrigated Corn per Acre

Total Revenue	*Per Acre*
1. Total Value Product	
(150 bushels/acre × $2.75/bushel)	$412.50
Variable Costs	
2. Labor (6 hours × $3.00)	$ 18.00
3. Seed (17 # × $.76)	13.00
4. Herbicide and Insecticides	15.00
5. Fertilizer (125 pounds × $0.20 per pound)	25.00
6. Fuel and oil	4.00
7. Machinery and equipment repairs	6.00
8. Crop Insurance	0.00
9. Drying ($.10/bu.)	15.00
10. Custom harvest ($.15/bu.) and hauling ($.10/bu.)	37.50
11. Miscellaneous	.50
12. Total variable costs	$134.00
Fixed Costs	
13. Real estate taxes	4.00
14. Taxes and insurance	
Machinery ($42,000 × 0.5%)/676	0.31
Irrigation equipment ($20,400 × 0.5%)/160	0.64
15. Depreciation	
Crop machinery ($42,000 × 12.5%)/676 acres	7.77
Irrigation equipment ($20,400 × 7%)/160 acres	8.93
16. Interest on capital investment	
Land ($500 per acre × 6%)	30.00
Machinery ($42,000 × 6%)/676	3.73
Irrigation equipment ($20,400 × 6%)/160	7.65
17. Total fixed costs	$ 63.03
Income	
18. Net cash income (line: 1–12–13–14)	$273.55
19. Net income per acre (Line: 18–15)	$256.85
20. Income to labor and management (line: 19–16)	$215.47

Source: Adapted from Bogle, T. Roy and Overley, Frank, *Irrigated Corn Production in Western Kansas*, KSU Farm Management Guides, Department of Economics, Cooperative Extension Service, Kansas State University, Revised, September 1974.

such as gasoline, labor, or grease, which do not last more than one production period.

Cash expenses can be determined directly from the farmer's account books. For the budget in Table 3–5, cash costs include all variable costs, real estate taxes, machinery taxes, and insurance. For each acre of corn, the farmer must pay out a total of $134 + $4.00 + $0.95 = $138.95 from cash flow.

Net cash income from the production period is determined by subtracting cash expenses from total cash receipts. The corn farmer thus has a net cash income of $412.50 − $138.95 = $273.55 per acre.

This is not "pure" profit, however, because noncash costs have yet to be considered.

Noncash costs consist of depreciation and payments to resources owned by the farmer. Depreciation represents an attempt to spread the investment cost or purchase price of durable inputs over their productive lifetime. A durable input, such as a tractor, can be used profitably for 15 years. But when purchased, the tractor must be purchased in one piece, not 15 pieces. To charge the total purchase price of the tractor as a cost during the year of its purchase would distort the cost structure of the farm for the first year and for the next 14 years. Profits would appear too low the first year and too high thereafter. To remedy this, an annual depreciation charge should be levied against the farm expenses each year of the tractor's productive lifetime. When the tractor is worn out, it should be completely "paid for" by depreciation, thus insuring that capital will be available to purchase another. When farmers are unable to make depreciation payments because of low farm earnings, they are said to be "living off their investment." They use the depreciation fund to pay annual living and farming expenses and are unable to replace durable equipment when it wears out.

Depreciation of durable resources used only on one enterprise can be charged directly to that enterprise; depreciation should be charged each time the production process is repeated. On a farm, durable assets such as tractors, trucks, and combines are commonly used in several enterprises; that is, they are involved in several production processes. In such cases allocation of depreciation among the enterprises is difficult, and the depreciation can only be charged against the farm business as a whole.

The corn farmer in the example has $42,000 in crop machinery and $20,400 invested in irrigation equipment. Depreciation on these items are estimated at 12.5 and 7 percent, respectively. The farmer is assumed to have 676 acres of cropland, 160 of it irrigated, so he will incur machinery depreciation costs of $7.77 per acre and irrigation equipment depreciation costs of $8.93 per acre (line 15 in Table 3–5). He must, in effect, save that much each year or reinvest it in machinery and equipment, or he will eventually find himself with worn-out equipment and no cash reserves to replace it. His actual cash income must be reduced by $16.70 per acre to determine what is ordinarily called net income per acre, $273.55 − $16.70 = $256.85. When this concept, cash income minus depreciation, is carried from a one-acre budget to a budget for an entire farm, it is called net farm income.

Opportunity Costs

Net farm income is the source of payments to the farmer's labor, management, and capital. Before considering these payments, we must digress to explain one of the most important ideas in economics—

opportunity costs. Every resource used in the productive process has but one true cost: its opportunity cost. The opportunity cost of a resource is the return the resource can earn when put to its best alternative use. Suppose a farmer has a ton of commercial fertilizer. Suppose further that spreading it on his wheat field will add $150 to the total revenue from wheat, while spreading it on his barley field will add $100 to the total revenue from barley. If he fertilized the barley, his opportunity cost is $150; he has foregone $150 to earn $100. If he fertilizes the wheat, his opportunity cost is $100; he has foregone $100 to earn $150. The most return from a unit of input is realized when the actual earned return is equal to or greater than the opportunity cost.

The best use of the fertilizer in the example just mentioned was to fertilize wheat. No mention was made of the cost of the fertilizer, and, indeed, none is needed. Once fertilizer is purchased and committed to the farm, its cost is irrelevant to the problem. It is a sunk cost. Whether the fertilizer cost $100 or $1000, its best use is on wheat. Suppose now that the purchase price of the fertilizer is $110 per ton, and the farmer has not yet purchased it. He may now keep the $110, buy and fertilize wheat, or buy and fertilize barley. If he fertilizes wheat, his opportunity cost is $110, for he could have had that much by not buying anything. If he fertilizes barley, his opportunity cost is $150.

Now consider payments to the farmer's owned resources. Farmers own many durable resources, such as machinery or land, used in the production process. Once purchased these resources are used over several production periods. Even though they do not have a purchase price each year, these resources do have an opportunity cost—the money they would earn from their best alternative use. The alternative uses of many of the items on a farm, such as plows, fences, or trucks, may be difficult to evaluate. Their opportunity costs could be evaluated directly, but a much easier convention is usually adopted. A dollar value is placed on all resources owned by the farmer; that is, the total amount of his investment in land, machinery, buildings, tools, and other items is determined. The opportunity cost of these items is then defined as the amount the total capital on the farm could earn if invested in its best alternative use. This opportunity cost is called interest on investment.

The corn farmer has $500 invested per acre of land, while his machinery investment is $42,000/676 = $62.13 per acre, and his irrigation equipment costs $20,400/160 = $127.50 per acre. At 6 percent, the opportunity cost of this investment is $30.00 + $3.73 + $7.65 = $41.38 per acre (line 16 of Table 3–5). Therefore, the acre of corn must return the farmer, in addition to other costs, $41.38 per acre. If it does not, he should consider alternative investments; the $41.38 is the opportunity cost of his capital.

Family labor represents another noncash cost. When budgeting, charges can be made for family labor used on the farm. These charges represent salaries the farm operator and his family could earn if employed off the farm. Just as with interest on investment, these off-farm salaries represent an opportunity cost. Farmers can also charge themselves the opportunity cost of their management services. If the opportunity cost of family labor and management is zero, then no cost for these items should be charged against the business. Rather than attempt to estimate the opportunity cost of the farmer's labor and management, the budgeter often labels the residual earnings "Labor and Management Income." The farmer can then decide for himself whether the return is satisfactory to him.

Allocation of Limited Capital Among Variable Inputs

Farmers who have a strong financial position, an "unlimited" access to capital, apply variable inputs to all fixed factors in quantities such that the added marginal returns from the last unit of the input just equal their cost ($VMP = P_X$). But many farmers do not have enough capital (cash plus credit) to purchase the quantities of the variable inputs needed to make the most profitable application for all technical units or enterprises. For them, the application rates of inputs is not a simple factor-product problem to be considered in isolation for each enterprise. Instead, the problem is to decide how the limited amount of variable input should be allocated to best enhance the growth of the farm business. In this situation, each input competes with the other inputs for the limited amount of capital. In this type of competition, the variable input is used in each enterprise in such a way that the marginal returns per dollar are the same for the input in each use, that is, $VMP_{Y_1} = VMP_{Y_2} = \ldots. = VMP_{Y_N}$, where the Y_i represent different enterprises. In this case, the marginal returns will be equal but must also equal or exceed input cost, just as in the case when capital is more plentiful.

As the farmer accumulates capital, additional funds will become available to enable the farmer to apply the optimum amount of variable input on all technical units or enterprises. For example, the optimum applications of fertilizer per acre will return greater net revenue to the farm when the total acreage is fertilized. But when the manager is unable to apply the optimum per acre to the total acreage, then he should apply less than the optimum on all acres but equate marginal returns among all acres. This assumes that input use takes place in Stage II. Allocation of limited inputs among competing enterprises is considered further in Chapter 5.

Profit

Profit is total revenue minus total cost. Care must be taken when interpreting accounts to determine the appropriate amount of profit. Because noncash items may be included in revenues or costs, profit is not the same as net cash receipts. Also, the income to labor and management listed in Table 3–5 is not profit to the farmer. To compute pure profit from this figure the farmer must subtract the value of labor he has contributed to the farm business as well as the value of his management. But this pure profit is not the farmer's personal income. His personal income is the return on all owned resources including interest on investment, returns to labor and management, and pure profit. When the farmer owns his land debt-free, the opportunity costs represent personal earnings; if the land were mortgaged, a portion of these returns would be used to meet principal and interest payments on the land, thus reducing income to the farmer.

What is pure profit? Total cost, you remember, included payments to all resources used in production. These payments were based on the opportunity costs of the resources and were therefore large enough to keep the resources committed to the farm business. Interest on investment, representing the opportunity cost of durable resources, depreciation, cash costs of nondurable resources, family labor, and management charges are included in costs so that even if pure profit is zero, the farmer is paid for the use of all of his resources and will have no incentive to shift them to other uses. Profit, then, represents the surplus earnings remaining after all resources are "paid."

There are several views on the function and meaning of profit. Some believe profit is the reward to the entrepreneur for visualizing and organizing a production process—that it is due him because of his enterprise and ability to innovate. Another view is that profit is a reward for risk taking. Profit is thus viewed as the entrepreneur's reward for the mental and physical wear and tear incurred by organizing and seeing through a production process. A third view is that profit is the reward for natural or contrived scarcity of resources. The farmer may earn a high profit because his soil is both highly productive and scarce relative to other agricultural soils. This type of scarcity is due to nature, and in a free-enterprise economic system those people either lucky or smart enough to own scarce resources are allowed to retain the profit earned by those resources. Profits from contrived scarcities occur when production is deliberately limited, that is, when a productive resource is held out of production and a shortage of the product occurs so that market values of products are thereby forced up and profits increased.

Problems and Exercises

3–1. Consider the production function $Y = X^{1/2}$. (a) Find the exact *MPP* equation. (b) What is the exact *MPP* where $X = 4$, 16, and 25 units? (c) At how many units of X are the net returns a maximum when $P_X = \$1$ and $P_Y = \$4$? (d) Derive and plot demand function for X when $P_Y = \$4$.

3–2. A potato grower has the following production function:

No. of 500-lb Units of Mixed Fertilizer per Acre	*Total Production Bushels per Acre*
0	0
1	103
2	174
3	223
4	257
5	281
6	298

If potatoes sell for $2.00 a bushel, derive the farmer's demand schedule for fertilizer.

3–3. The farmer has 10 acres with a production response as shown in Problem 3–2. He has another 5-acre field with a production response for potatoes as follows:

No. of 500-lb Units of Mixed Fertilizer per Acre	*Total Production Bushels per Acre*
0	220
1	380
2	460
3	494
4	506
5	510

Potatoes sell for $2.00 a bushel in the field. In a rationing program the farmer is allotted 27,500 pounds of fertilizer. How should he divide it between the two fields to maximize his output? His profit?

3–4. Consider the production function $Y = 70 + 2X - 0.02X^2$. (a) Find the level of X at which Y is a maximum. (b) Calculate elasticities of production, *APP*, and *MPP* when $X = 10, 20, 30, 40$, and 50. (c) Find levels of X maximizing net returns when $P_X = \$1, P_Y = \$1$; $P_X = \$1, P_Y = \$2$; $P_X = \$1, P_Y = \$4$; $P_X = \$1, P_Y = \$10$.

3–5. Consider the production function $Y = 2X^{1/3}$ and $P_X = \$8$. Relate costs to output (Y) and derive the total cost, marginal cost, and average cost functions.

3–6. Consider the total cost function $TC = 120Y - Y^2 + 0.02Y^3$. (a) What level of Y yields minimum cost per unit of Y? (b) What is the relationship between *MC* and *AVC* at the above level of output? (c) Does this level of output yield maximum net returns when P_Y is $108?

3–7. Consider the production function (Problem 2–5) $Y = 37 + 0.8X - 0.001X^2$. Determine the optimum use of X when P_X/P_Y ratio is 0.20, 0.40, 0.60, 0.80 and 1.00.

3–8. Consider two production functions for products Y_1 and Y_2 where $Y_1 = 10 + 2X - 0.1X^2$ and $Y_2 = 5 + 4X - 0.2X^2$; $P_{Y_1} = \$2$ and $P_{Y_2} = \$1$. How will you allocate 10 units of X among the two enterprises?

3–9. Consider the production function $Y = 68 + 0.8X - 0.002X^2$ where Y is grain sorghum in bushels per acre and X is nitrogen in pounds per acre. (a) Find optimum pounds of nitrogen per acre to apply under different crop and nitrogen prices and note the amount in appropriate blank columns in the following table.

Nitrogen Price per Pound of N ($)	*Sorghum Price per Bushel ($)*				
	1.50	*2.00*	*2.50*	*3.00*	*3.50*
0.00	—	—	—	—	—
0.20	—	—	—	—	—
0.40	—	—	—	—	—
0.60	—	—	—	—	—
0.80	—	—	—	—	—

(b) At what P_X/P_Y ratio would the net returns from nitrogen be zero?

3–10. Let "pure competition" be used to describe a situation in which there are many producers and a homogeneous product. (a) In the short run, how does a producer in a perfectly competitive industry maximize his profit? (b) Under what conditions will he produce at a loss? (c) Under what conditions will he stop producing temporarily?

3–11. Minimum unit cost of a product is not always an indicator of maximum net returns. Discuss and illustate.

3–12. Consider the *VMP* curve for the classical production function. Suppose the price of the variable input were to increase until $P_X = VMP$ in Stage I (actually, at two points in Stage I). Would it then be profitable for the manager to operate in Stage I? Why? Illustrate using the data in Table 3–2.

3–13. Referring to Figure 3–3, why does the distance *OK* represent the relative magnitude of profit (before fixed cost) while *KB* represents the relative magnitude of variable costs? Why are they relative, that is, relative to what? (Hint: Examine the general form of the family of price lines used in Figure 3–3, $Y = K + (P_X/P_Y)X$ where *K* is the parameter which, when changed, shifts the price line up and down.)

3–14. Show that P_Y = marginal revenue = average revenue in pure competition. Then refer to Figure 3–2C. (a) The area of what rectangle represents total revenue on the graph in Figure 3–2C? Why? (b) The area of what rectangle represents total cost? Total variable cost? Profit? Why? (c) Do rectangles falling under the *MC* curve [such as you found in the answer to (a) and (b)] have any interpretation? What would the area under *MC* represent?

3–15. Show that P_X = average cost per unit of input = marginal cost of a unit of input in pure competition. On a figure such as Figure 3–1C: (a) The area of what rectangle represents total variable cost? (b) If average value product, $VAP = P_X \cdot APP$, were placed on the graph, what would be the interpretation of the area of a rectangle with sides on the *X* and *Y* axes and with the northeast corner positioned exactly on *VAP*? (c) On a graph of *VMP*, *VAP*, and P_X, how could profit before fixed cost be depicted?

3–16. In what sense can P_X be interpreted as the opportunity cost of a unit of variable input? If all variable costs are held as cash at the beginning of the production process, should each input actually earn $P_X\ (1 + i)$ where i is the opportunity cost of capital over the time period of the production process?

Suggested Readings

Baumol, W. J. *Economic Theory and Operations Analysis.* Englewood Cliffs, N.J.: Prentice-Hall, Inc., 1961, Chapter 11.

Cohen, K. J. and Cyert, R. M. *Theory of the Firm.* Englewood Cliffs, N.J.: Prentice-Hall, Inc., 1965, Chapters 3 and 7.

Cramer, Gail L. and Jensen, Clarence W. *Agricultural Economics and Agribusiness,* Second Edition. New York: John Wiley and Sons, Inc., 1982, Chapters 4 and 6.

Dillon, John. *The Analysis of Response in Crop and Livestock Production.* Oxford: Pergamon Press, 1968, Chapters 1 and 2.

Ferguson, C. E. and Gould, J. P. *Microeconomic Theory,* Fourth Edition. Homewood, Ill: Richard D. Irwin, Inc., 1975, Chapters 8 and 13.

Heady, Earl O. *Economics of Agricultural Production and Resource Use.* New York: Prentice-Hall, Inc., 1952, Chapters 4 and 11.

Hicks, J. R. *Value and Capital,* Second Edition. Oxford: Oxford University Press, 1953, Chapter 6.

Leftwich, Richard H. *The Price System and Resource Allocation,* Third Edition. Hinsdale, Ill: Holt, Rinehart, and Winston, 1966, Chapter 7.

Mansfield, E. *Microeconomics,* Third Edition. New York: W. W. Norton and Company, Inc., 1979, Chapters 6 and 7.

Mueller, A. G. and Hinton, R. A. "Farmers' Production Costs for Corn and Soybeans by Unit Size." *American Journal of Agricultural Economics,* Volume 57, December 1975.

PRODUCTION WITH TWO OR MORE VARIABLE INPUTS

CHAPTER 4

The previous two chapters developed concepts of economic analysis basic to the factor-product relationship. This elementary production process results when one input is varied and all remaining inputs are held constant.

In this chapter the fundamental relationships among one output and two or more variable inputs will be developed. The principles presented will build on those developed earlier. As before, the assumption is made that the farm manager buys inputs and sells products in purely competitive markets. Prices of inputs and outputs are determined by economic forces beyond the control of the manager and can be regarded by him as given. Second, the assumption is made that at least one productive input is fixed in quantity so that all production processes discussed in this chapter are short-run situations to which the law of diminishing returns is applicable. If all inputs are variable, the law of diminishing returns no longer holds, the short run is replaced by the long run, and the economic interpretation of the relationship among inputs and output differs somewhat from that presented in this chapter. A development of the long run is presented in Chapter 6.

In the factor-product relationship a given level of output can be produced in only one way. Thus, in Table 2–2, 2 units of X, when combined with the fixed factors, produce 3.7 units of Y, 10 units of X produce 66.7 units of Y, and so on. When two or more inputs are variable, a given amount of output may be produced in more than one way. This is particularly true in agriculture when a product can be produced using many different combinations of inputs. Substitution possibilities among inputs or factors of production create what is often called the *factor-factor* relationship. The factor-factor relationship adds an extra dimension to decision making. As the number of alternative ways of producing a given product increases, the challenge to the manager to choose the combination of inputs that meet his or society's objectives also increases. In a developing economy, the number of inputs and alternative ways of production increase continually, increasing the difficulty of decision making.

To simplify the analysis in the beginning, the special case of one output resulting from two variable inputs will be discussed. The production function will be discussed conceptually and presented geometrically. The basic problem facing the manager when more than one variable input is used, that of finding the "right" combination of inputs, will be analyzed. The optimum or most profitable combination of inputs will be determined. Finally, the analyses developed in detail for two inputs will be generalized to include any number of variable inputs. Mathematical and empirical examples will be presented.

AN EXAMPLE FROM AGRICULTURE

Chapter 2 began with an example of a production function depicting corn yield response to fertilizer. To begin the study of yield response to two variable inputs, an empirical example will again be presented.

Table 4–1 contains the response of corn yields to varying amounts of applied nitrogen and plant population. These data were collected from a controlled experiment conducted in northern Missouri. Seven rates of nitrogen and four rates of plant population were applied to experimental plots located on Seymour silt loam near Spickard, Missouri. The resulting yields were measured and used to estimate the "smooth" production function contained in Table 4–1.

The production function in Table 4–1 can be regarded as a series of individual or "sub-production" functions. For example, the first row of the table depicts corn yield response to increasing numbers of plants when nitrogen is held fixed at zero. Other rows in the table represent yield response to plant populations for higher levels of applied nitrogen. In a like manner, each column represents corn yield response to nitrogen given a fixed number of plants. In total, ten sub-production functions are contained in Table 4–1. All yields represent the corn

Table 4–1 Corn Yield Response in Bushels per Acre to Two Inputs, Pounds of Applied Nitrogen and Number of Plants Per Acre, North Missouri Research Center, 1962

Pounds of Nitrogen Applied per Acre	*Number of Plants per Acre*				
	9,000	*12,000*	*15,000*	*18,000*	*21,000*
0	50.6	54.2	53.5	48.5	39.2
50	78.7	85.9	88.8	87.5	81.9
100	94.4	105.3	111.9	114.2	112.2
150	97.8	112.4	122.6	128.6	130.3
200	88.9	107.1	121.0	130.6	135.9

Source: Kroth, Earl M. and Doll, John P., *Response of Corn Yields to Nitrogen Fertilization and Plant Population in Missouri, 1962*, Progress Report No. 2, Missouri Agricultural Experiment Station Special Report 27, Columbia, Missouri, 1963. The treatment means for the experiment are contained in Table 3 of Special Report 27. The predicted values in Table 4–1 were computed using regression procedures.

yields that would result from the use of the combinations of nitrogen and plant population shown, one acre of land (the technical unit), along with a host of fixed inputs, including weather, not described here.

The yields resulting from two variable inputs are not as readily depicted on a graph as are those resulting from one variable input. In general, a three-dimensional drawing is needed. One way to picture this type of production situation is to consider only the input levels contained in Table 4–1 and draw a picture as in Figure 4–1A. For each level of nitrogen and plant population a block is drawn to represent corn yield. The height of the block is determined by the amount of the corn yield. By placing these blocks in tiers, the corn yield response to both inputs can be visualized. In total, 25 blocks would be needed to trace out response to the 25 input combinations. To avoid clutter, only a few of these are presented in Figure 4–1A.

Figure 4–1A represents yields resulting from 0, 50, 100, 150, and 200 pounds of nitrogen and 9000, 12,000, 15,000, 18,000, and 21,000 plants per acre. But these inputs can be used at any rate because they are perfectly divisible. The farmer may apply 47, 175, 20, 199, or any other rate of nitrogen; the same is true of plant population. Thus, the method of picturing yield response shown in Figure 4–1A is not only cumbersome but does not depict yields resulting from all possible input combinations. In reality, the yields in Table 4–1 represent but 25 points through which a smooth curve can be drawn. Such a "surface" is depicted in Figure 4–1B.

The base of Figure 4–1B measures the units of plants and nitrogen; these units mark off a grid similar to a checkerboard. The intersections of the lines on the base represent an input combination. The point immediately above each intersection represents the corn yield resulting from the input combination. For example, to find the corn yield resulting from 100 pounds of nitrogen and 15,000 plants per acre, locate in the base grid the intersection of the lines marked 100 and 15,000, respectively, and then place a perpendicular line segment (such as a flagpole) on the intersection so that the segment exactly touches the surface above. The height of this perpendicular line segment represents the resulting corn yield, in this case 111.9 bushels. In Figure 4–1B, this point occurs exactly where the two sub-production functions (one resulting from varying stand while nitrogen is fixed at 100 pounds per acre, and the other from varying nitrogen while stand is fixed at 15,000 plants per acre) intersect in space. By imagining that a perpendicular is raised from each point of intersection in the base grid and that a smooth sheet of, say, plastic is stretched over the tops of the perpendiculars (flagpoles), the resulting production surface can be developed. In Figure 4–1B, ten curves, each representing a sub-production function from Table 4–1, are interlaced to depict the general shape of the surface; points in between are not depicted but are imagined to exist. The surface thus displays the corn yields resulting

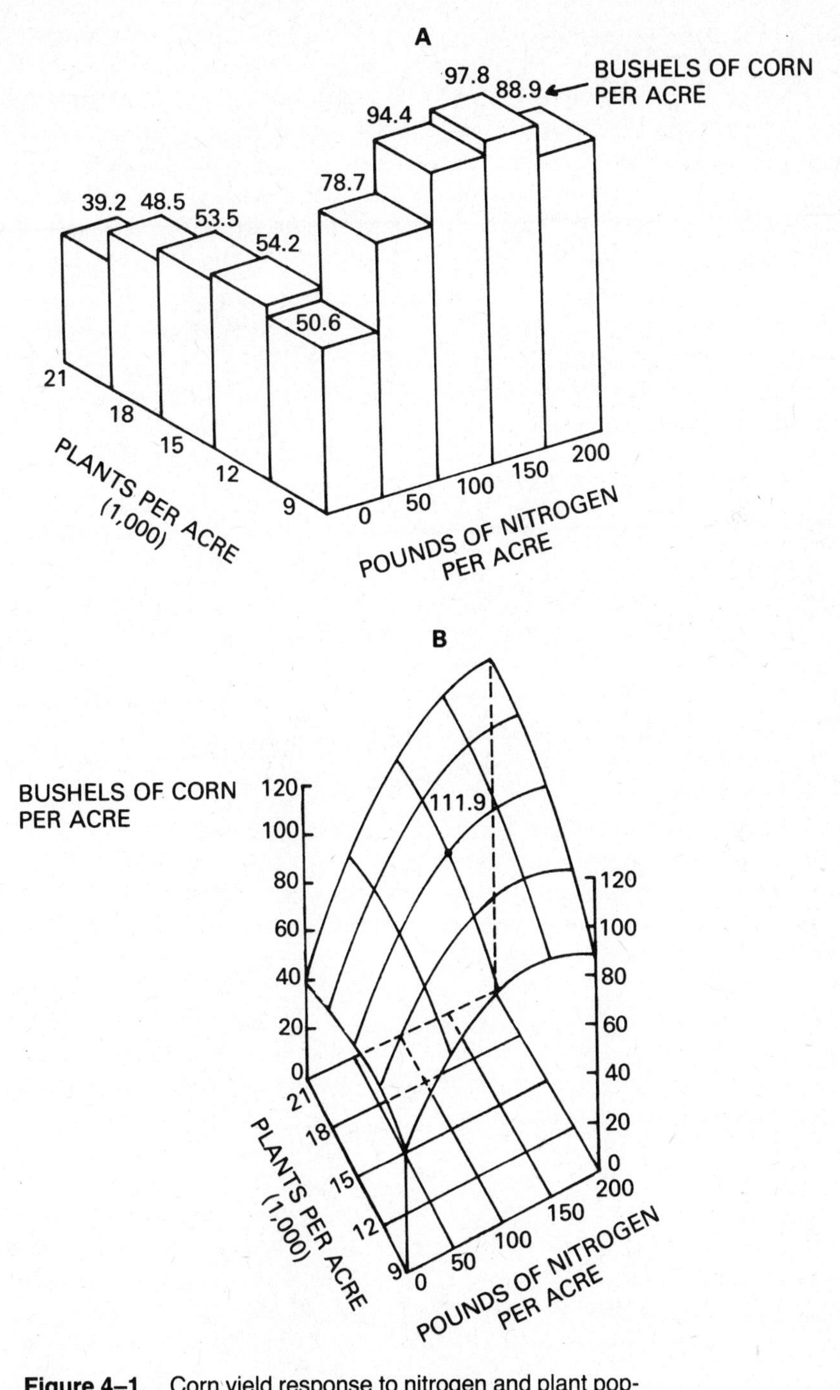

Figure 4–1. Corn yield response to nitrogen and plant population, North Missouri Research Center, 1962.

from any combination of nitrogen and plant population from 0–200 and 9000–21,000, respectively.

The surface shown in Figure 4–1B demonstrates the problems faced when more than one variable input is included in the production function. First, what is the most profitable corn yield? Clearly, of the many yields possible, one is the most profitable. Second, what combination of nitrogen and plant population should be used to produce the most profitable yield? The second question arises because the possibility now exists that a given corn yield can be produced using many different combinations of nitrogen and plant population. For example, from Table 4–1 it appears that 100 bushels of corn can be produced using 15,000 plants and slightly more than 50 pounds of nitrogen; 12,000 plants and slightly less than 100 pounds of nitrogen; 150 pounds of nitrogen and some planting rate between 9000 and 12,000 plants per acre, etc. In other words, a perpendicular line segment representing 100 bushels would fit snugly between the surface and the grid in many places in Figure 4–1B. The manager who first decided to produce 100 bushels must also decide which of these combinations to use. The next section of this chapter is directed at two questions: what input combination should be used to produce a given yield and what level of output meets the producers' objective?

THE PRODUCTION FUNCTION FOR TWO VARIABLE INPUTS

The production function for two variable inputs does not differ conceptually from that for one variable input. Each combination of the two inputs produces a unique amount of output. In symbolic notation, the production function for two variable inputs is often written as follows:

$$Y = f(X_1, X_2 | X_3, \ldots, X_n)$$

Or, suppressing the notation representing the fixed inputs, it is more simply written as

$$Y = f(X_1, X_2)$$

where Y is the amount of product and X_1 and X_2 are amounts of the two variable inputs. This expression says that the amount of output, Y, depends in a unique way on the amounts of the two variable inputs, X_1 and X_2, used in the production process along with fixed inputs.

The Production Surface

A hypothetical production function for two variable inputs is presented in Table 4–2. Amounts of the two inputs, X_1 and X_2, are listed along

Table 4–2 Output Resulting from Various Combinations of Two Inputs

X_1											
10	80	93	104	113	120	125	128	129	128	125	120
9	81	94	105	114	121	126	129	130	129	126	121
8	80	93	104	113	120	125	128	129	128	125	120
7	77	90	101	110	117	122	125	126	125	122	117
6	72	85	96	105	112	117	120	121	120	117	112
5	65	78	89	98	105	110	113	114	113	110	105
4	56	69	80	89	96	101	104	105	104	101	96
3	45	58	69	78	85	90	93	94	93	90	85
2	32	45	56	65	72	77	80	81	80	77	72
1	17	30	41	50	57	62	65	66	65	62	57
0	0	13	24	33	40	45	48	49	48	45	40
	0	1	2	3	4	5	6	7	8	9	10
					X_2						

the left side and bottom of the table. The body of the table presents, in appropriate units of bushels, tons, bales, etc., the amount of output resulting from each combination of inputs. Thus, zero output results when no inputs are used; 30 units result from one unit each of X_1 and X_2; 56 units of output from two units of each input; and so on. The maximum possible output, 130, results from the use of 9 units of X_1 and 7 units of X_2.

The data in Table 4–2 are regarded as continuous, and the input-output combinations presented represent only a few of all possible combinations. Thus, for 2 units of X_1 and 2.5 units of X_2, some amount of output, between 56 and 65, will be produced. The same is true for other input combinations not represented by whole numbers. All of these combinations are not presented in Table 4–2 because the table, already large, would thereby expand without limit. The input-output data in Table 4–2 were derived from the following quadratic production function:

$$Y = 18X_1 - X_1^2 + 14X_2 - X_2^2 \tag{4.1}$$

The output from any combination of inputs can be computed by substituting the selected values for inputs X_1 and X_2 into (4.1).

The output or the total physical product is increasing at a diminishing rate for low levels of X_1 and X_2. When $X_1 = 0$ and $X_2 = 0$, then $Y = 0$. The output reaches a maximum when the marginal physical products, MPP_{X_1} and MPP_{X_2} are zero. This can be determined by

$$\frac{\partial Y}{\partial X_1} = MPP_{X_1} = 18 - 2X_1 = 0 \quad \text{or} \quad X_1 = 9 \tag{4.2}$$

$$\frac{\partial Y}{\partial X_2} = MPP_{X_2} = 14 - 2X_2 = 0 \quad \text{or} \quad X_2 = 7 \tag{4.3}$$

When $X_1 = 9$ and $X_2 = 7$, the output reaches a maximum of 130 units of Y. For input levels $X_2 > 9$ and $X_2 > 7$, both MPPs would be negative and the resulting output would drop below 130.

The above production function represents an expected relationship that is useful for planning purposes. It predicts the amount of product forthcoming if a particular combination of inputs is used. It is subject to all the restrictive interpretations placed on the production function for one variable input in previous chapters. Changes in the fixed factors or in the technology used will, in turn, affect the production function.

Isoquants

Factor-factor relationships and the resulting substitution possibilities among variable inputs permit a given level of output to be produced with different combinations of inputs. With the exception of the minimum output, zero, and the maximum output, 130, all output levels can be produced using several different input combinations. For example, Table 4–2 shows that 105 units of output can be produced using the following input combinations:

X_1	X_2
9	2
6	3
5	4
4	7
5	10

As pointed out above, the input-output combinations in Table 4–2 represent only a few of all possible combinations, the rest of which are not presented because of space limitations. Because inputs are assumed to be divisible, there must exist many other input combinations that also produce 105 units of output. For example, when X_2 is 5, some amount of X_1 greater than 4 but less than 5 would produce 105 units of product. Similarly, slightly more than 4 units of X_1 would, when used with 6 units of X_2, also produce 105 units.

The curve representing all combinations of X_1 and X_2 that produce a given level of output is called an *isoquant—iso* meaning equal and *quant* meaning quantity. It is also referred to as an *isoproduct* or *equal-product* curve. The isoquant for 105 units of output is shown in the three-dimensional drawing of Figure 4–2A. The isoquant traces out all input combinations that produce 105 units of output. If a perpendicular line is dropped from any point on the surface exactly 105 units above the input base, it will always touch the base grid at a point on the isoquant for 105 units of output.

The isoquant is more conventionally drawn in two dimensions. If the production surface in Figure 4–2A were viewed directly from above, it would appear as in Figure 4–2B. This diagram contains all the information in Figure 4–2A and has the additional advantage of simplicity. In fact, the isoquant for 105 units of output can be determined directly from Table 4–2 without recourse to the production surface. The input combinations from Table 4–2 can be located on the graph (denoted by the heavy dots in Figure 4–2B) and connected by a smooth curve. The result will be the same isoquant depicted in Figure 4–2A. All points falling on the curve are those which will produce 105 units of product. For example, in addition to the points already mentioned, $X_1 = 7$ and $X_2 = 2.4$, and $X_1 = 4.4$ and $X_2 = 5$, $X_1 = 4.1$ and $X_2 = 6$ will all produce 105 units of product.

The accuracy of the isoproduct curve or isoquant depends on the number of points available for plotting. If the production function is expressed in algebraic form, the isoquant equation can also be deter-

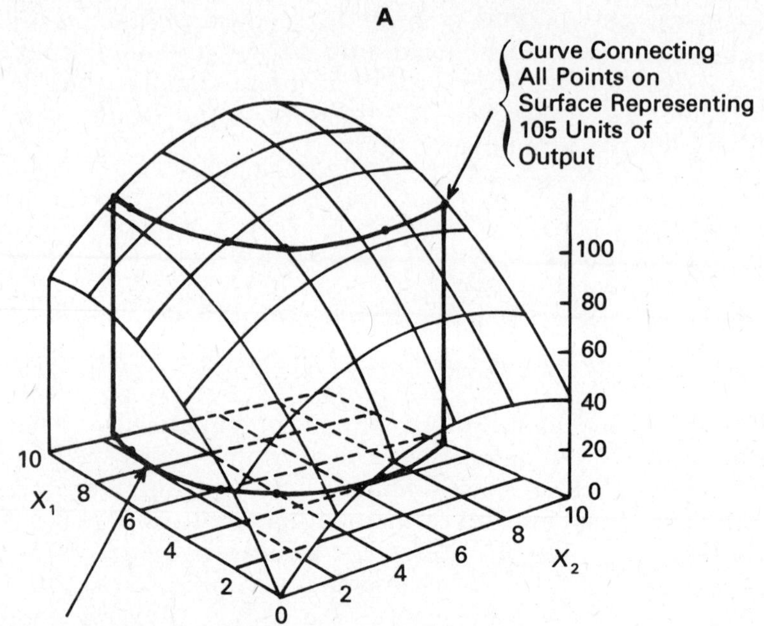

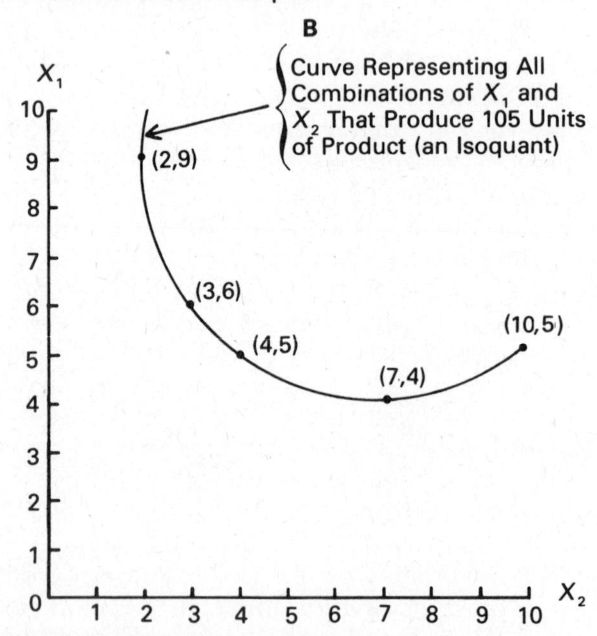

Figure 4–2. Relationship of isoquants to the production surface.

mined. In the present example, equation (4.1) can be solved for X_1 and as a function of X_2 and Y by using the quadratic formula. To do so, the production function (4.1)

$$Y = 18X_1 - X_1^2 + 14X_2 - X_2^2$$

is rewritten as

$$-X_1^2 + 18X_1 + (14X_2 - X_2^2 - Y) = 0$$

which is substituted into the quadratic equation to obtain

$$X_1 = \frac{18 - (324 + 56X_2 - 4X_2^2 - 4Y)^{1/2}}{2}$$

or

$$X_1 = 9 - (81 + 14X_2 - X_2^2 - Y)^{1/2} \qquad (4.4)$$

Thus, substituting $Y = 105$ and $X_2 = 6$ into equation (4.4) will determine the value $X_1 = 4.1$. Solutions obtained when the sign on the radical is positive are outside the area of economic relevance and therefore ignored.

Isoquants can be determined for any particular output level or, conversely, every output level has an isoquant. For the example, an isoquant exists for every output between 0 and 130. Several isoquants for this example are shown in Figure 4–3. This isoquant map was drawn using points derived from the isoquant equation (4.4). The isoquants presented are for 26-unit increments in output: 0, 26, 52, 78, 104, and 130.

The isoquant map has the same interpretations as a contour map used to demonstrate the topography of a countryside. On a contour map, the contour lines represent the altitude above sea level; by following the contours, the lay of the land can be determined. The isoquant map in Figure 4–3 indicates the shape of the production surface, which, in turn, indicates the nature of the output response to the inputs. In Figure 4–3, each isoquant measures the output above zero; the shape of the isoquants indicate that output increases smoothly to the maximum point, 130.

Isoquants derived from a production function have a number of important properties. They never cross because to do so would violate a basic definition underlying the production function—each combination of inputs can produce one, and only one, amount of output. Isoquants must be either convex to the origin or linear. A concave isoquant would lead to solutions that are not optimal because the condition for economic efficiency would be violated. (An isoquant is strictly

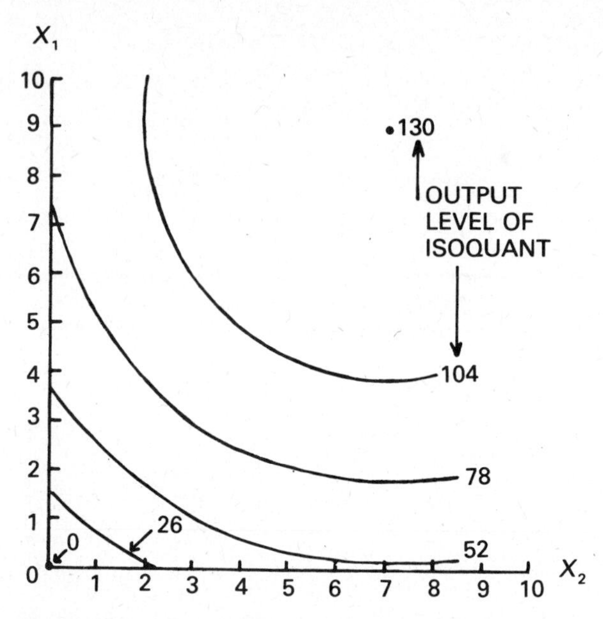

Figure 4–3. Output isoquants for a hypothetical production function.

convex to the origin if any segment of the isoquant connected by two points on the isoquant always lies below a straight line connecting the same two points. On a concave isoquant, the segment would be above the straight line. The straight line is itself *weakly* convex—or *weakly* concave.)

Marginal Rate of Input Substitution

The marginal rate of input substitution is as important to the factor-factor relationship as the marginal physical product is to the factor-product relationship. It is represented by the slope of the isoquant. The 105-unit isoquant is presented in Figure 4–4. When $X_2 = 2, X_1 = 9$, but when $X_2 = 3, X_1 = 6$. Thus, if output is to be held constant at 105 units when X_2 is increased from 2 to 3 units, X_1 must decrease from 9 units to 6 units. This type of problem is analogous to the mountain climber, two miles above sea level on the southwest slope of a mountain, who asks himself the question, "If I walk south one mile, how far east must I go to remain two miles above sea level?" In this case, the marginal rate of substitution of X_2 for X_1 is defined as the amount by which X_1 must be decreased to maintain output at a constant amount when X_2 is increased by one unit. Between the points ($X_2 = 2, X_1 = 9$) and ($X_2 = 3, X_1 = 6$), the marginal rate of substitution of X_2 for X_1, abbreviated *MRS* of X_2 for X_1, is

$$MRS \text{ of } X_2 \text{ for } X_1 = \frac{\Delta X_1}{\Delta X_2} = \frac{6 - 9}{3 - 2} = \frac{-3}{1} = -3$$

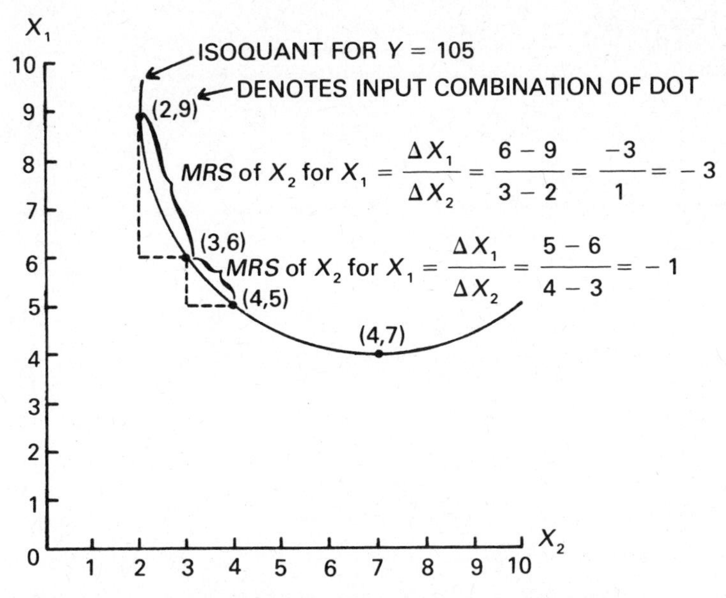

Figure 4–4. Measuring the marginal rate of substitution between two points on an isoquant.

where ΔX_1 is the change in X_1 and ΔX_2 is the change in X_2.[1] The *MRS* is negative because the isoquant slopes downward and to the right; that is, the isoquant has a negative slope.

MRS of X_2 for X_1 varies depending on the points on the isoquant considered. Between the points ($X_2 = 3, X_1 = 6$) and ($X_2 = 4, X_1 = 5$)

$$MRS \text{ of } X_2 \text{ for } X_1 = \frac{5-6}{4-3} = \frac{-1}{1} = -1$$

The addition of one unit of X_2 will permit a simultaneous reduction of X_1 by one unit if output is to be held constant at 105.

Marginal rates of substitution of X_2 for X_1 are computed in Table 4–3 for the 105-unit isoquant. Because the isoquant is curved, the slope changes continuously and, as the amount of X_2 increases, the *MRS* of X_2 for X_1 decreases absolutely (ignoring the negative sign). This type of substitution is known as a "decreasing marginal rate of substitution." As the amount of X_2 increases, its ability to increase output decreases, and one unit of X_2 replaces successively smaller and smaller amounts of X_1. Whenever *MRS* is decreasing, the isoproduct curve is convex to the origin.

If a straight line were drawn between the bracketed points in

[1] We should warn the student that some authors define *MRS* exactly opposite. That is, $\Delta X_1/\Delta X_2$ is called the *MRS* of X_1 for X_2. The concept remains the same, but the terminology is reversed. If you understand the ideas, the names are unimportant.

Table 4–3 Computing the Marginal Rates of Substitution X_2 for X_1 for an Output of 105

Units of X_2	*Units of* X_1	ΔX_2	ΔX_1	*MRS of* X_2 *for* X_1
2	9.0			
		1	−3.0	−3.0
3	6.0			
		1	−1.0	−1.0
4	5.0			
		1	−0.6	−0.6
5	4.4			
		1	−0.3	−0.3
6	4.1			
		1	−0.1	−0.1
7	4.0			
		1	0.1	0.1
8	4.1			

Figure 4–4, the slope of that line would represent the *MRS* computed between the points. As with the marginal physical product, the *MRS* between two points represents the approximate slope of the isoquant; it is the average of all possible slopes between the two points. This approximation can be improved by moving the two points together until, in a limiting sense, the slope of the tangent at any point on an isoquant represents the exact *MRS* at that point. This is illustrated in Figure 4–5, where a straight line is drawn tangent to the isoquant at the point (X_2 = 3, X_1 = 6). The slope of this straight line is equal to the *MRS* of X_2 for X_1 at the point (3,6) and, from measurement on the graph, is equal to −3.2/2.4 = −4/3. This *MRS* is different from the average *MRS* between the points.

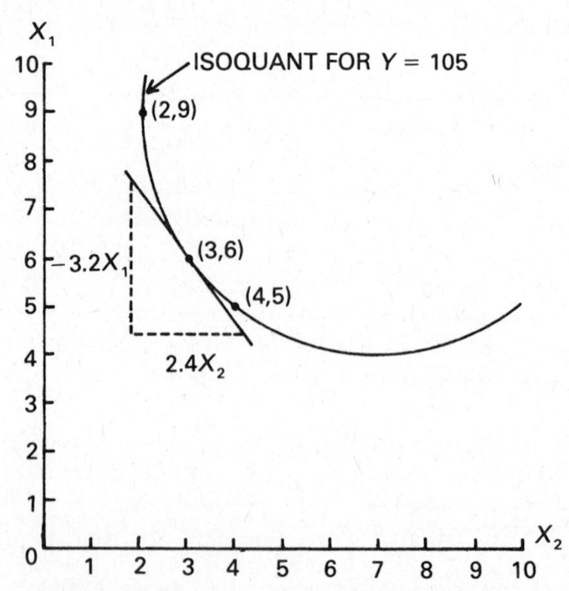

Figure 4–5. Measuring the exact marginal rate of substitution at a point.

Thus far, little has been said about the marginal products of the inputs. This may seem unusual; marginal analysis was singularly emphasized in preceding chapters. Actually, the *MRS* is a convenient way to examine two marginal products simultaneously.

A small portion of the 105 unit isoquant is reproduced in Figure 4–6. The *MRS* of X_2 for X_1 on the isoquant between *B* and *C* is $\Delta X_1/\Delta X_2 = (6 - 9)/(3 - 2) = -3$. But the output at point *A*, where $X_1 = 6$ and $X_2 = 2$, is 96 units (Table 4–2). Thus, between *A* and *B*,

$$MPP_{X_1} = \frac{105 - 96}{9 - 6} = \frac{9}{3} = 3$$

while, between *A* and *C*

$$MPP_{X_2} = \frac{105 - 96}{3 - 2} = \frac{9}{1} = 9$$

Thus, it is apparent that

$$-\frac{MPP_{X_2}}{MPP_{X_1}} = -\frac{9}{3} = -3 = MRS \text{ of } X_2 \text{ for } X_1 \text{ between } B \text{ and } C.$$

The negative sign must be added because the isoquant slopes downward to the right. The *MRS* therefore represents the ratio of the marginal physical products.

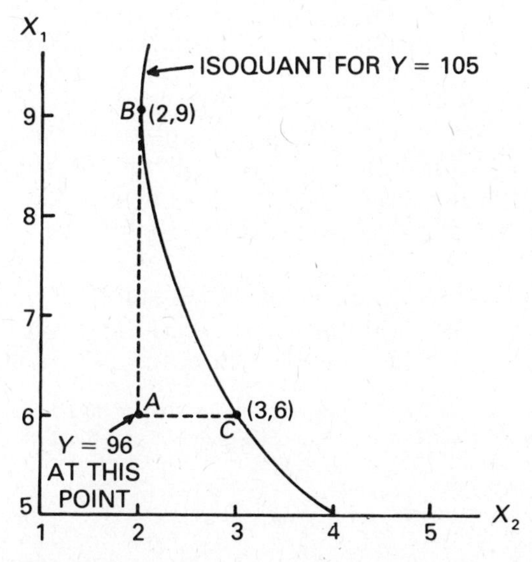

Figure 4–6. Computing the marginal rate of substitution along an isoquant.

This interpretation of the *MRS* provides a direct method for computing the exact *MRS* discussed above and depicted in Figure 4–5. Using the *MPP* equations derived from the production function (4.1), this exact *MRS* will be

$$MRS \text{ of } X_2 \text{ for } X_1 = -\frac{MPP_{X_2}}{MPP_{X_1}} = -\frac{\partial Y/\partial X_2}{\partial Y/\partial X_1} = \frac{dX_1}{dX_2}$$

$$= -\frac{14 - 2X_2}{18 - 2X_1} \qquad (4.5)$$

Thus, the exact *MRS* on the isoquant for $X_1 = 6$ and $X_2 = 3$ is

$$-\frac{14 - 2(3)}{18 - 2(6)} = -\frac{8}{6} \text{ or } -\frac{4}{3};$$

one unit of X_2 replaces 1.33 units of X_1. The exact *MRS* can be calculated for any point on any of the isoquants by substituting the appropriate X_1 and X_2 values into (4.5).[2]

The production function is a relationship between input and output that can be used when planning business activity. Isoquants are derived from production functions and have the same interpretations. Movement along an isoquant can be interpreted in a planning sense only; *if* a certain combination of inputs were to be used, *then* the *MRS* would take a specific value. In practice, once a combination of inputs has been selected and used, that combination cannot be immediately altered.

RELATIONSHIPS BETWEEN INPUTS

Inputs are technical substitutes when an increase in one input permits a decrease in the other input while maintaining the level of output. When resources or inputs are technical substitutes, they are often said to compete with each other. The marginal rate of substitution for inputs that are technical substitutes is negative.

Decreasing Rates of Substitution

Decreasing rates of substitution occur when the input being increased substitutes for successively smaller amounts of the input being re-

[2]The slope of the isoquant can be derived using calculus. The production function is $Y = f(X_1, X_2)$, but along an isoquant, Y is constant and X_1 can be regarded as a function of X_2, say, $X_1 = g(X_2)$. Then, the total derivative of Y with respect to X_2 is

$$\frac{dy}{dX_2} = \frac{\partial Y}{\partial X_1}\frac{dX_1}{dX_2} + \frac{\partial Y}{\partial X_2} = 0$$

The total derivative is zero because output does not change along an isoquant. Solving this expression for dX_1/dX_2 gives the expression in (4.5).

placed. Thus, if the MRS of X_2 for X_1 decreases absolutely (ignoring the negative sign) as X_2 is increased along an isoquant, then a decreasing rate of substitution is exhibited.

Decreasing rates of substitution are caused by the law of diminishing returns. The MRS of X_2 for X_1 is defined as MPP_{X_2}/MPP_{X_1}. When diminishing returns are present, MPP_{X_2} decreases as X_2 increases and MPP_{X_1} increases as X_1 decreases. Thus, the ratio decreases numerically. Decreasing rates of substitution among inputs are quite common.

The production function in Table 4–2 is characterized by decreasing rates of substitution. Therefore, further illustrative tables will not be presented here. Figure 4–7A contains some hypothetical isoquants illustrating decreasing rates of substitution. In the left diagram, the isoquant intersects each input axis; thus, the output represented by the isoquant can be produced using only X_1, only X_2, or some combination of X_1 and X_2, depending on the choice indicator. In the center figure, the isoquant does not intersect either axis, signifying that the output cannot be produced unless a minimum amount of both inputs

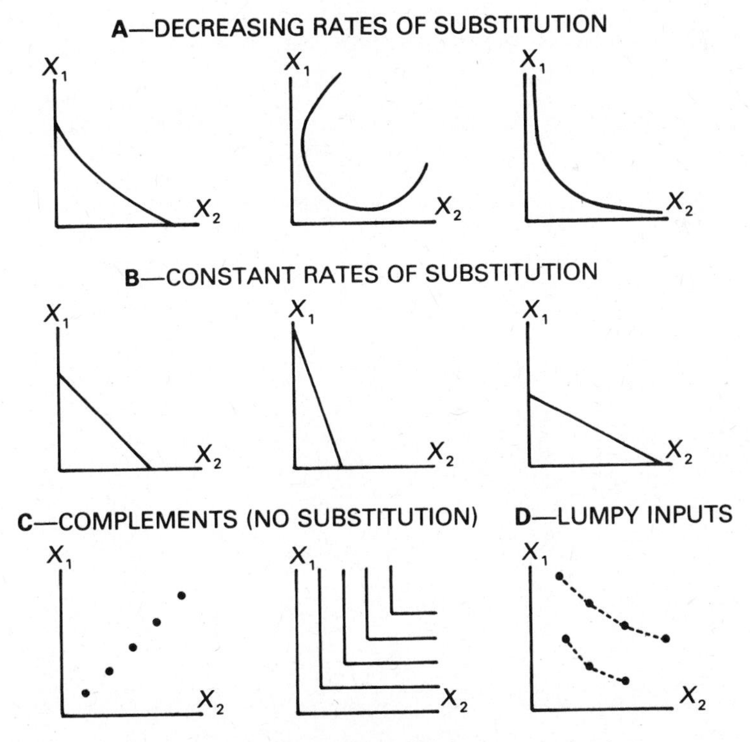

Figure 4–7. Isoquants representing different types of substitution between inputs.
A. Decreasing rates of substitution.
B. Constant rates of substitution.
C. Complements (no substitution).
D. Lumpy inputs.

are present. Also, the isoquants attain a positive slope, suggesting that if too much of one input is used, increasing amounts of the other must be applied to maintain output. The zone of economic relevance is, of course, that portion which has negative slope. The isoquant in the right diagram becomes parallel to each input axis as the amount of either input is increased indefinitely; in this case the inputs substitute within limits, but after one input decreases to a certain low level, the second can be added in extremely large amounts without causing significant changes in either output or the amount of the first input.

Constant Rates of Substitution

Constant rates of substitution occur when the amount of one input replaced by the other input does not change as the added input increases in magnitude. Thus, the *MRS* of X_2 for X_1 is a constant and the isoquant is a straight line.

Examples of production functions with inputs that substitute at constant rates are presented in Table 4–4. In Table 4–4A, one unit of X_2 substitutes for one unit of X_1 at each output level (except 12). In

Table 4–4 Examples of Production Functions with Inputs that Substitute at Constant Rates

A

X_1					
	3	6	8	10	12
	2	4	6	8	10
	1	2	4	6	8
	0	0	2	4	6
		0	1	2	3
			X_2		

B

X_1					
	3	3	5	7	9
	2	2	4	6	8
	1	1	3	5	7
	0	0	2	4	6
		0	1	2	3
			X_2		

C

X_1					
	3	18	21	23	24
	2	14	18	21	23
	1	8	14	18	21
	0	0	8	14	18
		0	1	2	3
			X_2		

this case, the production function displays constant marginal returns to the inputs but, in general, this is not a requirement. When one unit of an input replaces exactly one unit of another input, the inputs are the same from an economic point of view. The isoquant forms 45° interior angles with the input axes (see the figure on the left in Figure 4–7B).

Table 4–4B illustrates a production function with the property that one unit of X_2 replaces two units of X_1 at each output level. In this particular case, the isoquant would not form a 45° interior angle with the axes. In general, two cases are possible: 1) A unit of X_2 replaces more than one unit of X_1. This is illustrated in Table 4–4B and the center diagram in Figure 4–7B. 2) A unit of X_2 replaces less than one unit of X_1. This situation is illustrated in the right diagram in Figure 4–7B. Constant rates of substitution require that the slope of a given isoquant remains unchanged, not that isoquants be parallel or equidistant. Table 4–4C contains a production function with linear isoquants but diminishing returns to each input. In some economic literature, inputs that substitute at constant rates are termed "perfect" substitutes.

Complementary Inputs

Inputs that increase output only when combined in fixed proportions are called technical complements. Complements represent the opposite of substitutes; "complements" imply that only one exact combination of inputs will produce the specified output.

Two examples of production functions showing complementary relationships between inputs are presented in Table 4–5. In Table 4–

Table 4–5 Examples of Production Functions with Inputs that Are Complements

A

	X_1				
	3	0	0	0	12
X_1	2	0	0	9	0
	1	0	5	0	0
	0	0	0	0	0
		0	1	2	3
			X_2		

B

	X_1				
	3	0	2	4	6
X_1	2	0	2	4	4
	1	0	2	2	2
	0	0	0	0	0
		0	1	2	3
			X_2		

5A, the inputs must be combined in a 1:1 ratio; any deviation from this ratio causes output to drop to zero. If one unit of X_2 is used, one unit of X_1 must also be used. Any other amount of X_1 results in no output. In Table 4–5B, if one unit of X_2 is used, then at least one unit of X_1 must also be used to attain an output of two. If more than one unit of X_1 is used, output is not diminished, but neither is it increased.

Figure 4–7C illustrates two possible complementary relationships. The diagram on the left is comparable to Table 4–5A. Isoquants degenerate to single points. The center diagram portrays the production function in Table 4–5B. In this case, the isoquants appear as right angles set within the input axis.

When the isoquant degenerates to a dot, only one possible combination of inputs may be used to produce a given output level regardless of cost. The inputs in combination can be considered as one input. Either it pays to use both in combination or none is used. When the isoquant appears as a right angle (center, Figure 4–7C), the input combination at the vertex is used; other combinations on the isoquant would cost more but produce no more. Decision making in this case is relatively simple because only one input combination is relevant for consideration.

Complementary inputs are common in agriculture. A trivial example would be all the parts that make up a machine—tractor tires and other components of tractors. Other common examples are tractors and tractor drivers, fuel and grease, fence wire and posts, cylinders and spark plugs. Chemical compounds used in agriculture are also examples; water is composed of two parts hydrogen and one part oxygen, etc.

Often complements are regarded as single inputs, such as a horse and a man versus a tractor and a man. In each case, the man and the power source are complements; however, each combination may be regarded as a single input that substitutes for the other combination. Thus, the same amount of work may be accomplished using many different combinations of tractors and men as compared to horses and men.

Lumpy Inputs

The discussion presented up to this time, along with the graphs drawn, has assumed that each of the variable inputs is completely divisible. If both of the inputs are not completely divisible but exist only in discrete units, then the isoquants appear as dots which, when connected, appear as linear segments (Figure 4–7D). Decision making with lumpy inputs, because of fewer alternatives, is less complex than when substitution relationships are continuous.

Elasticity of Factor Substitution

The elasticity of factor substitution in factor-factor relationships is defined in a manner similar to the elasticity of production, ϵ_P, in the factor-product relationship. Elasticities are ratios of proportionate (or percentage) changes; in the factor-product relationship, ϵ_P is defined as the proportionate change in output divided by the proportionate change in input. In factor substitution, two elasticity concepts have been proposed. Both are commonly found in production economics literature, and we will discuss each in turn. The discussion below assumes input use in Stage II where isoquants are either convex or linear.

The elasticity of input substitution, as traditionally defined in the theory of the firm, is

$$\sigma = \frac{\Delta\left(\frac{X_1}{X_2}\right) \Big/ \frac{X_1}{X_2}}{\Delta\ MRS/MRS}$$

Thus, σ is the proportionate change in the ratio of the input amounts divided by the proportionate change in the *MRS*. Because the proportionate change in the *MRS* appears in the denominator, this elasticity could be called the elasticity of the marginal rate of substitution. σ is always zero or positive and measures the ease with which the inputs can be substituted for each other. It is, of course, a pure number and is inversely proportional to the curvature of the isoquant. When inputs must be used in fixed proportions, and hence cannot be substituted for each other, σ will be zero. For example, σ is zero for the production functions shown in Table 4–5. When the inputs substitute at constant rates, that is, are perfect substitutes, σ is infinite. (More appropriately, σ is undefined because there can be no change in the *MRS* along a linear isoquant.) Ease of substitution increases as σ increases; as the curvature of the isoquant decreases, the isoquant approaches linearity and the elasticity of substitution increases.

Finally, σ is symmetrical. At a given point on the isoquant, it will have the same value when X_1 is substituted for X_2 as when X_2 is substituted for X_1.

The elasticity of substitution, σ, is difficult to compute from tabular data. The reason is that the *MRS* is the slope of the isoquant and determining changes in the *MRS* involves estimating the rate of change of a slope, which is difficult when graphs or tables represent the only available data. Use of calculus simplifies the problem. Consider the problem of estimating σ along the isoquant for 105 units of output between the input quantities ($X_2 = 3$, $X_1 = 6$) and ($X_2 = 4$, $X_1 = 5$) as shown in Figure 4–4. To obtain an estimate of ΔMRS over the arc of the isoquant requires estimating the exact slope at each point and subtracting. This can be done graphically, but a simpler and more accurate method is to use derivatives when an equation for the production function is available. Substitution into equation (4.5), the exact equation for the *MRS*, gives an exact *MRS* of -1.33 and (3,6) and

-0.75 at (4,5). Substituting these into the formula for σ gives the arc estimate of

$$\sigma = \frac{\left(\frac{5}{4} - \frac{6}{3}\right) \Big/ \frac{5 + 6}{4 + 3}}{(-0.75 - (-1.33)) \Big/ \frac{-0.75 + (-1.33)}{2}} = \frac{0.48}{0.56} = 0.86$$

between the two input combinations on the isoquant.[3]

An alternative measure of the elasticity along an isoquant was introduced by Heady in 1952 (page 144). Heady's measure, denoted E_s, is defined as the percentage change in one input, X_1, divided by the percentage change in the other input, X_2. Thus,

$$E_s = \Delta X_1/X_1 \div \Delta X_2/X_2 = \frac{(\Delta X_1)X_2}{(\Delta X_2)X_1} = (MRS_{X_2 \text{ for } X_1})\frac{X_2}{X_1}$$

For example, between the input quantities (3,6) and (4,5), E_s is $[(-1)(3.5/5.5)]$ or -0.64. The -0.64 in this case is an arc elasticity of substitution between the points on the isoquant for $Y = 105$.

The exact estimate of E_s at the point $X_1 = 6$, $X_2 = 3$ can be calculated from MRS equation (4.5) by substituting $X_1 = 6$ and $X_2 = 3$ into the equation and multiplying the quotient by 3/6. In this case E_s is -0.67. At the point $X_1 = 5$, $X_2 = 4$, the exact elasticity of substitution is -0.60. Thus the average of the two exact E_s, -0.64, equals the arc elasticity of substitution between the two points.

E_s is a pure number, independent of units of measure. It will be zero for technical complements (as in Table 4–5) but will be negative for inputs that are substitutes because the slope of the isoquants for such inputs will be negative. E_s measures the rate at which the slope of the isoquant changes as X_2 is increased. The maximum value E_s can attain is zero, and it can decrease from zero without limit (to "minus infinity").

E_s is much easier to calculate than σ when calculus is not used, but the two elasticities are not the same. E_s is zero or negative while σ is

[3]When the production function is known, an exact measure of σ is given by

$$\sigma = [f_1 f_2(X_1 f_1 + X_2 f_2)]/[X_1 X_2(f_{11} f_2^2 - 2 f_{12} f_1 f_2 + f_{22} f_1^2)]$$

where X_1 and X_2 are the input quantities; f_1 and f_{11} are the first and second partial derivatives of the production function taken with respect to X_1 and evaluated at the point X_1 and X_2; f_2 and f_{22} have the same interpretation for X_2; and f_{12} is the second-order crosspartial derivative evaluated at X_1 and X_2. For the example, this formula determines a value of $\sigma = 0.82$ for $X_1 = 5.5$ and $X_2 = 3.5$. The formula is derived in Ferguson C. E., *Neoclassical Theory of Production and Distribution*, Cambridge: Cambridge University Press, 1969, p. 91.

zero or positive. E_s is not necessarily symmetrical and may vary depending on whether X_1 is substituted for X_2 or vice versa. Headley has compared the two elasticities and suggested that E_s might be called the "isoquant elasticity."[4]

The Meaning of Input Substitution

Input substitution in production processes is possible when the inputs both increase output; the marginal physical products must be positive. In farming, land can be substituted for labor, machinery can be substituted for labor, fertilizer can be substituted for land, etc. This is possible because many different combinations of land, labor, and capital can be used to produce a certain amount of farm output. The fact that the inputs are substitutes does not mean they perform the same technical or physiological function in the production process but only that their function, whatever it is, is output-increasing.

For example, isoquants could be derived for plant population and nitrogen in corn production (review the example at the beginning of this chapter). This suggests only that a given corn yield can be obtained using many different combinations of nitrogen and plant population. It does not imply that nitrogen and plant population increase corn yield biologically in the same manner. In fact, they are known to affect yield differently. Nitrogen is a chemical that stimulates growth within the plants while an increase in plant population increases the number of plants growing per acre of soil. Despite their completely different biological effects, however, these inputs are technical substitutes in an economic sense because they have a positive effect on corn yield.

ISOCOST LINES

Each combination of inputs has a cost associated with it. The cost is variable because the inputs considered are variable. (See Chapter 2 for discussion of variable and fixed costs). Denoting the cost per unit of X_1 as P_{X_1} and the cost per unit of X_2 as P_{X_2}, the total variable cost, TVC, is given by

$$TVC = P_{X_1}X_1 + P_{X_2}X_2$$

The input prices are assumed known, and as a result TVC can be computed for each input combination. If $P_{X_1} = \$2$ and $P_{X_2} = \$3$, then the cost of 5 units of X_1 and 2 units of X_2 is $\$2 \cdot 5 + \$3 \cdot 2 = \$16$. TVC is a function of the amount of X_1 and X_2 and can be graphed in a manner similar to a production surface. A total variable cost surface is shown in Figure 4–8A. The surface is linear and touches the base

[4]Headley, J.C., "Elasticity of Substitution: Some Confusion," *The Canadian Journal of Agricultural Economics,* Vol. 30, March 1982, pp. 71–74.

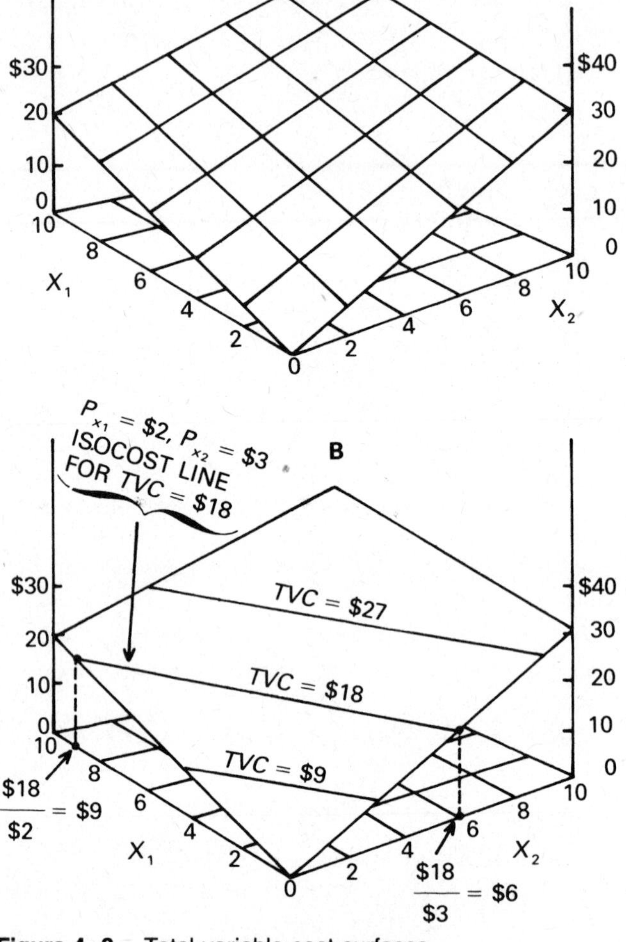

Figure 4–8. Total variable cost surfaces.

grid only at $X_1 = X_2 = 0$ because no variable costs are incurred at that point. The slope of the surface is determined by the input prices. The slope of TVC on a line parallel to the X_2 axis (X_1 held constant) is equal to P_{X_2}, and vice versa. The surface is linear because input prices are constant in pure competition.

Just as production surfaces are characterized by isoquants, total variable cost surfaces can be described using isocost lines—*iso* again meaning equal. An isocost line traces a set of points on the cost surface that are an equal distance above the input base or grid. For example, suppose $TVC = \$18$; that is, the farmer has \$18 to spend on variable inputs. He may then purchase $TVC/P_{X_1} = \$18/\$2 = 9$ units of X_1 or $TVC/P_{X_2} = \$18/\$3 = 6$ units of X_2. The appropriate isocost line for

\$18 can then be located on the *TVC* surface (Figure 4–8B). For perspective, the isocost lines for total variable costs \$9 and \$27 are also shown in Figure 4–8B.

Isocost lines determine all combinations of the two inputs that cost the same amount. Each point on the isocost line represents a combination of inputs that can be purchased with the same outlay of funds. As an isoquant, the isocost line can be placed on a two-dimensional graph. If the *TVC* surface in Figure 4–8B were viewed directly from above, an isocost map would be the result. Such a map is shown in Figure 4–9A. (To avoid clutter, only isocost lines for $TVC = \$9$ and $TVC = \$18$ are included in Figure 4–9A.) The isocost lines in Figure 4–9A can be derived directly by locating the endpoints ($X_2 = 6$ and $X_1 = 9$ for $TVC = \$18$) and connecting them with a straight line.

The equation of the isocost line can be found by solving the *TVC* equation for X_1 as an explicit function of X_2. For example,

$$P_{X_1}X_1 = TVC - P_{X_2}X_2$$

and

$$X_1 = \frac{TVC}{P_{X_1}} - \frac{P_{X_2}}{P_{X_1}} X_2$$

From this expression it can be seen that the slope of the isocost line is $-P_{X_2}/P_{X_1}$ while the intercept on the X_1 axis is given by TVC/P_{X_1}. For the given prices, the isocost line for $TVC = \$18$ in Figure 4–9A is $X_1 = 9 - (3/2)\,X_2$.

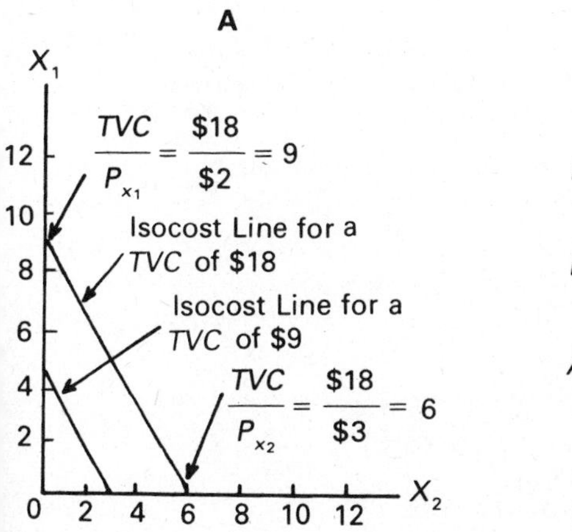

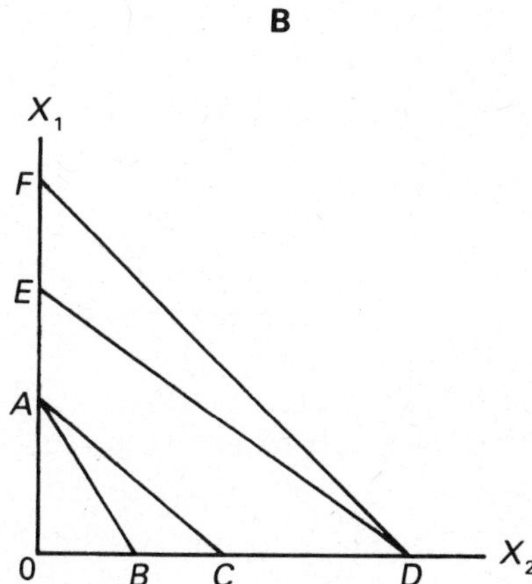

Figure 4–9. Constructing isocost lines.

The two important aspects of the isocost line are its distance from the origin and its slope. When prices do not change, each possible total variable cost has a different isocost line. As total variable cost increases, the ratio TVC/P_{X_1} increases and the isocost line moves to higher points on the cost surface, located further from the origin. An isocost line for $9 is drawn in Figure 4–9A; it lies halfway between the origin and the $18 cost line. Both lines have the same slope, however, because input prices are the same in both cases.

Changes in the input price change the slope of the isocost line. A decrease in the input price means that more of that input can be purchased with the same total variable cost; an increase means that less can be purchased. In Figure 4–9B, if *AB* were the original isocost line and the price of X_2 dropped, then *AC* would represent the new isocost line. If *AC* were the original isocost line and the unit price of X_2 increased, then *AB* would represent the new isocost line. The lines *DE* and *DF* represent similar change for the price of X_1, given that the price of X_2 is constant.

THE LEAST COST CRITERION

As in factor-product relationship (Chapter 3), economic efficiency in the factor-factor relationship is attained when the necessary and sufficient conditions are met. The necessary condition in the factor-factor case is met when the marginal rate of substitution is equal to or less than zero. Table 4–3 and Figure 4–2B show combinations of inputs X_1 and X_2 that produce 105 units of output, some of which meet the necessary conditions. To select the combination that would meet the individual or social objective, a criterion is needed. As always the objective may differ with the wishes of the manager; for example, he may wish to produce a given level of output with minimum effort. In a market economy he may choose to produce a given level of output with minimum cost. It is usually the latter criterion, cost minimization, that is employed in economic analyses. The least cost criterion is used in the examples that follow in this chapter.

The concepts needed to determine the combination of inputs that will produce a given output at a minimum of cost are available. As seen, a given level of output can be produced using many different combinations of the two variable inputs. Usually, no two of these input combinations will have the same cost (a special case where they do will be discussed later); therefore, one combination must necessarily be cheaper than all others. The problem of cost minimization is to determine the one combination of the two inputs that will produce a given output at the least possible cost.

One possible way to determine the least cost combination is to compute the cost of all possible combinations and then select the one with minimum cost. This method is satisfactory when only a few combinations produce a given output. Table 4–6 contains seven combinations of inputs that produce 105 units of output; these points were read from Figure 4–2B. If the price of X_1 is \$2 per unit and the price of X_2 is \$3 per unit, the cost of each combination can be determined by multiplication. For example, the cost of 2 units of X_2 and 9 units of X_1 is equal to $(\$3 \cdot 2) + (\$2 \cdot 9) = \$24$.

Of the seven combinations in Table 4–6, 3 units of X_2 and 6 units of X_1 represent the least cost combination of inputs, \$21. For many purposes, this may be as accurate an estimate of the least cost combination as is needed. However, many input combinations not listed in Table 4–6 will also produce 105 units of output. Some of these combinations may cost less than the combination (3, 6). The exact location of the least cost combination of inputs can be determined geometrically, but to do so, concepts associated with the marginal rate of input substitution and the isocost line must be utilized.

Least Cost–Geometric Determination

The isoquant in Figure 4–2B has an infinite number of points—only one will represent the cost minimizing combination. At this point the following criterion, called the *least cost criterion,* will hold:

$$MRS \text{ of } X_2 \text{ for } X_1 = -\frac{P_{X_2}}{P_{X_1}}$$

Because of the definition of *MRS*, the criterion can be written

$$\frac{\Delta X_1}{\Delta X_2} = -\frac{P_{X_2}}{P_{X_1}}$$

The left side of this expression represents the slope of an isoquant; the right side the slope of the isocost line. Thus, the least cost com-

Table 4–6 Computing the Minimum-Cost Combination of Inputs for an Output of 105 ($P_{X_1} = \$2$, $P_{X_2} = \$3$)

Units of X_2	*Units of* X_1	*Cost of* X_2	*Cost of* X_1	*Total Variable Cost*
2	9.0	\$ 6	\$18.0	\$24.0
3	6.0	9	12.0	21.0
4	5.0	12	10.0	22.0
5	4.4	15	8.8	23.8
6	4.1	18	8.2	26.2
7	4.0	21	8.0	29.0
8	4.1	24	8.2	32.2

bination of inputs occurs at the point where the isocost line is tangent to the isoquant, given that the isoquant is convex to the origin.

Figure 4–10 depicts the least cost solution for the 105-unit isoquant. In this case the minimum total variable cost needed to produce 105 units of product is unknown. The solution can be found, however, by finding the point where a straight line with a slope of $-3/2$ is tangent to the isoquant, that is, finding the point where the isoquant has a slope of $-3/2$. Once located, this straight line can be interpreted as the isocost line and can be extended to the input axes. As shown in Figure 4–10, the point of tangency occurs at $X_1 = 6.2$ and $X_2 = 2.8$. The total variable cost for this combination is $(6.2 \cdot \$2) + (2.8 \cdot \$3) = \$20.80$, slightly less than the cheapest combination in Table 4–6 when only discrete (whole) units of inputs were considered.

An intuitive interpretation of the least cost combination is as follows: If the finite changes in X_1 and X_2 could be made sufficiently small, then the criterion implies that

$$-P_{X_1}\,(\Delta X_1) = P_{X_2}\,(\Delta X_2)$$

because either ΔX_1 or ΔX_2 will be negative for movements along the isoquant in the economically relevant range. If between two points on

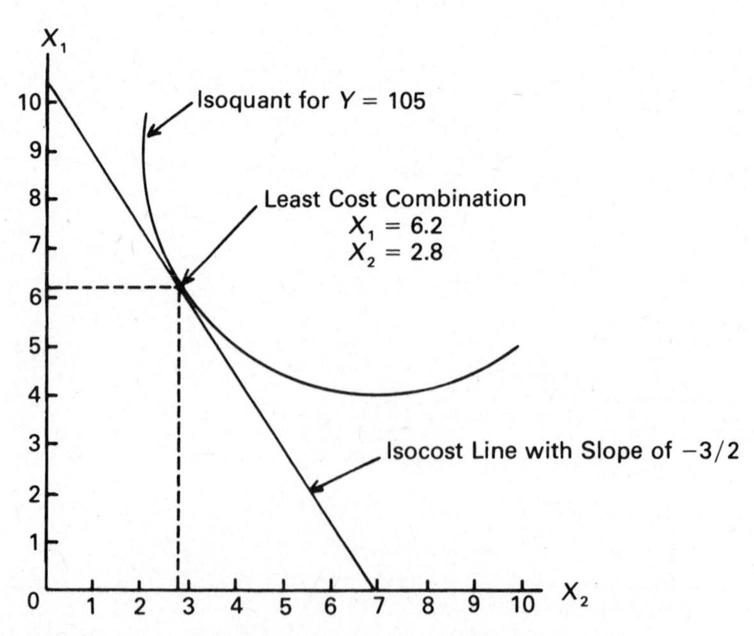

Figure 4–10. Determining the combinations of inputs to produce 105 units of output at a minimum cost ($P_{x_1} = \$2$, $P_{x_2} = \$3$).

the isoquant,

$$-P_{X_1}\,(\Delta X_1) > P_{X_2}\,(\Delta X_2)$$

then the cost of producing the given output amount could be reduced by increasing X_2 and decreasing X_1 because the cost of an added unit of X_2 is less than the cost of the units of X_1 it replaces. On the other hand, if between two points on the isoquant,

$$P_{X_1}\,(\Delta X_1) < -P_{X_2}\,(\Delta X_2)$$

then the cost of producing the specified quantity of output can be reduced by using less X_2 and adding X_1. The equality of the least cost criterion signifies that a change in the input combination would increase the cost of producing the output.

The least cost criterion ensures that returns to the inputs are equal per dollar of cost. At the least cost combination, $X_1 = 6.2$ and $X_2 = 2.8$, a dollar invested in X_1 gives the same return as a dollar invested in X_2.

The nature of the solution obtained by the least cost criterion can be also illustrated using a three-dimensional drawing (Figure 4–11). The isocost line defines the set of inputs that can be purchased with the given total cost in the example, \$20.80. The manager can buy any amounts of the inputs up to and including combinations that cost

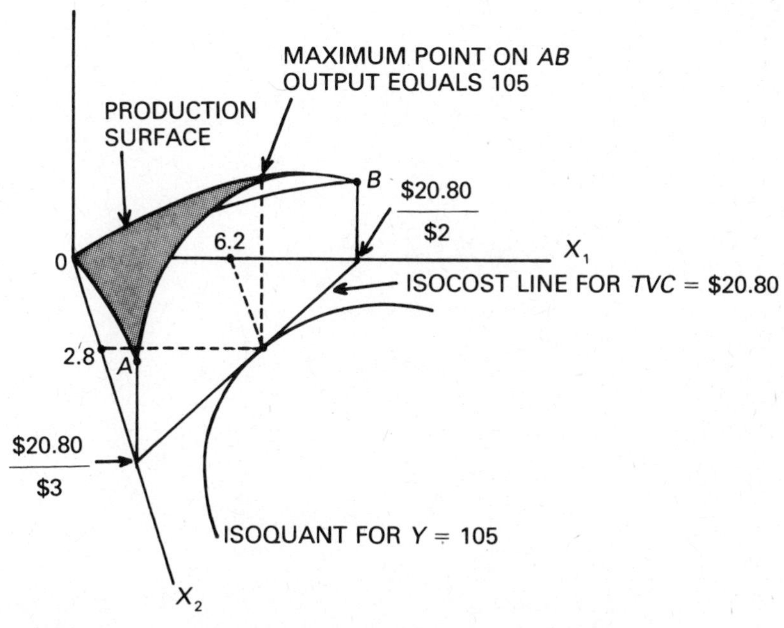

Figure 4–11. Schematic presentation of least cost criterion.

\$20.80. For each of these "feasible" input combinations, there exists an associated output. The production surface is vertically sliced above the isocost line; these outputs and all outputs resulting from less expensive input combinations make up the segment of the production surface shown in Figure 4–11. The curve *AB* represents all outputs that could result from an expenditure of \$20.80—the maximum of the outputs on the line *AB* is 105, resulting from 2.8 units of X_2 and 6.2 units of X_1.

The analysis in Figure 4–11 depicts a situation in which yield is maximized subject to TVC = \$20.80. But this results in exactly the same answer that would have been obtained from minimizing the total variable cost of producing 105 units of Y. One problem is the "dual" of the other. This type of problem occurs often in microeconomic theory (see Chapter 9) and, in this case, the graphic solution of the dual of the original problem seems clearer.

Least Cost–Algebraic Determination

Algebraic determination of the least cost combination follows the same procedure as that discussed for a geometric case. The principle involved is to equate the isoproduct and isocost slopes. For example,

$$MRS \text{ of } X_2 \text{ for } X_1 = -\frac{P_{X_2}}{P_{X_1}}$$

From the equations (4.1) and (4.5) and using the given prices,

$$-\frac{7 - X_2}{9 - X_1} = -\frac{3}{2} \tag{4.6}$$

and

$$X_2 = \frac{3X_1 - 13}{2}$$

Values for X_1 and X_2 must be calculated. Substituting the above expression for X_2 into the production function (4.1) gives a quadratic expression in Y and X_1. Setting $Y = 105$ in that equation and solving using the quadratic formula determines the value $X_1 = 6.2$.[5] Then,

$$X_2 = \frac{3(6.2) - 13}{2} = 2.8$$

[5]In the factor-product relationship, when solving the quadratic equation it was noted that the second derivative of the profit equation must be negative for the profit to be maximum. In this case, the second derivative of *TVC* equation must be positive for costs to be minimum. The alternative solution of X_1 does not meet this objective.

Thus the least cost combination for $Y = 105$, given production function (4.1) and the prices, is 6.2 units of X_1 and 2.8 units of X_2. This solution is identical to that obtained in Figure 4–10.

The marginal physical product equations can be used to determine the returns per dollar spent at the least cost point. Rewriting the least cost criterion as

$$\frac{MPP_{X_1}}{P_{X_1}} = \frac{MPP_{X_2}}{P_{X_2}}$$

and substituting the derivatives (4.2 and 4.3) and prices gives

$$\frac{14 - 2X_2}{3} = \frac{18 - 2X_1}{2} = \frac{14 - 2(2.8)}{3} = \frac{18 - 2(6.2)}{2} = 2.8$$

Thus, the last dollar spent on X_1 and X_2 is returning 2.8 units of output.

Some Special Cases

The preceding sections have shown how to determine least cost combinations of inputs when (1) the inputs are infinitely divisible (the isoquants are smooth curves rather than dots), (2) the solution always requires the use of both inputs, and (3) the *MRS* is decreasing rather than constant. Cases in which these requirements are not met will be discussed in this section. Although called "special," they certainly should not be regarded as uncommon.

The first exception to be considered is often called the *lumpy* input. In this case, inputs are not infinitely divisible but come in discrete units: 1, 2, 3, etc. If both inputs are lumpy, the isoquant map might appear as in Figure 4–12A. The three points in Figure 4–12A represent an isoquant because the inputs can be used in differrent combinations to produce a given output level. Such a map might arise when capital items of differing size and embodying differing amounts of input *services* are available for use in the production process. An isocost line is imposed on this isoquant map to touch the isoquant at point E. A tangency is not possible because the isoquant has no slope, but point E is nonetheless the least cost combination. This solution will be quite stable; a relatively large shift in relative input prices would be required to cause the least cost ratio to shift from E.

The second special case leads to the *corner* solution—a least cost combination that involves the use of a zero amount of one input. In this case, the least cost combination falls on one of the axes. This situation is depicted in Figure 4–12B. The isoquants are continuous and convex but intersect the axes. Neither input is limiting; production from one is possible in the absence of the other. When such an intersection occurs, it is possible for the isocost line to have a slope that is steeper than any slope on the isoquant. Again, tangency will not occur,

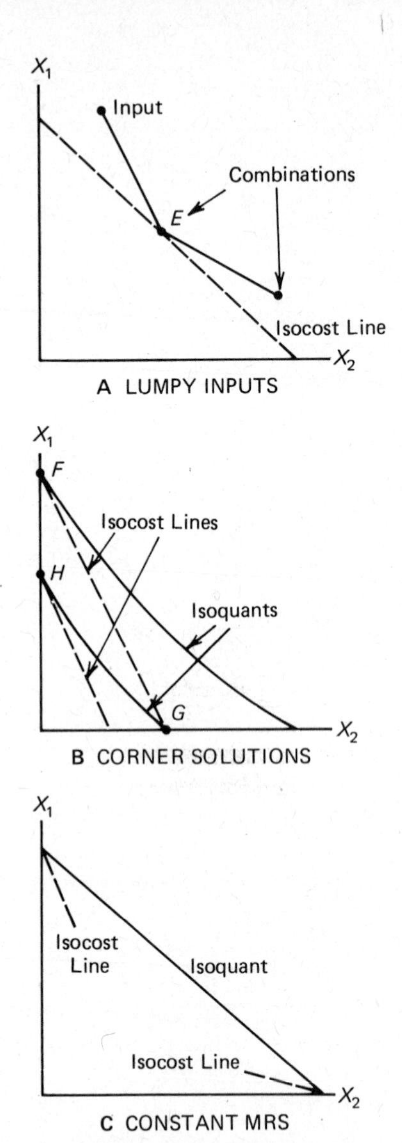

Figure 4–12. Least-cost combinations: special cases.
A. Lumpy inputs.
B. Corner solutions.
C. Constant *MRS*.

but the least cost combination will exist. When a given output level can be produced with more than one combination of inputs, there will always be at least one combination that costs the least. In this case, the least cost combination will involve either all X_1 and no X_2 or vice versa. In Figure 4–12B, point F represents the least cost combination for the higher isoquant. On the lower isoquant, point H represents the least cost combination; if the manager selected point G, he would maximize the cost of producing the yield level represented by the lower isoquant. (Note that the corner solution could occur if the tangency happened to fall exactly in the axis. This would seem unlikely.)

The corner solution will also exist when the *MRS* is constant and, hence, the isoquants are linear. This situation is shown in Figure 4–12C. The inputs in this case might be two grades or sources of fuel. Depending on the relative prices, either one fuel or the other will be used. If the isocost line were to be coincident (have the same slope) with the isoquant, all combinations of X_1 and X_2 would be least cost, and the manager would be indifferent among them.

Corner solutions usually appear to be stable, but this, of course, can be quite misleading. If the slopes of the isoquant and isocost line are close and the isoquant has a very gradual slope, a small shift in price could cause a large change in the least cost combination.

In the special cases just presented, the least cost criterion

$$MRS \text{ of } X_2 \text{ for } X_1 = -\frac{P_{X_2}}{P_{X_1}}$$

generally does not hold (the exceptions were noted). For lumpy inputs, the isoquant had no slope, while for corner solutions this equality was not attainable. For corner solutions, the least cost solution did involve minimizing the difference between the two slopes even though equality could not be achieved. Thus, in Figure 4–12B, the slopes of the isoquant and isocost line differ less at *F* than at *G*. (Corner solutions are discussed in more detail in Chapter 9 and Appendix II.)

ISOCLINES, EXPANSION PATHS, AND PROFIT MAXIMIZATION

Isoclines are lines or curves that pass through points of equal marginal rates of substitution on an isoquant map. That is, particular isocline will pass through all isoquants at points where the isoquants have a specified slope. There are as many different isoclines as there are different slopes or marginal rates of substitution on an isoquant.

The expansion path, too, is an isocline. However, an expansion path is a special isocline that connects the least cost combinations of inputs for all yield levels. In other words, the expansion path connects points on an isoquant map that, in addition to meeting the necessary condition of economic efficiency, also meets the sufficient condition. The expansion path is a unique isocline that is selected from the set of all isoclines. The choice indicator is the input price ratio: On the expansion path, the marginal rate of substitution must equal the input price ratio. If resources are technical substitutes and the marginal rate of substitution is decreasing, any change in the relative prices of the inputs also changes the expansion path. The change in input prices shifts the expansion path to a new isocline that assumes the role of the expansion path.

Expansion paths and isoclines have important economic implications. If the expansion path is a straight line emanating from the origin, then the inputs will be used in the same proportions at all output levels. When the expansion path is curved, the proportion of the inputs that must be used to achieve the least cost combination will vary among yield levels. For example, feeding rations will likely include more protein relative to carbohydrates at lower weight levels of hogs and relatively more carbohydrates and less protein at heavier weights. Similarly, fertilizer nutrient ratios may vary depending upon the planned level of corn yield.

Expansion paths can be constructed from an isoquant map or derived mathematically. In Figure 4–13, an isocost line with a slope of $-3/2$, representing $P_{X_2} = \$3$ and $P_{X_1} = \$2$, is drawn tangent to each isoquant. Then a line, the expansion path, is drawn connecting the points of tangency. At all points on this expansion path

$$MRS \text{ of } X_2 \text{ for } X_1 = -3/2$$

which also equals the slope of the isocost line, for example, $-3/2$.

The expansion path for the price ratio $-3/2$ intersects the X_1 axis at $X_1 = 4.3$, where output is 59. Outputs below this amount would be

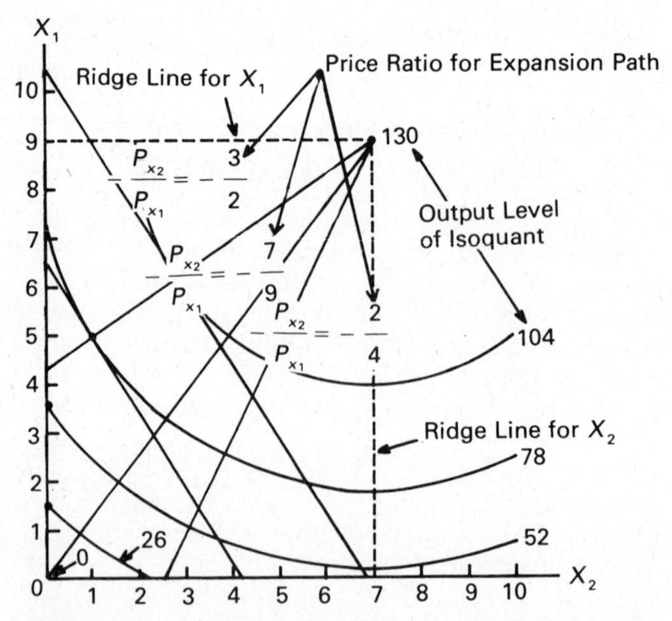

Figure 4–13. Isoquant map with expansion paths for three different input price ratios.

produced most cheaply using only X_1—the corner solutions prevail; outputs above this amount can be produced at least cost using some combination of the two inputs. For 52 units of output, the least cost combination is $X_1 = 3.6$ and $X_2 = 0$; for 78 units, $X_1 = 5$ and $X_2 = 1$; for 130 units, the least cost and only possible combination is 9 units of X_1 and 7 units of X_2.

Just as there is an isoquant for each output level, there is an expansion path for each price ratio. Two additional expansion paths are included in Figure 4–13; the corresponding isocost lines are omitted for clarity. Note that as the price of X_2 drops relative to the price of X_1, increasing amounts of X_2 are included in the least cost combination for any given output. For 78 units of output, the least cost combination is $X_1 = 5$ and $X_2 = 1$ when the price ratio is $-3/2$, but it is 3.3 and 2.6 when the price ratio is $-7/9$, and 2.6 and 3.8 when the price ratio is $-2/4$. All expansion paths converge at the point of maximum output.

A general expression for all isoclines can be derived from the production function equation. The criterion for the least cost combination of inputs is

$$\frac{MPP_{X_2}}{MPP_{X_1}} = \frac{P_{X_2}}{P_{X_1}}$$

Substituting the derivatives of the production function into this expression and solving for X_1 as a function of X_2 and the prices gives

$$X_1 = 9 - [(P_{X_1}/P_{X_2})(7 - X_2)] \tag{4.7}$$

which will give the isocline equation for any particular slope, represented by (P_{X_1}/P_{X_2}). [Thus, (4.7) represents a family of curves with the price ratio as the determining parameter.]

The expansion path equation is determined by substituting the relevant input prices into (4.7). When X_1 costs \$2 and X_2 costs \$3, the resulting expansion path is

$$X_1 = \frac{13}{3} + \frac{2}{3}X_2$$

when output exceeds 59 units. Below yields of 59, only X_1 would be used. The equations for the other two expansion paths in Figure 4–13 are

$$X_1 = \frac{9}{7}X_2 \text{ when } \frac{P_{X_1}}{P_{X_2}} = 9/7$$

and

$$X_1 = -5 + 2X_2 \text{ when } \frac{P_{X_2}}{P_{X_1}} = 2/4$$

The dashed lines in Figure 4–13 represent special types of isoclines called ridge lines. Ridge lines represent the limits of economic relevance, the boundaries beyond which the isocline and isoquant maps cease to have economic meaning or the boundaries beyond which the necessary condition of economic efficiency is not met. The horizontal ridge line represents the points where MPP_{X_1} is zero; the vertical line, the points where MPP_{X_2} is zero. The slope of the isoquant was shown to be

$$MRS \text{ of } X_2 \text{ for } X_1 = -\frac{MPP_{X_2}}{MPP_{X_1}}$$

On the ridge line for X_1, MPP_{X_1} is zero, the tangent to the isoquant is vertical and has no defined slope. On the ridge line for X_2, MPP_{X_2} is zero and the isoquant has a zero slope and thus $MRS = 0$. Ridge lines are so named because they trace the high points up the side of the production surface, much like mountain ridges that rise to the peak of the mountain.

Ridge lines represent the points of maximum output from each input, given a fixed amount of the other input. When $X_1 = 1$, output can be increased by adding X_2 up to the amount denoted by the ridge line (7 units). At that point, output from X_2 is a maximum, given one unit of X_1 and MPP_{X_2} is zero. Past $X_2 = 7$, MPP_{X_2} is negative while MPP_{X_1} is positive; the inputs have an opposite effect on output and are no longer substitutes. Thus, the ridge lines denote the limits of substitution. Within the ridge lines, in Stage II, the isoquants have a negative slope and the inputs are substitutes. Outside the ridge lines in Stage III, the inputs do not substitute in an economically meaningful way and production in this area would be irrational. Output is a maximum, 130, where the ridge lines, and all other isoclines, converge.[6]

Expansion Path and Profit Maximization

The expansion path traces out the least cost combination of inputs for every possible output level. The question now arises, which output level is the most profitable? Conceptually this question is answered by proceeding on the expansion path, that is, increasing output until the value of the product added by increasing the two inputs along the

[6]The stages of production are discussed in more detail when homogeneous functions are analyzed in Chapter 6.

expansion path is equal to the combined cost of the added amounts of the two inputs. Viewed from the input side, this is the same as saying that the *VMP* of each input must equal the unit price of that input; viewed from the output side, it is the same as saying marginal cost must equal marginal revenue. Thus, while all points on an expansion path represent least cost combinations, only one point represents the maximum profit output.

Analytically, several methods of determining the optimal input combination are available. They are all based on the same principle; only the methods differ. One method is to maximize profits directly. For one output and two variable inputs, the profit equation is

$$Profit = P_Y Y - P_{X_1} X_1 - P_{X_2} X_2 - TFC$$

where $Y = f(X_1, X_2)$. Maximizing this function with respect to the variable inputs gives two equations in two unknowns

$$\frac{\partial Profit}{\partial X_1} = P_Y \frac{\partial Y}{\partial X_1} - P_{X_1} = 0$$

$$\frac{\partial Profit}{\partial X_2} = P_Y \frac{\partial Y}{\partial X_2} - P_{X_2} = 0 \qquad (4.8)$$

The unknowns are X_1 and X_2—the solution of (4.8) will determine the profit-maximizing amounts of the two inputs. Equations (4.8) can be written as follows:

$$VMP_{X_1} = P_{X_1}$$

$$VMP_{X_2} = P_{X_2}$$

Thus, the profit-maximizing criterion requires that the marginal earnings of each input must be equal to its cost; this must be true for both inputs simultaneously. Profit will be a maximum at this point if, for each input, $MPP < APP$ and both are decreasing, that is, input use is in Stage II and the isoquants are convex. This criterion must be fulfilled at the point of maximum profit; it is the familiar requirement of Diminishing Returns.[7]

For production function (4.1), the marginal physical products, equations (4.2) and (4.3), when multiplied by $P_Y = \$0.65$, result in

$$VMP_{X_1} = \$(18 - 2X_1)0.65$$

$$VMP_{X_2} = \$(14 - 2X_2)0.65$$

[7]Again, the sufficient conditions for a maximum would require a second derivative test. We will not present those conditions here, but the reader should be aware of the need for such a test. See Appendix I.

Equating the *VMP*s with the input prices of \$9 and \$7 for P_{X_1} and P_{X_2}, respectively, gives two equations and two unknowns:

$$\$(18 - 2X_1)0.65 = \$9$$
$$\$(14 - 2X_2)0.65 = \$7$$

The solutions are 2.08 for X_1 and 1.6 for X_2, respectively. Substituting these values in the production function predicts a value of Y equal to 53 units. Profit at that point would be

$$\begin{aligned} Profit &= \$0.65 \cdot 53 - \$9 \cdot 2.08 - \$7 \cdot 1.60 - TFC \\ &= \$4.53 - TFC \end{aligned}$$

The optimum criterion for two variable inputs is often expressed in somewhat different forms. By the appropriate manipulations, (4.8) can be rewritten

$$\frac{VMP_{X_1}}{P_{X_1}} = 1, \quad \frac{VMP_{X_2}}{P_{X_2}} = 1$$

or, because both equal one,

$$\frac{VMP_{X_1}}{P_{X_1}} = \frac{VMP_{X_2}}{P_{X_2}} = 1$$

All variable inputs must be earning as much as they cost on the "margin." Rewriting (4.8) in a different form,

$$MPP_{X_1} = \frac{P_{X_1}}{P_Y}$$
$$MPP_{X_2} = \frac{P_{X_2}}{P_Y}$$

This last form results in a convenient method of depicting the optimum on a graph of the expansion path. Solving the first equation ($18 - 2X_1 = 13.85$) gives the optimal amount of X_1, 2.08. This line, $X_1 = 2.08$, when graphed, will trace out all points representing the optimal amount of X_1, given X_2. Solving the second equation gives the optimal amount of X_2, given X_1. This is $X_2 = 1.6$. Profit will be maximized where the two lines cross—and that point will always be on the expansion path. This solution is shown in Figure 4–14.

Figure 4–14 depicts a graphical solution to the two simultaneous equations in (4.8). The solution is particularly simple in this case be-

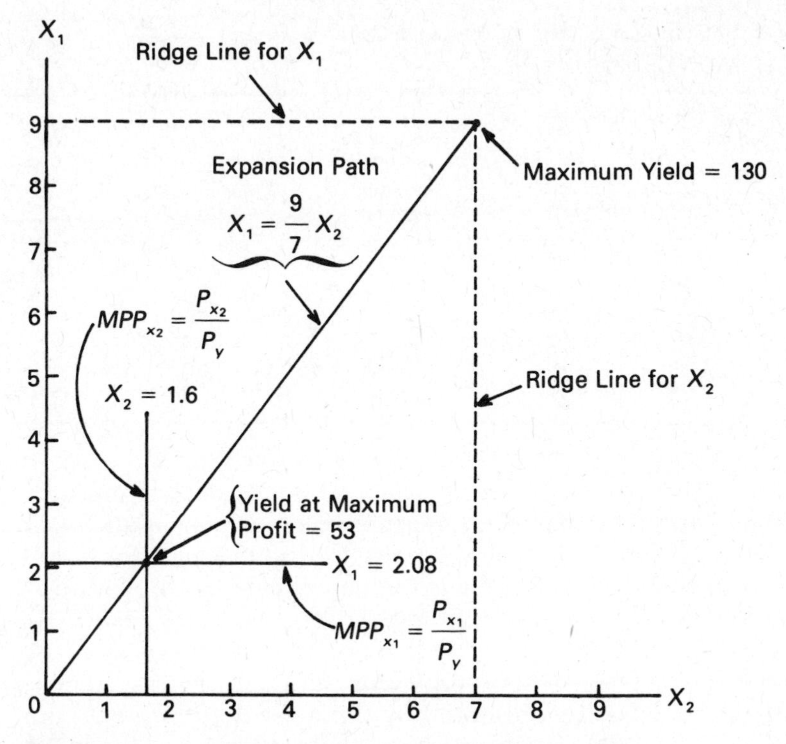

Figure 4–14. Locating the optimum amounts of two variable inputs.

cause of the algebraic form of the equations. When the production function is more complex (such as when the two inputs interact), the marginal product of one input will be a function of the quantities of both inputs, and the lines representing optimal input use will not necessarily be parallel to the input axes. The concept is valid in general, however, and the optimum combination will occur where the lines (or curves) intersect.

Cost Curves and Profit Maximization

Another method that can be used to determine the optimum output on a graph is to develop the appropriate cost curves. Inputs are usually measured in different physical units and have different prices; any difficulties caused by differences in physical measures can be avoided by converting to dollar measures and computing costs. Once the appropriate curves are derived, the optimum output can be found by equating MC to P_Y. This method is illustrated in Table 4–7 and Figure 4–15.

The left side of Table 4–7 contains points from the sub-production function determined by the expansion path for the input prices $P_{X_1} = \$9$ and $P_{X_2} = \$7$. These points were read from Figure 4–13 by deter-

Table 4–7 Computing Costs for the Least Cost Combination of Inputs (P_{X_1} = \$9, P_{X_2} = \$7)

Least Cost Combinations of X_1 and X_2 for Indicated Outputs			*Minimum Total Outlay for Output*	*Average Variable Cost*	*Marginal Cost*
X_1	X_2	Y	*TVC*	*AVC*	*MC*
0.00	0.00	0	0	...	
					\$0.53
0.96	0.75	26	\$ 13.89	\$0.53	
					0.58
2.00	1.55	52	28.85	0.56	
					0.72
3.30	2.55	78	47.55	0.61	
					0.95
5.00	3.90	104	72.30	0.70	
					2.22
9.00	7.00	130	130.00	1.00	

Source: Least cost combinations from Figure 4–13.

mining where the expansion path intersects each isoquant. *TVC*, *AVC*, and *MC* were then computed for each output level. Fixed costs are omitted in this example. *TVC* is the combined cost of X_1 and X_2; for example, in line 2, Table 4–7, \$13.89 = \$9 · 0.96 + \$7 · 0.75. *AVC* and *MC* are computed as always.

AVC and *MC* are graphed in Figure 4–15. If the price per unit of output, P_Y, is \$0.65, represented by the horizontal line 0.65 units up the dollar axis, then the optimum output is 53 units of output, determined by dropping a perpendicular from the intersection of *MC* and

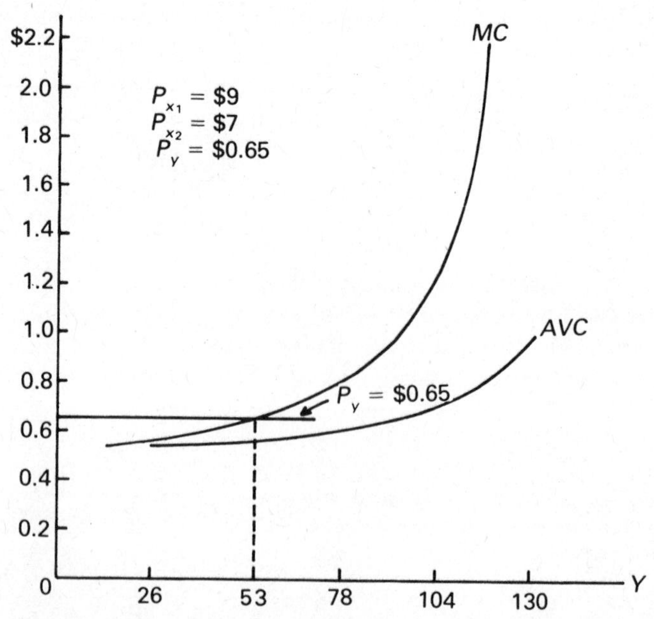

Figure 4–15. Determining optimum output for a production process utilizing two variable inputs.

P_Y. Reference to Figure 4–13 shows that 53 units of output would result from the use of a combination of inputs on the expansion path slightly past the 52-bushel isoquant.

This method also lends itself to an analytical solution. The equation for one variable input as expressed by the expansion path is substituted directly into the production function. The resulting sub-production function expresses output along the expansion path as a function of one input. (This is called a *parametric* representation.) The same expression for one variable input as given by the expansion path equation is also substituted into the *TVC* equation; variable costs are thus also restricted to the least cost combination. The inverse of the sub-production function, expressing input as a function of output, is found and substituted into *TVC* to express costs as a function of output.

For the present example, the expansion path is $X_1 = (9/7)\ X_2$. When this expression for X_1 is substituted into the production function,

$$Y = 18X_1 - X_1^2 + 14X_2 - X_2^2$$

the result is

$$Y = \frac{260}{7} X_2 - \frac{130}{49} X_2^2$$

For total variable cost,

$$TVC = \$9X_1 + \$7X_2$$

the substitution is

$$TVC = \$9\left(\frac{9}{7} X_2\right) + \$7X_2$$

$$= \frac{\$130}{7} X_2$$

Using the quadratic formula, the inverse production function for outputs between zero and 130 is found to be

$$X_2 = 7 - \frac{49}{260} \sqrt{\left(\frac{260}{7}\right)^2 - \frac{520}{49} Y}$$

Total variable cost along the expansion path as a function of output is given by

$$TVC = \$130\left(1 - \frac{7}{260} \sqrt{\left(\frac{260}{7}\right)^2 - \frac{520}{49} Y}\right)$$

Marginal cost is, of course, the derivative of TVC with respect to Y:

$$MC = \frac{dTVC}{dY} = \frac{130}{7}\left[\left(\frac{260}{7}\right)^2 - \frac{520}{49}Y\right]^{-1/2}$$

Equating $MC = P_Y = \$0.65$, and solving results in the optimum yield of 53.08. Substituting $Y = 53.08$ in the inverse production function gives $X_2 = 1.61$, from which the expansion path equation predicts X_1 to be 2.08. This is the least cost combination of inputs that would produce the most profitable level of output. Cost and profit are as shown above.

PROFIT MAXIMIZATION—TWO MATHEMATICAL EXAMPLES

Examples that incorporate the least cost and profit-maximizing principles for two variable inputs are presented in this section. They utilize all the concepts discussed in this chapter but use different production functions. While the production functions might vary, the solution to the factor-factor and factor-product problems will be the same.

The first example will be a quadratic equation similar to (4.1) but with an interaction term. Input interaction is common in agriculture; this example will illustrate the effects of such a term. The second example will utilize the Cobb-Douglas function, a famous form that has had wide application in economics.

Quadratic Function with Interaction

The quadratic equation (4.1) represented a production process where the yield response to one input was independent of the other input. In effect, the response to each input was determined by sub-production functions for each input because the yield response to X_1 and X_2 were added to obtain total yield. Another form of the quadratic function often found in agricultural research includes input interaction. For example, adding an interaction term to (4.1) gives

$$Y = 18X_1 - X_1^2 + 14X_2 - X_2^2 + X_1X_2$$

The interaction term, X_1X_2, has a coefficient value of one for this example but could take any value in general. The effect of the interaction term is best illustrated by examining the marginal products

$$\frac{\partial Y}{\partial X_1} = MPP_{X_1} = 18 - 2X_1 + X_2$$

$$\frac{\partial Y}{\partial X_2} = MPP_{X_2} = 14 - 2X_2 + X_1$$

The marginal product of each input is a function of the quantities of both inputs. For example, the marginal response to X_1 depends not only on the amount of X_1 but also on the amount of X_2 being used. In this example, increases in one input increase the marginal product of the other; such inputs are called *complements*. When one input does not affect the marginal product of the other, as in (4.1), the inputs are called *independent*. When the interaction is negative, they are called *competitive*.[8]

In our example, the addition of the interaction term shifts the marginal product equations upward. That is, increasing X_2, for example, amounts to shifting the intercept of MPP_{X_1} up the Y axis. Compared to equation (4.1), the interaction term increases the yield resulting from any combinations of X_1 and X_2, for input combinations not on the axes.

As before, setting the marginal product equations to zero and solving will determine the values of X_1 and X_2 that produce the maximum yields. Equated to zero and rewritten, the marginal product equations are

$$-2X_1 + X_2 = -18$$
$$X_1 - 2X_2 = -14$$

which require simultaneous solution. [Equations (4.2) and 4.3) could be solved individually.] For example, from the first equation, $X_2 = 2X_1 - 18$; this equation can be substituted into the second equation to determine the numerical value of X_1. When X_1 is known, X_2 can be determined. The values $X_1 = 16.66$ and $X_2 = 15.33$ result in the maximum yield level, 257.33. For comparison, recall that with the interaction term omitted, the quadratic equation (4.1) had a maximum yield of 130 when $X_1 = 9$ and $X_2 = 7$. Thus, the positive interaction increases yield response substantially.

The isoquant equation is found in a manner analogous to equation (4.4) and is only slightly more complex:

$$X_1 = 9 + 0.5 \pm 0.5(324 + 92X_2 - 3X_2^2 - 4Y)^{1/2}$$

Isoquants for 50-unit increments in output are presented in Figure 4–16. Comparison with Figures 4–13 and 4–15 shows the effect of the interaction term.

[8]Frisch, Ragnar, *Theory of Production*, Chicago: Rand McNally and Company, 1965, p. 59. Frisch examines the signs of $\partial MPP_{X_1}/\partial X_2$ and $\partial MPP_{X_2}/\partial X_1$ to determine the relationship between inputs. Examination of Figure 4–1 suggests that nitrogen fertilizer and plant population were complements in the Missouri experiment. For the milk experiment to be presented later in this chapter, hay and grain are competitive. Note that Frisch's definition of complements is weaker than the one presented above on page 105.

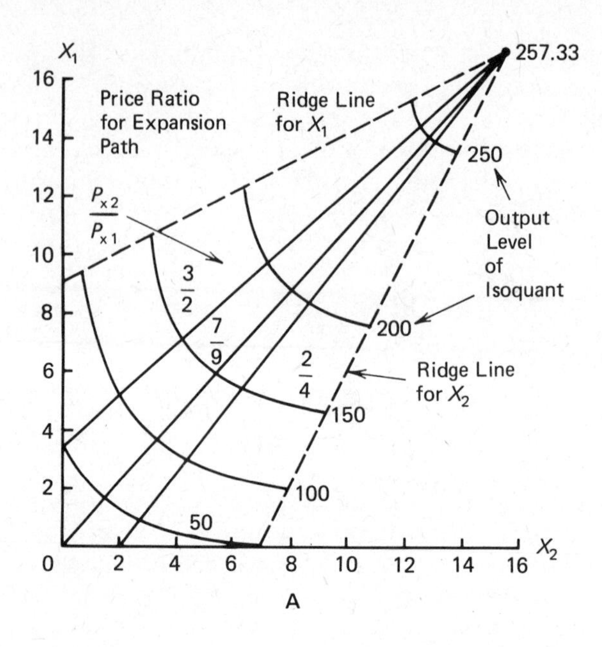

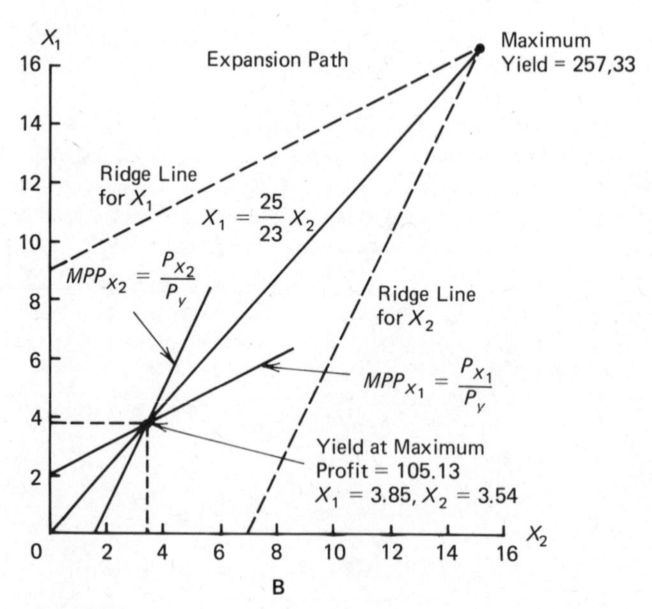

Figure 4–16. Isoquant map and determination of the optimum input amounts, quadratic function with input interaction.

The exact marginal rate of substitution is now given by

$$MRS \text{ of } X_2 \text{ for } X_1 = \frac{MPP_{X_2}}{MPP_{X_1}} = \frac{14 - 2X_2 + X_1}{18 - 2X_1 + X_2}$$

The interaction term causes a change in the slope of the isoquants. For example, when X_2 is constant, increasing X_1 simultaneously decreases the denominator of the *MRS* while increasing the numerator. In contrast, when interaction is not present, the numerator is constant [see equation (4.5)]. Thus, on a given vertical line in Figure 4–16, say, $X_2 = 6$, the isoquants become more steeply inclined than on a similar vertical line in Figure 4–13. In general, the effect of interactions on isoquant slopes will depend on the nature of the particular type of interaction present.

The equation for the family of isoclines is found by equating the marginal rate of substitution to the price ratio. For this example,

$$\frac{MPP_{X_2}}{MPP_{X_1}} = \frac{14 - 2X_2 + X_1}{18 - 2X_1 + X_2} = \frac{P_{X_2}}{P_{X_1}} = r$$

so that the isocline equation is given by

$$X_1 = \frac{(18r - 14) + (2 + r)\,X_2}{2r + 1}$$

which defines a family isoclines depending on the price ratio, r. Expansion paths for three different price ratios are presented in Figure 4–16A; the expansion path for the ratio, $P_{X_2}/P_{X_1} = 7/9$ passes through the origin.

Ridge lines fall on the set of input combinations for which the marginal products of the inputs are zero. Setting the marginal product equations equal to zero and expressing X_1 as a function of X_2 gives the ridge line for X_1,

$$X_1 = 9 + \tfrac{1}{2}\,X_2$$

and the ridge line for X_2,

$$X_1 = -14 + 2X_2$$

Inside the ridge lines, both inputs have positive marginal products; Stage II is the region within the ridge lines. Stage III lies outside the ridge lines. The ridge lines intersect at the point of maximum yield. Ridge lines for this example are plotted in Figure 4–16; when compared to Figure 4–13, it can be seen that the positive interaction extends the region of substitution, Stage II, to higher yield levels.

The profit-maximizing criterion requires that the value of the marginal product of an input equals the price of that input; this condition must hold simultaneously for all inputs. Using $P_Y = 0.65$, $P_{X_1} = \$9$ and $P_{X_2} = \$7$, the resulting equations are

$$\$0.65\ (18 - 2X_2 + X_1) = \$9$$
$$\$0.65\ (14 - 2X_1 + X_2) = \$7$$

which is a system of two equations and two unknowns. Each variable appears in each equation; the solution is simultaneous as explained above for the marginal product equations. The solutions are 3.85 for X_1 and 3.54 for X_2, and the yield that maximizes profits is 105.13 units. Profit, computed as before, is \$8.90 minus fixed costs.

The profit-maximizing solution is shown on a graph in Figure 4–16B. The two equations that specify the necessary conditions for profit maximization, $MPP_{X_1} = P_{X_1}/P_Y$ and $MPP_{X_2} = P_{X_2}/P_Y$ are, in the same order,

$$X_1 = 2.08 + 0.05X_2$$
$$X_1 = -3.23 + 2X_2$$

These intersect at $X_1 = 3.85$ and $X_2 = 3.54$ where profit is a maximum. This point is located on the expansion path. (Notice that these two lines could be interpreted as ridge lines for the profit surface, given the assumed prices.) Comparison with Figure 4–14 suggests the interaction effect nearly doubles the most profitable yield level.

The equation for *TVC* can be derived by proceeding exactly as illustrated above for the quadratic equation without interaction. In this case, the expansion path equation is $X_1 = (25/23)\ X_2$; this would be substituted directly into the production function and total variable cost function. Because no new interpretations are required, determination of the cost functions will be left as an exercise for the student.

The Cobb-Douglas Function

Assume a Cobb-Douglas function with prices as follows:

$$Y = X_1^{1/5} X_2^{3/5};\ P_{X_1} = \$3,\ P_{X_2} = \$1, \text{ and } P_Y = \$10 \qquad (4.9)$$

The objective is to find the least cost combination of inputs, X_1 and X_2, as well as the level of output at which net returns or profits are maximized.

The equation for the marginal rate of substitution is

$$\frac{\frac{\partial Y}{\partial X_2}}{\frac{\partial Y}{\partial X_1}} = \frac{\frac{3}{5} X_2^{-2/5} X_1^{1/5}}{\frac{1}{5} X_1^{-4/5} X_2^{3/5}} = \frac{3X_1}{X_2} \tag{4.10}$$

and thus the expansion path equation will be given by

$$\frac{3X_1}{X_2} = \frac{1}{3}$$

or

$$X_2 = 9X_1 \tag{4.11}$$

The expansion path equation (4.11) defines the coordinates of the least cost combinations of the inputs. For example, if $X_1 = 1$, then X_2 must equal 9 if the combination is to be a least cost combination. If $X_1 = 3$, then $X_2 = 27$, etc. For the least cost combinations, input X_1 can be expressed in terms of X_2 or vice versa.

The expansion path equation can be viewed as a restriction; out of all possible input combinations, it will select only those that are cost-minimizing. This restriction can be substituted directly in the production function. If X_2 in (4.9) is expressed in terms of X_1, then (4.9) can be written as

$$Y = X_1^{1/5} (9X_1)^{3/5}$$
$$Y = 9^{3/5} X_1^{4/5} \tag{4.12}$$

The least cost input combination for any specified level of output, say, Y_0, can be found by solving (4.12) for X_1. That value for X_1 can then be substituted into the expansion path equation (4.11) to determine the appropriate value of X_2. To determine the level of output at which profits are maximized, a profit equation is needed.

$$Profit = P_Y Y - X_1 P_{X_1} - X_2 P_{X_2} - TFC$$
$$Profit = \$10Y - \$3X_1 - \$1X_2 - TFC \tag{4.13}$$

But, from (4.11),

$$X_2 = 9X_1$$

so that

$$Profit = \$10Y - \$12X_1 - TFC$$

If costs are related to inputs rather than output, then output, Y, in (4.13) has to be expressed in terms of X_1. Y is defined in terms of X_1 in (4.12). Thus, the profit equation can be expressed as a function of X_1 as follows

$$Profit = \$10 \cdot 9^{3/5}\, X_1^{4/5} - \$12X_1 - TFC$$

where the expression

$$\$10 \cdot 9^{3/5}\, X_1^{4/5}$$

represents total value product and $\$12X_1$ represents total variable cost. Their respective slopes will represent the value of the marginal product and marginal factor cost for outputs and costs restricted to the set of points falling on the expansion path. Because the marginal factor cost will be constant and equal to input prices (assuming pure competition), the derivative of profit will be

$$\frac{d(Profit)}{dX_1} = \$\,\frac{(4/5)(10)\, 9^{3/5}}{X_1^{1/5}} - \$12 \tag{4.14}$$

or

$$\$\,\frac{8 \cdot 9^{3/5}}{X_1^{1/5}} = \$12$$

which has the solution

$$X_1 = \frac{2^5\, 9^3}{3^5} = 96$$

From the expansion path (4.11),

$$X_2 = 9 \cdot 96 = 864$$

In this example, the least cost and profit-maximization point is achieved when $X_1 = 96$ and $X_2 = 864$ units. The level of output can be calculated by substituting the input values into either the original production function, (4.9), or in (4.12). If (4.12) is used,

$$Y = 9^{3/5}\, 96^{4/5} = 144$$

Principles discussed earlier can be tested. At the point where profits are maximized, it must be true that

$$\frac{VMP_{X_1}}{P_{X_1}} = \frac{VMP_{X_2}}{P_{X_2}} = 1$$

$$\frac{P_Y MPP_{X_1}}{P_{X_1}} = \frac{P_Y MPP_{X_2}}{P_{X_2}} = 1$$

From (4.10),

$$MPP_{X_1} = \frac{X_2^{3/5}}{5X_1^{4/5}} = \frac{864^{3/5}}{5 \cdot 96^{4/5}} = 0.3$$

$$MPP_{X_2} = \frac{3X_1^{1/5}}{5X_2^{2/5}} = \frac{3 \cdot 96^{1/5}}{5 \cdot 864^{2/5}} = 0.1$$

Multiplying MPP_{X_1} and MPP_{X_2} by the price of Y,

$$P_Y MPP_{X_1} = 0.3 \cdot \$10 = \$3$$

$$P_Y MPP_{X_2} = 0.1 \cdot \$10 = \$1$$

and finally

$$\frac{VMP_{X1}}{P_{X1}} = \frac{VMP_{X_2}}{P_{X_2}}$$

or

$$\frac{\$3}{\$3} = \frac{\$1}{\$1} = 1$$

If costs are related to output rather than inputs, the profit-maximizing principle $MC = P_Y$ can be used. In this case, the most profitable level of output, Y, must be determined first, and then the least cost combination of inputs producing the level of output must be determined. From (4.13),

$$TVC = \$12X_1$$

and because

$$Y = 9^{3/5}\, X_1^{4/5}$$

then

$$X_1 = \frac{Y^{5/4}}{9^{3/4}}$$

This equation is the inverse production function for input and output points restricted to the expansion path.

Total variable cost can be related to output by substituting output, Y, for input, X_1, in the cost equation

$$TVC = \frac{\$12\ Y^{5/4}}{9^{3/4}} \tag{4.15}$$

$$MC = \frac{dTVC}{dY} = \frac{\$15\ Y^{1/4}}{9^{3/4}} \tag{4.16}$$

Using the principle that $MC = P_Y$, the optimum amount of output is

$$\frac{\$15\ Y^{1/4}}{9^{3/4}} = \$10 \tag{4.17}$$

$$Y = \frac{2^4\ 9^3}{3^4} = 144$$

Thus, the solution for the optimum amount of Y of 144 units is the same as that obtained when profit was expressed as a function of inputs. The amount of X_1 and X_2 can be determined by substituting 144 units of Y in equation (4.12), solving for X_1, and then using the expansion path equation, (4.11), to determine the amount of X_2 needed for the least cost combination.

In the above example, profits can be calculated by substituting the level of Y and required inputs X_1 and X_2 in the profit equation (4.13) [Profit $= \$1440 - 96 \cdot \$3 - 864 \cdot \$1 - TFC = \$288 - TFC$]. Any other combination of X_1 and X_2 that produces 144 units of Y would cost more and thus would result in lower profits.

SUBSTITUTION AND EXPANSION EFFECTS

Any change in relative prices among variable inputs changes the slope of the isocost line and results in a new least cost combination of the variable factors used to produce a given output. This change in factor mix is due to the substitution effect. A relatively cheaper input is substituted for a more expensive input. The degree of change will depend on the rate at which resources substitute for each other. If the marginal rate of substitution among inputs is constant (corn and sorghum grain), a decline in the price of sorghum may cause sorghum to replace all of the corn used in a feeding ration or vice versa. If the marginal rate of substitution is diminishing, the substitution effect among inputs will tend to be less dramatic.

In the example from the previous section, if price of X_1 decreases from \$3 to \$2 per unit, and the price of X_2 remains at \$1 per unit, the slope of the new isocost line is 1/2 and the new expansion path equation would be $X_2 = 6X_1$. The new least cost input combination for producing 144 units of Y consists of 131 units of X_1 and 788 units of X_2. Thus, the new X_1 and X_2 mix calls for 35 more units of X_1 (relatively cheaper input) and 76 fewer units of X_2 (relatively more expensive input). In Figure 4–17, the substitution effect is illustrated by moving from point A to point B on the 144-unit isoproduct curve.

Because the price of Y of \$10 has not changed and the cost of producing 144 units of Y is lowered due to a decrease in the price of X_1, the values of the marginal products at point B, Figure 4–17, are greater than their respective input prices. From equation (4.10), VMP_{X_1} and VMP_{X_2}, when $X_1 = 131$ and $X_2 = 788$, are \$2.2 and \$1.1, respectively. While the least cost criterion is still satisfied at point B, Figure 4–17, the profit-maximizing condition is not.

The profit-maximizing condition is met on a higher isoquant. Following the same procedure as that outlined for equations (4.11) through (4.14), the new isoquant represents 216 units of Y. The expansion of

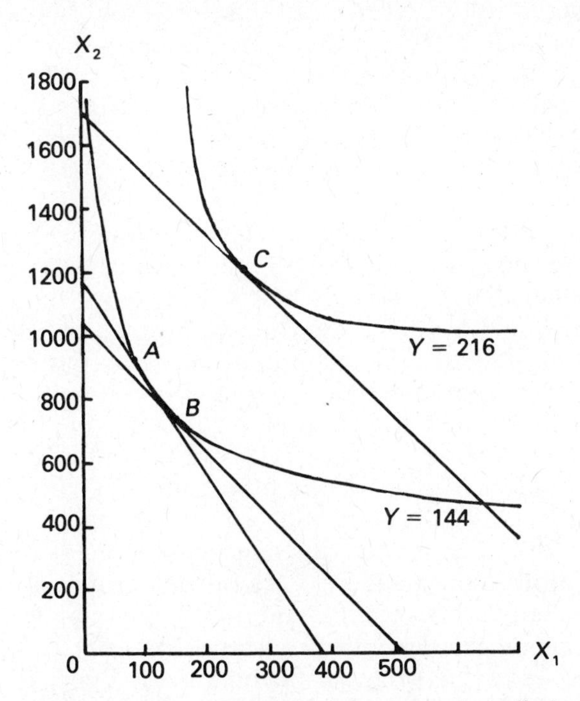

Figure 4–17. Substitution and expansion effects among variable inputs.

Y from 144 to 216 calls for 216 units of X_1 and 1296 units of X_2. This change to point C in Figure 4–17 is due to the expansion effect. At that point, the $VMP_{X_1} = P_{X_1}$ and $VMP_{X_2} = P_{X_2}$ or

$$\frac{VMP_{X_1}}{P_{X_1}} = \frac{VMP_{X_2}}{P_{X_2}} = 1; \frac{\$2}{\$2} = \frac{\$1}{\$1} = 1$$

The total value product is \$2160, the total cost of the two variable inputs, X_1 and X_2, is \$1728, and the profit is \$432 (before fixed costs).

Relative Strengths of Substitution and Expansion Effects

For the relatively more expensive input, X_2 in our example, the two effects work in opposite directions. The substitution effect tends to reduce the amount of the input that becomes relatively more expensive and substitute the cheaper input for it; the expansion effect, on the other hand, tends to increase the use of the expensive input. This increase is due to imperfect substitutability among inputs. If inputs substitute at a constant rate, the power of the expansion effect in restoring the use of the expensive input is nil. If inputs substitute for each other at a diminishing rate, the expansion effect will increase the use of the more expensive input. If the increase in the expensive input is greater than the decrease in that input due to the substitution effect, the expansion effect outweighs the substitution effect and more of the expensive input will be used.

If the expansion effect simply restores the use of the expensive input to its original level, the two effects are neutral. If less of the expensive input is used after the price change, the substitution effect is stronger.

In the example, the expansion effect is stronger than the substitution effect because at C, Figure 4–17, 1296 units of X_2 are used; this amount is greater than the 864 units of X_2 used in the original position, point A. In this case, even though the two inputs are technical substitutes on a given isoquant, the changing level of output causes them to become economic complements.

In cases where the expansion effect is stronger than the substitution effect, a decrease in the price of an input may result in an increase in the market price of the other input. That is, because the price of an input falls, farmers will use more of both. But when *all* farmers buy more of the input, the aggregate effect is to increase the price of that input to all farmers. One farmer cannot affect the price of an input by buying more; all farmers acting in concert can. Therefore, although one farmer faces constant input prices in pure competition, prices may rise when all farmers make the same adjustment in production practices.

In cases where the substitution effect predominates, a decrease in the price of one input will ordinarily result in a corresponding decline in the market price of other inputs. For example, a decrease in corn price usually causes the prices of other feed grains to drop. In this case, farmers will use less of the other feed grains and the aggregate effect is to cause their market price to fall.

Derived Demand for Inputs

In Chapter 3, it was determined that the derived demand curve for one variable input is the *VMP* curve in Stage II. But when the production process involves two (or more) variable inputs, the value of the marginal product curves for each input, such as those given in (4.8), are not the demand curves for the inputs. Essentially, this is because of the existence of interrelationships between the inputs as described by the substitution and expansion effects. Another, more mathematically astute way to visualize the problem would be to say that the demands for X_1 and X_2 depend on the simultaneous solution of the two equations in (4.8); solving either one alone will not provide a satisfactory solution.

When the price of one input changes, the optimal quantities of both inputs will change because of the simultaneous impact of the substitution effect and the expansion effect, or the readjustment to a new profit-maximizing output. Demand for one input, say, X_1, is thus a function of P_{X_1} but also of P_{X_2} and P_Y. The result is a demand "surface" that cannot be adequately depicted on a graph. For normal inputs, the function does possess the properties usually associated with demand functions in Stage II.

The derived demand functions for inputs can be illustrated using the Cobb-Douglas function from the previous section. The production function was

$$Y = X_1^{1/5} X_2^{3/5}$$

where $P_Y = \$10$, $P_{X_1} = \$3$, and $P_{X_2} = \$1$. The equations that determine the optimum amounts of input are

$$VMP_{X_1} = P_{X_1}$$
$$VMP_{X_2} = P_{X_2}$$

or

$$P_Y\left(\frac{1}{5} X_1^{-4/5} X_2^{3/5}\right) = P_{X_1}$$

$$P_Y\left(\frac{3}{5} X_1^{1/5} X_2^{-2/5}\right) = P_{X_2}$$

The assumed values of the prices are not substituted into the equations at this time—to do so would give the specific solution. Rather, the equations must be solved simultaneously to find X_1 and X_2 as functions of the prices. Solving by the method of substitution will, with patience, determine the following derived demand functions:

$$X_1 = \frac{3^3 P_Y^5}{5^5 P_{X_1}^2 P_{X_2}^3}$$

$$X_2 = \frac{3^4 P_Y^5}{5^5 P_{X_1} P_{X_2}^4}$$

These functions give the optimal amounts of X_1 and X_2 for all input and output prices. Substituting the assumed prices for the example will result in $X_1 = 96$ and $X_2 = 864$, the answers obtained before.

Examination of the equations shows that demand for the inputs will increase as output prices increase and will vary inversely with input prices. They are obviously too complex to illustrate on a graph. Again, they are based on the assumption that the manager wishes to maximize profit.

ALTERNATIVE PRODUCTION FUNCTION AND ISOCLINE FORMS

A discussion of the different types of substitution between inputs is actually a discussion of the shape or form of the production surface; the two are different ways of looking at the same thing. Also, the type of substitution (or shape of the production surface) determines the shape of the isoclines. All these are interrelated and are determined by the technical relationship among the inputs and outputs.

Several algebraic forms of production functions are used in agricultural economics research. Two common types, the quadratic function without an interaction term and the Cobb-Douglas function, have been discussed previously in this chapter. Two other common forms are the quadratic with an interaction term and the square root function. The equations for these functions are:

Quadratic: $Y = aX_1 - bX_1^2 + cX_2 - dX_2^2$

Quadratic with Interaction: $Y = aX_1 - bX_1^2 + cX_2 - dX_2^2 + eX_1X_2$

Cobb-Douglas: $Y = aX_1^bX_2^c$

Square Root with Interaction: $Y = aX_1^{1/2} - bX_1 + cX_2^{1/2} - dX_2 + eX_1^{1/2} X_2^{1/2}$

where the coefficients a through e are intended to represent numerical values that need not be specified for our purposes. This list is illustrative but not exhaustive. Other forms have also been used in agricultural research; included among these are the Spillman function, the Mitcherlich function, and additional variations on the polynomial and exponential functions.

Isocline maps for the four equations listed are contained in Figure 4–18. The upper two are for the quadratic function—the upper left diagram showing the effects of an interaction term and the upper right without interaction. The lower left isocline map is typical of the square root function while the lower right displays the isocline map for the Cobb-Douglas function. Each function assumes a somewhat different technical relationship between the inputs, and thus each may be appropriate depending on the nature of the production process.

GENERAL CRITERIA FOR TWO OR MORE INPUTS

Thus far, production processes with one and two variable inputs have been discussed. While these production processes are relevant and encountered in agricultural situations (feeding rations, balancing of primary inputs), most production processes involve many inputs. Some inputs are readily available and free, such as sunlight and the gases in the air, these are not of economic concern. Other inputs, that are scarce and therefore have a cost, are the ones of economic importance.

The physical production function for many inputs is written

$$Y = f(X_1, X_2, X_3, \ldots, X_n)$$

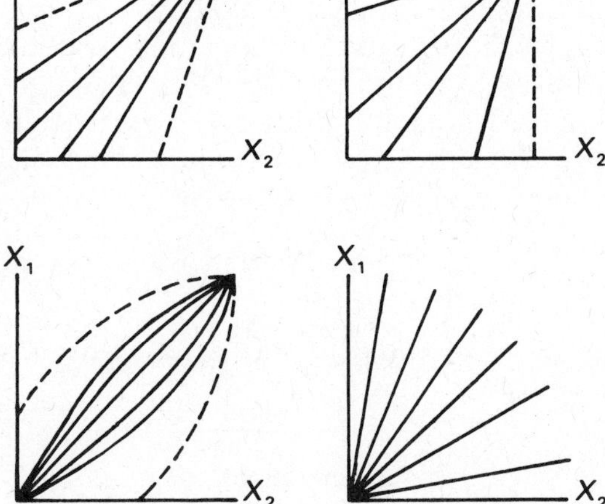

Figure 4–18. Alternative forms of isocline maps.

where n signifies the number of variable inputs and can be any number, four, five, six, etc. As usual, although it does not affect the criteria presented below in any way, the assumption is made that there is at least one fixed input. Thus, a short-run production period is specified.

Minimizing Cost

To produce any output at a minimum of cost, the following criterion must be fulfilled:

$$\frac{MPP_{X_1}}{P_{X_1}} = \frac{MPP_{X_2}}{P_{X_2}} = \frac{MPP_{X_3}}{P_{X_3}} = \ldots = \frac{MPP_{X_n}}{P_{X_n}}$$

The ratio of the price to the marginal product must be equal for all inputs. Thus, if one input costs twice as much as the other inputs, it must produce twice as much on the "margin." For two inputs, the minimum cost criterion as shown earlier is

$$\frac{MPP_{X_1}}{P_{X_1}} = \frac{MPP_{X_2}}{P_{X_2}}$$

This criterion can be expanded to three or more inputs by adding additional terms to the equality.

The ratios of the marginal product to the unit input cost need not be equal to any particular value, but should be positive and equal to each other. For example, if X_1, X_2, and X_3 cost \$1, \$2, and \$3, respectively, then a least cost combination would exist if the marginal products of the inputs were 1, 2, and 3, in the same order. However, a least cost combination would also exist if the marginal products of X_1, X_2, and X_3 were 2, 4, and 6 or 0.3, 0.6, and 0.9, or even 0, 0, and 0. The latter combination would exist at the maximum output.

When n variable inputs are used, a geometrical interpretation no longer exists for the expansion path. However, for a given set of input prices, there will exist for each output one combination of inputs that minimizes cost. For this combination of inputs, total variable cost, TVC, is computed as

$$TVC = P_{X_1} X_1 + P_{X_2} X_2 + P_{X_3} X_3 + \ldots + P_{X_n} X_n$$

Other cost curves are computed as always to derive a complete set of costs for each output.

Profit Maximization

Profits from the production process are maximized when the value of the marginal product of each input equals the unit cost of that input.

In symbolic notation,

$$
\begin{aligned}
VMP_{X_1} &= P_{X_1} \\
VMP_{X_2} &= P_{X_2} \\
VMP_{X_3} &= P_{X_3} \\
&\vdots \\
VMP_{X_n} &= P_{X_n}
\end{aligned}
$$

To attain the optimum, all these equalities must be satisfied simultaneously. Thus, the marginal earnings of each input must be equal to its cost. Again, if an input costs twice as much as another input, its marginal earnings must be twice as much (see the algebraic example in this chapter). To have an optimum, decreasing marginal and average products are required for all inputs.

Dividing each equality in the above statement by the input price on the right and rearranging, the optimum criterion can also be expressed as

$$\frac{VMP_{X_1}}{P_{X_1}} = \frac{VMP_{X_2}}{P_{X_2}} = \frac{VMP_{X_3}}{P_{X_3}} = \ldots = \frac{VMP_{X_n}}{P_{X_n}} = 1$$

Because the ratios are all equal to one at the optimum, they also equal each other. The ratios in the expression would never be equated to a number less than one because to do so would require input use above the optimum; added cost would exceed added returns. When capital available to buy inputs is limited, the ratios may equal some value larger than one. For example, when the ratios equal two, each dollar used to purchase variable inputs is earning a marginal return of two dollars. In this case, the combination being used is a least cost combination but not an optimum combination. Thus, an alternative way of expressing the least cost criteria is to equate the above ratios to each other and to a value greater than or equal to one.

The profit-maximizing combination of inputs is always a least cost combination of inputs. The converse is not necessarily true, of course.

AN EMPIRICAL EXAMPLE

Milk production represents an important agricultural example of a production process involving two variable inputs. Regarding a dairy cow as the fixed input, many different combinations of hay and grain can be used to produce milk. The economic problem of feeding a dairy cow can be divided into the two basic problems of input combination: (1) What combination of hay and grain should be used to produce a given amount of milk at a minimum of cost and (2) what combination of hay and grain should be used to maximize profits per cow? To simplify discussion of this example, possible limitations on capital available to purchase feed and alternative earnings of inputs in other enterprises will not be considered. That is, the example presented will be a simplified enterprise analysis, not an entire farm budgeting plan.

Research workers at Iowa State University investigated the hay-grain relationships in dairy cow feeding.[9] An experiment using 36 Holstein cows was conducted over a 17-month period. After calving, the cows were placed under experimental conditions and fed the same ration during an adjustment period. For the experiment, four hay-to-concentrate ratios were selected, ranging from a ration in which 75 percent of the energy came from hay and 25 percent from concentrates, to one in which 15 percent of the energy came from hay and 85 percent from concentrates. The four hay-to-concentrate ratios used were 75:25, 55:45, 35:65, and 15:85. For convenience, the concentrate mix will hereafter be called "grain." Pastures were not used in the experiment.

Milk production per cow was measured and used to estimate a production function for milk. In symbols, milk represents the output, M, for a four-week period, while pounds of hay and pounds of grain consumed for the four-week period represent the variable inputs, H and G, respectively. The dairy cow represents the fixed input, along with other inputs such as management, buildings, labor, and so forth.

The experimental data were used to estimate production functions for milk using standard statistical techniques. Several different equations were presented by the Iowa researchers. The one selected for discussion here is a quadratic function with an interaction term for hay and grain:

$$M = -340.10 + 1.5437H + 2.9740G - 0.001192G^2 - 0.00388H^2 - 0.001056HG$$

This particular equation represents the milk production response of the average cow for the first four-week period of the experiment. It displays diminishing returns to both inputs with a negative interaction between them. That is, increasing the amount of grain in the ration will decrease the marginal physical product of hay, even when the quantity of hay is held constant.

The estimated milk production surface is presented in Figure 4–19A. Pounds of hay fed per cow during the four-week period and

[9]The example presented is from Heady, E. O., Schnittker, J. A., Jacobson, N. L., and Bloom, Solomon, *Milk Production Functions, Hay/Grain Substitution Rates and Economic Optima in Dairy Cow Rations*. Iowa State Agricultural Experiment Station Research Bulletin 444, Ames, Iowa, October 1956. The discussion included here is quite general; the bulletin should be examined for details.

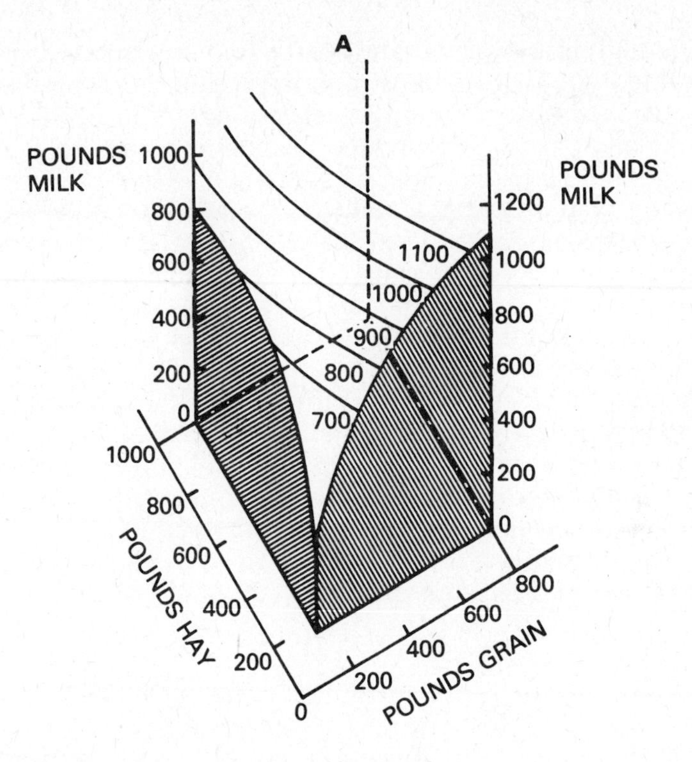

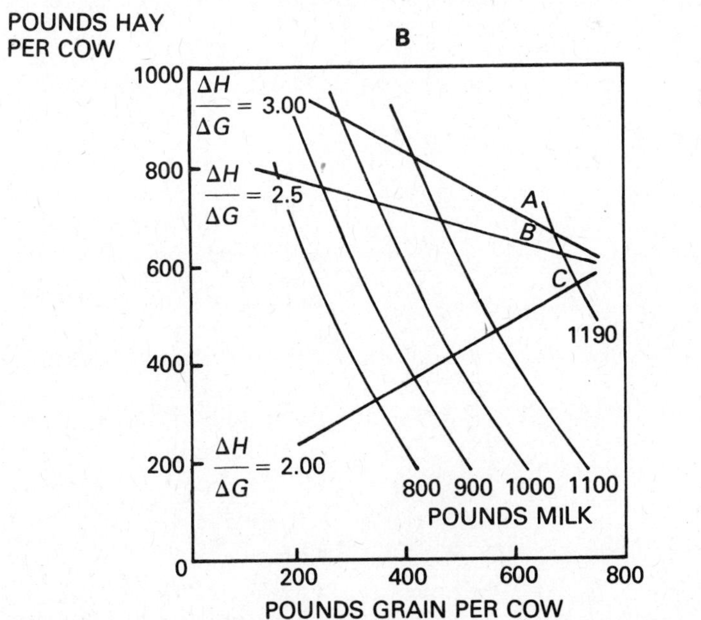

Figure 4–19. Milk production surface, isoquants and isoclines, Iowa State Experiment, 1956.

pounds of grain fed per cow during the four-week period are measured along the input axes; pounds of milk produced during the four-week period are measured along the vertical axis. The surface is indented from the input origin to represent the minimum amount of feed needed to ensure animal maintenance. Examination of the surface shows that increasing amounts of hay and grain increase total milk production, but at a decreasing rate. The marginal products of hay and grain are positive but decreasing. Because of this, the milk production isoquants, representing the various combinations of hay and grain that produce a given amount of milk during the production period, are curved with a negative slope.

A more detailed picture of the relationships between milk production and feed input can be gained by examining the isoclines and isoquants. The isocline-isoquant map for the milk production surface is presented in Figure 4–19B, and selected points from the 800, 1,000, and 1,190 pound milk isoquants are contained in Table 4–8.

The milk isoquant equation was derived from the production function and has the following form:

$$H = 1{,}989.36 - 1.3608G \pm 1{,}288.66\sqrt{1.8553 + 0.001355G - 0.000000736G^2 - 0.001552M}$$

The marginal rates of technical substitution can be computed from the equation

$$MRS \text{ of } G \text{ for } H = \frac{dH}{dG} = -\frac{\dfrac{\partial M}{\partial bG}}{\dfrac{\partial M}{\partial H}}$$

$$= -\frac{2.9740 - 0.002384G - 0.001056H}{1.5437 - 0.00776H - 0.001056G}$$

The general isocline equation can be found by setting the equation for the MRS equal to the grain/hay price ratio and solving for H as a function of G. Thus, the equation for the family of isoclines is

$$H = \frac{(-2.9740 + 1.5437r) + (0.002384 - 0.001056r)G}{(-0.001056 + 0.000776r)}$$

where $r = P_G/P_H$

The data in Table 4–8 were derived using the isoquant and MRS equations. When 250 pounds of grain are used, 577 pounds of hay must be used to attain a production of 800 pounds of milk during the four-week period. At this point on the 800-pound isoquant (grain =

Table 4–8 Combinations of Hay and Grain Needed to Produce Specified Amounts of Milk for a Four-Week Period

Pounds of Grain	*Pounds of Hay Required to Maintain Milk Output at*			*Pounds of Hay Replaced by One Added Pound of Grain on Isoquant*		
	800	*1,000*	*1,190*	*800*	*1,000*	*1,190*
150	815	—	—	−2.54	—	—
250	577	1,032	—	−2.24	−3.41	—
350	365	744	—	−2.02	−2.53	—
450	172	513	—	−1.85	−2.13	—
550	—	314	—	—	−1.87	—
650	—	138	—	—	−1.66	—
750	—	—	575	—	—	−2.09

Source: Heady, E. O., Schnittker, J. A., Jacobson, N. L., and Bloom, Solomon, *Milk Production Functions, Hay/Grain Substitution Rates and Economic Optima in Dairy Cow Rations.* Iowa State Agricultural Experiment Station Research Bulletin 444, Ames, Iowa, October 1956.

250, hay = 577), the marginal rate of substitution of grain for hay, dH/dG, is − 2.24. If one pound of grain is added, hay can be reduced 2.24 pounds and output over the four-week period will remain at 800 pounds. However, if 1000 pounds of milk are to be produced with 250 pounds of grain, 1032 pounds of hay will be required. The marginal rate of substitution at that point on the 1000-pound isoquant is − 3.41. Thus, as relatively more hay is used, a pound of grain will replace larger amounts of hay.

The milk isoclines are linear and converge toward the point of maximum output per cow. As discussed earlier in this chapter, the marginal products of the feeds are constant, relative to each other along an isocline. Thus, one pound of grain substitutes for three pounds of hay on isocline *A*. Isocline *A* thus traces out all least cost feed combinations for a grain/hay price ratio of three. In a like manner, isoclines *B* and *C* trace out least cost combinations for possible grain/hay price ratios of 2.5 and 2.0, respectively.

The isoclines are linear but do not pass through the origin. As milk production is increased, the ratio of grain to hay is not proportional but must change if least cost combinations of feed are to be used. Feeding a constant ratio and increasing production by increasing amounts fed would not be the most economical way of producing milk; only if the isocline were a straight line passing through the origin would such a recommendation be valid. Because of the slight curvature of the isoquants, a relatively small shift in the grain/hay price ratio can cause a large change in the least cost feed ratio for a given milk production level. Compare, for example, the input combination needed to produce 800 pounds of milk on isocline *B* to that of isocline *C*.

Optimum amounts of hay and grain and the resulting milk production for a four-week period are presented in Table 4–9. For pur-

Table 4–9 Estimated Optimum Feed Quantities and Milk Production for Selected Price Ratios, Milk Production for a Four-Week Period

Milk per cwt.	*Grain per cwt.*	*Hay per ton*	*Pounds of grain*	*Pounds of hay*	*Pounds of milk*
$ 8	$4	$30	560	710	1143
8	6	50	476	662	1080
8	8	70	390	618	988
10	4	30	612	686	1163
10	6	50	544	649	1121
10	8	70	476	613	1064

Source: Heady, E. O., Schnittker, J. A., Jacobson, N. L., and Bloom, Solomon. *Milk Production Functions, Hay/Grain Substitution Rates and Economic Optima in Dairy Cow Rations.* Iowa State Agricultural Experiment Station Research Bulletin 444, Ames, Iowa, October, 1956.

poses of illustration, several different price combinations are presented. These optima were derived using the theoretical conditions for maximum profit; the marginal product for each input must be equated to the input/output price ratio. The equations

$$\frac{\partial M}{\partial G} = 2.9740 - 0.002384G - 0.001056H = \frac{P_G}{P_M}$$

$$\frac{\partial M}{\partial H} = 1.5437 - 0.000776H - 0.001056G = \frac{P_H}{P_M}$$

were used, where P_G is the cost per pound of grain, P_H is the cost per pound of hay, and P_M is the price per pound of milk. The marginal products are also expressed in pounds.

Examination of Table 4–9 reveals some important economic considerations involved in feeding problems. First, none of the production rates represent the maximum possible production per cow, which was over 1190 pounds for the four-week period. Secondly, as the price of feed rises relative to the price of milk, the optimum feed quantities and milk production decrease. Third, the ratio of grain to hay in the optimum feed combination varies with the price ratio. For a milk price of $10 per hundredweight, 612 pounds of grain and 686 pounds of hay—almost a one-to-one ratio—represents the optimum combination when grain is $4 per hundredweight and hay is $30 per ton. However, when grain increases to $8 per hundredweight and hay to $70 per ton while milk remains at $10 per hundredweight, the optimum ration is 476 pounds of grain and 613 pounds of hay, approximately a 2:3 ratio. Also, when grain and hay are $4 and $30, respectively, the ratio of the optimum feed changes when milk increases from $8 to $10 per hundredweight.

Feeding a dairy cow is more complex than is represented in this example. Production varies during the lactation period. Inherent differences exist in the productive ability of cows. Nutrient requirements and stomach capacities must be considered. The experimental data presented in this section are not meant to be used for recommendations, but rather are illustrative of the principles developed in this chapter.

Problems and Exercises

4–1. A given product can be produced with the following combinations of resources:

	Resources	
Combination	X_1	X_2
1	39	0
2	28	1
3	20	3
4	14	5
5	9	8
6	5	12
7	1	17
8	0	25

Which combination meets the least cost objective when prices of the factors are: (*a*) P_{X_2} = \$6.40; P_{X_1} = \$8.00; (*b*) P_{X_2} = \$0.75; P_{X_1} = \$0.25; and (*c*) P_{X_2} = \$1.60; P_{X_1} = \$0.40.

4–2. Show graphically and discuss the resource adjustment that would be suggested when:

(*a*) $\frac{\Delta X_1}{\Delta X_2} < \frac{P_{X_2}}{P_{X_1}}$

(*b*) $\frac{\Delta Y_1}{\Delta X_1} > \frac{P_{X_1}}{P_Y}$

(*c*) $\frac{\Delta Y_1}{\Delta X_2} > \frac{\Delta Y_2}{\Delta X_2}$

where X_1 and X_2 are resources; Y_1 and Y_2 represent the same product of two technical units; and P_{X_1}, and P_{X_2}, and P_Y are the prices of the two resources and the product.

4–3. Consider the production function $Y = X_1^{1/2} X_2^{1/4}$; P_{X_1} = \$4, P_{X_2} = \$2. Find the least cost combinations of X_1 and X_2 to produce 8 units of Y.

4–4. Consider the production function $Y = X_1^{3/4} X_2^{1/4}$. Find the least cost combination of X_1 and X_2 to produce 12 units of Y when $P_{X_1} = 3$, $P_{X_2} = 1$; $P_{X_1} = 48$, P_{X_2} = \$1.

4–5. Consider the production function $Y = X_1^{1/5} X_2^{4/5}$; $P_{X_1} = 3$, $P_{X_2} = 3$. (*a*) What is the MPP_{X_1} when $X_1 = 1$ and $X_2 = 1$? (*b*) Find the expansion path equation.

4–6. Assume that a firm is operating in a purely competitive market and that its input-output relationship is of the following form:

$$Y = X_1^{1/3} X_2^{1/3} \text{ and } P_{X_1} = \$1 \text{ and } P_{X_2} = \$1$$

There is no fixed cost. At how many units of output, Y, will the firm maximize the net returns when the price of the product, P_Y is \$27?

4–7. Consider the production function $Y = X_1^{2/5} X_2^{1/5}$; $P_{X_1} = \$6$, $P_{X_2} = \$3$, and $P_Y = \$15$. (*a*) At which levels of X_1 and X_2 will the net returns be maximized? (*b*) Find the *TVC*, *AVC*, and *MC* equations.

4–8. In problem 4–7, which effect predominates (expansion or substitution effect) when P_{X_1} is reduced from \$6 to \$3, *ceteris paribus?*

4–9. Consider the production function $Y = 100 - 3X_1^2 + 4X_1 + 2X_1X_2 - 5X_2^2 + 48X_2$. Find the values of X_1 and X_2 at which the total physical product, *TPP*, is maximum.

4–10. Consider the production function $Y = X_1X_2$; $P_{X_1} = \$2$ and $P_{X_2} = \$1$. Which combination of X_1 and X_2 will produce maximum output when the manager has only \$1000 to spend on X_1 and X_2?

4–11. Consider the production function $Y = X_1^{1/4} X_2^{2/4}$. $P_{X_1} = \$1$ and $P_{X_2} = \$2$. How many units of X_1 and X_2 would be needed to produce 1 unit of Y?

4–12. Discuss and illustrate the conditions under which acreage control programs (*a*) will reduce output of single-product farms, (*b*) will not reduce output of single-product farms.

4–13. Discuss and illustrate the concept of economic efficiency as it is related to factor-product and factor-factor relationships.

4–14. Assume an isoproduct curve with decreasing *MRS* and that the price of one resource decreases, other things being equal. Discuss and illustrate (*a*) the situation where substitution effect predominates, and (*b*) the situation where expansion effect predominates.

4–15. "Technical substitutability between factors of production may lead to economic complementarity among the same factors of production." Discuss the meaning of this statement.

4–16. Assume two inputs are technical complements. How would the least-cost combinations be derived; that is, would the least-cost criterion be applicable? Show your answer on an isoquant graph. What would the isocline map look like? Can the production function be expressed in equation form? Consider these questions for the examples in Table 4–5.

4–17. Using the production function in problem 4–9 and prices of your choosing, derive the equations needed and plot the solution to the most profitable output as shown in Figure 4–14.

4–18. Explain why a concave isoquant would lead to an "optimal" solution that would in fact maximize the total variable cost of producing a given output level.

4–19. Find equations for the type of production functions contained in Table 4–4. Is a simple equation representing the response depicted in Table 4–4C possible?

4–20. Find the derived demand functions for X_1 and X_2 for the production function (4.1). How will they change when an interaction term is added to the production function?

4–21. For the Cobb-Douglas function, find the derived demand functions for the inputs in terms of P_{X_1}, P_{X_2}, P_Y, a, b and c where $Y = aX_1^bX_2^c$, given $0 < a$, $0 < b$, $0 < c$, $0 < c + b < 1$. What would happen when $c + b = 1$, when $c + b > 1$?

4–22. Explain what happens to production function (4.1) when the interaction term, $2X_1X_2$, is added to it. (Hint: Attempt to find the maximum yield. Can you maximize profits? Find least cost combinations? Explain.)

4–23. Given the production function

$$Y = 10X_1 - 1/2\ X_1^2 + 12X_2 - X_2^2 + X_1X_2$$

Find the marginal product equations, the isoquant equation expressing X_2 as a function of X_1 and Y, and the equation for the isocline family (as in our example).

4–24. For the production function in problem 4–23, draw an isoquant map for $Y = 0$, 100, 200, and 292 units. Plot the ridge lines and draw expansion paths for $r = 2/1$, $2/9$, $12/10$ where $r = P_{X_2}/P_{X_1}$.

4–25. For the production function in problem 4–23, show that $\sigma = 0.73$ on the isoquant for 100 units of output when $X_1 = 6.05$ and $X_2 = 4.16$.

4–26. Show that σ always equals one for the Cobb-Douglas function.

4–27. For the production function in problem 4–23, when $P_{X_1} = \$10$ and $P_{X_2} = \$12$, show that the cost-minimizing amounts of X_1 and X_2 and the resulting output can be expressed parametrically as

$$X_1 = X_1$$

$$X_2 = \frac{22}{32}X_1$$

$$Y = \frac{1}{32}\left(584X_1 - \frac{146}{16}X_1^2\right)$$

Depict this solution on a three-dimensional graph. Explain.

4–28. Continuing problem 4–27, show that along the expansion path in the region of substitution

$$X_1 = 32 - \frac{100}{1825}K$$

$$X_2 = 22 - \frac{22}{584}K$$

and

$$TVC = \$(584 - K)$$

where $K = (341056 - 1168Y)^{1/2}$

4–29. Given the Cobb-Douglas function,

$$Y = X_1^{1/6}\,X_2^{3/6}\,X_3^{2/6}$$

and the prices $P_{X_1} = \$4/3$, $P_{X_2} = \$1$, and $P_{X_3} = \$1/3$, show the expansion path is given parametrically as

$$X_1 = X_1$$

$$X_2 = 4X_1$$

$$X_3 = 8X_1$$

and output from any such least cost combination would be $Y = 4X_1$. Show also $VC = \$2Y$.

4–30. Repeat problem 4–29, changing the exponent on X_3 to 1/6. Why does the expansion path remain linear while output and variable cost become nonlinear?

4–31. Discuss the validity of the following statements:

Stage II implies convex isoquants. A convex isoquant implies Stage II.
Diminishing marginal products imply Stage II.
Stage II implies diminishing marginal products.

(Hint: See page 365 in Chiang, A.C., *Fundamental Methods of Mathematical Economics*, Second Edition, New York: McGraw-Hill, Inc., 1974.)

Suggested Readings

Allen, Clark Lee. *The Framework of Price Theory*. Belmont, California: Wadsworth Publishing Co., Inc., 1967, Chapter 4.

Baumol, W. J. *Economic Theory and Operations Analysis*. Englewood Cliffs N.J.: Prentice-Hall, Inc., 1961, Chapter 11.

Cohen, K. J. and Cyert, R. M. *Theory of the Firm*. Englewood Cliffs, N.J.: Prentice-Hall, Inc., 1965, Chapters 3 and 7.

Cramer, Gail L. and Jensen, Clarence W. *Agricultural Economics and Agribusiness*, Second Edition. New York: John Wiley and Sons, Inc., 1982, Chapters 5 and 6.

Dillon, John. *The Analysis of Response in Crop and Livestock Production*. Oxford: Pergamon Press, 1968, Chapters 1 and 2.

Ferguson, C. E. and Gould, J. P. *Microeconomic Theory*, Fourth Edition. Homewood, Ill.: Richard D. Irwin, Inc., 1975, Chapters 8 and 13.

Heady, Earl O. *Economics of Agricultural Production and Resource Use*. New York: Prentice-Hall, Inc., 1952, Chapters 4 and 11.

Hicks, J. R. *Value and Capital*, Second Edition. Oxford: Oxford University Press, 1953, Chapter 6.

Leftwich, Richard H. *The Price System and Resource Allocation*, Third Edition. Hinsdale, Ill.: Holt, Rinehart, and Winston, 1966, Chapter 7.

Mansfield, E. *Microeconomics*, Third Edition, New York: W. W. Norton and Company, Inc., 1979, Chapters 6 and 7.

PRODUCTION OF TWO OR MORE PRODUCTS

CHAPTER 5

The economic analysis of the production process as presented in the last three chapters has emphasized the allocation of inputs. Chapter 2 dealt with production functions that describe possible relationships between input and output. Chapter 3 presented a discussion of the allocation of one variable input, the Chapter 4 dealt with the combination of two or more variable inputs in the production of one output.

This chapter presents a different view of the production process. Rather than emphasizing the allocation of variable inputs within an enterprise, this chapter contains a discussion of enterprise combination, often called product-product relationships. The question asked is not "How should these inputs be allocated within an enterprise?" but rather "What combination of enterprises should be produced from a given bundle of fixed and variable inputs?" The two questions lead ultimately to the same answer. Only the vantage point differs.

The logic presented in this chapter represents a formalization of procedures used in budgeting or planning the farm business. The assumptions of this chapter remain unchanged from the previous chapter, including perfect certainty and pure competition in the input and output markets.

THE PRODUCTION POSSIBILITY CURVE

The production possibility curve is a convenient device for depicting two production functions on one graph. To begin, suppose that one input, X, can be used to produce two products, Y_1 and Y_2, and that all other inputs used to produce Y_1 or Y_2 are fixed or highly specialized so that their use cannot be diverted. Thus, the farm manager must determine how much of input X to use on each product. The relevant questions in "How much input is available?" The two possible situations are termed *unlimited* and *limited.*

Unlimited. When the amount of available input is unlimited, resource allocation is determined by equating the price of an input to the *VMP* of the input. No new problem arises. The manager can use the optimum amount in both enterprises. Increasing input use on one production process will not reduce the amount available for use in the other. Thus, other than the fact that the enterprises are on the same farm and are under the direction of the same manager, they are not related to each other.

The term *unlimited* means that the manager has a sufficient quantity of the input to use the optimum amount in all enterprises. It does not intend to suggest that the supply of the input is unlimited; if that were true, the input would be a free good. Often the term *unlimited capital* is used instead of *unlimited input*. The two terms have the same meaning, the assumption being that the variable input may be readily purchased if not already owned.

Limited. When the amount of input is limited, the optimum amount cannot be used in each enterprise. Thus, by definition, *limited* means that the total amount of input available is less than that amount needed to apply the optimum to each enterprise.

Limited input situations are also referred to as *limited capital* situations. Again, the implication is that the variable input can be purchased if it is not already owned and that the amount purchased will be limited by the amount of capital available. Limited capital therefore means that the capital available is not sufficient to allow the manager to use the optimum amount of input in each enterprise.

When inputs are limited in quantity, enterprises on the farm become uniquely related. No longer can they be considered independently. The degree of interdependence among enterprises depends on their technical and economic relationships. In some cases, if output in one enterprise is to be expanded, resources must be diverted to that enterprise, and output of other enterprises must be reduced. In other cases, expansion in one enterprise may lead to expansion in another enterprise as well. The objective of product-product relationships is to determine the combination of enterprises that best meets management objectives, given the resource limitation.

The primary use of the production possibility curve is to determine the most profitable combination of enterprises for a limited amount of input.

Deriving Production Possibility Curves from Production Functions

Production possibility curves portray the combinations of products that can be produced with a given set of inputs. Production possibility curves are also called isoresource curves because each point on the curve represents combinations of outputs produced using equal (iso) amounts of input. In a sense, a production possibility curve can be viewed as a boundary delineating the combinations of outputs that can be produced using the available inputs and technology from the combinations of outputs that cannot be produced. Thus, the combinations of outputs that fall on the isoresource curve represent the maximum amount of output that can be attained given the manager's resource situation.

The purpose of this section is to show how production possibility curves can be derived from production functions. Two production functions, one for Y_1 and one for Y_2, are presented in Table 5–1A. These production functions use the same limited input, X.

Suppose that four units of input were available. Production functions represent planning curves and, before any input is actually used, the manager has the opportunity to consider all possible ways the input can be used. By using all four units of input on Y_1, he can produce 22 units of Y_1, or if all four are used in Y_2, 36 units of Y_2 can be produced. Many other combinations are possible within these two extremes. One unit applied to Y_1 and three to Y_2 will produce 7 and 30 of these two

Table 5–1 Deriving Production Possibility Curves from Production Functions

A

X	Y_1	MPP_{XY_1}	X	Y_2	MPP_{XY_2}
0	0		0	0	
1	7	7	1	12	12
2	13	6	2	22	10
3	18	5	3	30	8
4	22	4	4	36	6
5	25	3	5	40	4
6	27	2	6	42	2
7	28	1	7	43	1
8	27	−1	8	42	−1
9	25	−2	9	40	−4

B

Production Possibilities for X = 4	
Y_2	Y_1
36	0
30	7
22	13
12	18
0	22

C

Production Possibilities for X = 7	
Y_2	Y_1
43	0
42	7
40	13
36	18
30	22
22	25
12	27
0	28

products, respectively. Dividing the four inputs evenly between the two outputs will produce 13 units of Y_1 and 22 units of Y_2. Three units of input to Y_1 and one unit to Y_2 will produce 18 units of Y_1 and 12 of Y_2. These combinations represent some of the production possibilities for four units of input; they are presented as a group in Table 5–1B. Each combination in Table 5–1B has one common feature—four units of input are required in total to produce each combination.

Graphs of the production functions and the resulting production possibility curve are presented in Figure 5–1. The production possibility curve for four units of input is the counterpart of the production possibilities presented in Table 5–1B. The output combinations in Table 5–1B were graphed and a smooth curve drawn through the points. Thus, the graph presents additional output combinations not included in Table 5–1B.

The production possibility curve in Figure 5–1B presents all possible combinations of the two products that can be produced using four units of input. It must be regarded as a planning curve because obviously one, and only one, combination of outputs can be produced

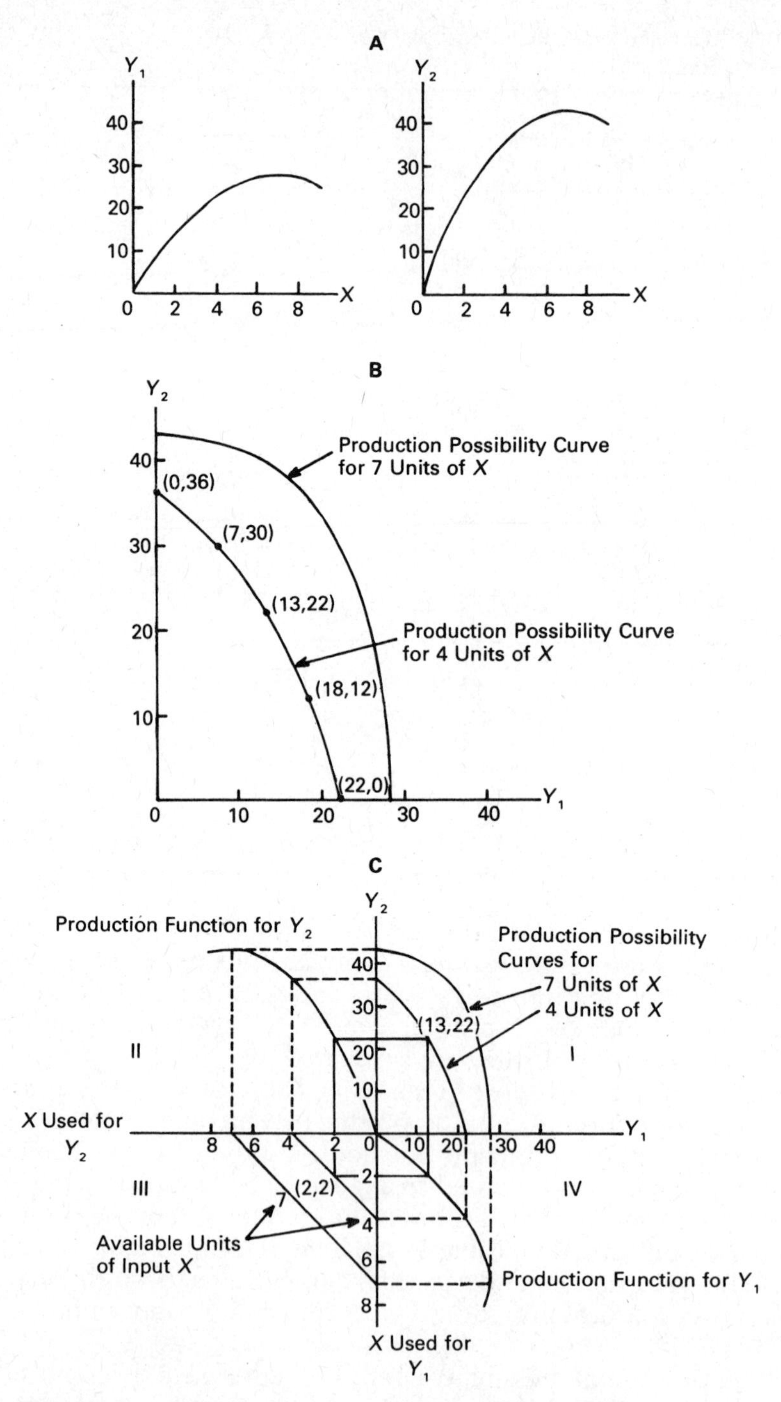

Figure 5–1. Production functions and production possibility curves.

when the four units of input are actually committed. Thus, while all combinations should be considered when planning production, only one combination will eventually be produced. The production possibility curve is affected by all the same factors that affect the production function. A change in technology or in the productivity of fixed inputs will cause a shift in the production functions and also a shift in the production possibility curve.

The production possibility curve is a convenient method of comparing two production functions simultaneously, but only within the limits determined by the available input. That is, the production possibility curve for four units of input (Figure 5–1B) compares the production functions for Y_1 and Y_2 only within the limits allowed for four units of input. If more, or less, than four units of input are available, a different production possibility curve is produced. Suppose seven units of input were available. Then combinations of Y_1 and Y_2 ranging from 43 of Y_2 and zero of Y_1 to zero of Y_2 and 28 of Y_1 are possible; production possibilities for $X = 7$ are presented in Table 5–1C. The corresponding production possibility curve is presented in Figure 5–1B (to avoid clutter, the output combinations are not labeled on this curve).

The production possibility curve for seven units of input represents a higher level of production of both products and is located farther from the origin than the production possibility curve for four units of input. Also, it has a somewhat different shape than the production possibility curve for four units of input. If a production possibility curve were drawn for less than four units of input, say, three, this production possibility curve would be located closer to the origin than the one for four units.

Figure 5–1C presents an alternative method of viewing the production possibility curve graphically. All axes of the graph are assumed to measure positive quantities. The straight line in the third quadrant depicts the allocation of the limited amount of input between the two products. All of the input may be used to produce Y_2, all may be used to produce Y_1, or some could be allocated to each of the two outputs. In any case, each combination chosen must be on the straight line in the third quadrant. The production functions in quadrants II and IV are used to determine the outputs resulting from the combination of X selected. The resulting outputs can then be plotted in the first quadrant to determine the production possibility curve. For example, when four units of input are available and two are used on each product, a vertical movement from the point (2,2) in the third quadrant to the production function for Y_2 in the second quadrant shows that 22 units of Y_2 will result. A horizontal movement into quadrant IV shows that 13 units of Y_1 will result. Plotting the point (13,22) in quadrant I determines a point on the production possibility curve for $X = 4$. Similarly, all points on the production possibility curve for $X = 4$ result from some combination of X falling on the straight line in the third quadrant. The production possibility curve for $X = 7$ is derived in a similar fashion.

Production possibility curves that are based on the same "bundle" of limited inputs and the same technology do not cross when input use is in Stage II. They are either concave to the origin or linear, depending on the input-output relationships described by the underlying production functions.

The meaning of the terms *production possibility curve* or *isoresource curve* is intuitively clear from their derivation. A third name often used in economics is *opportunity curve*. The production possibility curve, such as the one for 4 units of X in Figure 5–1B, delineates all the production opportunities available; further, it leads directly to the concept of opportunity cost, introduced in Chapter 3. For example, if 0 units of Y_1 and 36 units of Y_2 are being produced with the four units of X_1, production of Y_1 can be increased to 7 units only if production of Y_2 is reduced to 30. The opportunity cost of the additional 7 units of Y_1 is 6 units of Y_2. This concept is of great importance and will be developed in more detail later in this chapter.

As mentioned, the production possibility curve represents the maximum amounts of Y_1 and Y_2 that can be produced from a limited quantity of input. Outputs requiring lesser input amounts can also be produced. The production possibility curve for X equals 7 in Figure 5–1B represents the maximum amount of output that can be produced from the 7 units of X. However, the farmer possessing 7 units of X could produce any output amounts on or within the production possibility curve, including the axes. The entire area bounded by and including the axes and the production possibility curve is often called the *feasible set* or the *attainable set* of outputs. This area restricts outputs to those that are possible from the use of the available input amounts.

RELATIONSHIPS AMONG PRODUCTS

Production possibility curves illustrate the relationships among enterprises on the farm. These relationships may take different forms depending on the particular situation. In general, enterprises can be competitive, complementary, or supplementary.

Competitive Products

Products are termed competitive when the output of one product can be increased only by reducing the output of the other product. Outputs are competitive because they require the same inputs at the same time. Often, when planning the farm business, the manager can expand production of one output only by diverting inputs—labor, capital, land, management—from one enterprise to the other.

When the production possibility curve has a negative slope, the products concerned are competitive. Two examples are presented in Figure 5–2A. The curved production possibility curve results when the production functions for the product show decreasing marginal returns. As increasing amounts of input are transferred to one product, the marginal product of the input in that use becomes increasingly smaller, while the marginal product of the input in the other use becomes increasingly larger. Thus, each one unit increment of one product requires a successively larger sacrifice of the competing product.

Curved production possibility curves might exist when allocating labor among enterprises such as hogs and cattle. If all labor were used on cattle, no hogs could be grown. However, the last few hours spent on cattle would not be very productive—would have a small marginal product relative to the alternative marginal earnings on hogs. This would occur because more could be earned by fulfilling the basic labor requirements of the hog enterprise, thus allowing hogs to be produced, than could be earned by accomplishing the last few details on the cattle enterprise.

Also, when decreasing returns to fertilizer exist, production possibilities for applying commercial fertilizer on two crops would assume a curved shape. Many other examples can be found in animal feeding, etc.

A straight line or linear production possibility curve for competitive products is also shown in Figure 5–2A. In this case, diminishing returns do not exist for either production function. This type of production possibility curve would be appropriate when a reasonably homogeneous field of soil is being diverted to either of two crops. For example, suppose a field of soil will produce 80 bushels of corn or 25 bushels of soybeans per acre. Each acre diverted from corn to soybeans will reduce corn production by 80 bushels and increase soybean production by 25 bushels. This is true for the first acre diverted from corn and is also true for all successive acres. Thus, the production possibility curve is linear.

Complementary Products

Two products are complementary if an increase in one product causes an increase in the second product, when the total amount of inputs used on the two are held constant. Production possibility curves for complementary products are shown in Figure 5–2B. In the left figure, Y_1 is complementary with Y_2 until the point A is reached. To the right of A, the two products are competitive. Thus, if the farmer wishes to produce the maximum possible amount of Y_2, he should grow at least OB amount of Y_1. When larger quantities of Y_1 are produced, the two products become competitive, and the manager will select the combination that maximizes revenue or satisfies some other objective. This will be some point between A and the Y_1 axis, depending on the prices of the outputs.

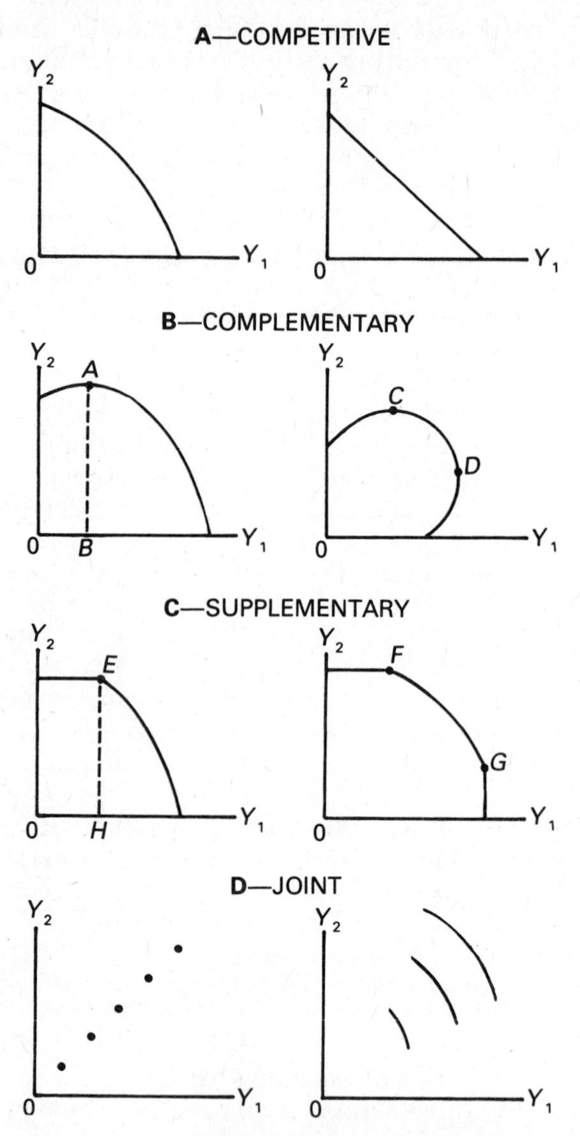

Figure 5–2. Production possibility curves showing possible relationships among enterprises.
A. Competitive.
B. Complementary.
C. Supplementary.
D. Joint.

The production possibility curve on the right in Figure 5–2B shows a situation where each output is complementary to the other output over a certain range. Thus, Y_1 is complementary to Y_2 up to C, and Y_2 is complementary to Y_1 up to D. Between C and D, the two products are competitive.

Complementarity usually occurs when one of the products produces an input used by the other product. An example of this is the use of a legume in a rotation with cash crops. The legume may add nitrogen and improve soil structure or tilth and improve weed and insect control. These factors, in turn, serve as "inputs" for the cash crops, thus causing, over the period of time required by the rotation, an increase in the production of cash crops. For example, in a four-year rotation with three years of corn and one of alfalfa, the inputs supplied by alfalfa may increase corn yields to such an extent that more corn is produced in three years production than could otherwise be produced in four.

Complementarity often occurs over time. Alfalfa and corn are competitive in a single year, giving rise to a linear production possibility curve. Over a period of years, however, the legume contributes to the production of corn. Some products, such as nurse crops, may give rise to complementarity in a single year.

Complementary products produce byproducts that are in fact inputs in the production process of the other product. In a sense, they change the "fixed" bundle of inputs. Of course, extensive use of an input, such as overgrazing, use of excessive amounts of irrigation water, or fertilizers, may cause one of the enterprises to be operated in Stage III and thus cause an apparent complementary relationship among products. That is, output in that enterprise can be increased simply by using less input. (Notice that if one enterprise were operated on Stage III, the production possibility curves for different amounts of input could cross—implying irrational resource allocation.)

The complementary products eventually must become competitive. That is, when a large amount of the resource in question is devoted to one product, production of the other product must decrease. For example, while one year of alfalfa in a four-year rotation may be complementary, two, three, or four years of alfalfa could be produced only by successive reductions in the cash crops. The quantities of cash crops and alfalfa would then be determined according to their relative profitability.

A change in technology may also change the complementary relationship among products. For example, commercial fertilizer has apparently caused the complementary relationship between corn and alfalfa to change to a competitive one.

Supplementary Products

Two products are called supplementary if the amount of one can be increased without increasing or decreasing the amount of the second.

Production possibility curves for supplementry enterprises are shown in figure 5–2C. For the diagram on the left, Y_1 is supplementary to Y_2. Y_1 can be increased from zero to OH amount without affecting the amount of Y_2 produced. Past E on the production possibility curve, the two inputs become competitive. For the diagram on the right in Figure 5–2C, each enterprise is supplementary to the other and competitive between FG.

Supplementary enterprises arise through time or when surplus resources are available at a given point in time. Once purchased, a tractor is available for use throughout the year. Its use in one month does not preclude its use in another month. Thus, a tractor purchased to plow and plant may be put to a lesser use during the off-season. Or a combine may be used to harvest wheat in June and soybeans in the fall; clearly the hours of use in June have no bearing on the hours of use in the fall. If the two crops were harvested at the same time, however, the relationship would be competitive—use on one could come about only by reducing the amount of use on the other.

The supplementary relationship between products depends on amount of use left in the resource. If the combine is completely worn out harvesting wheat in June, it will not be available for use in the fall. In this case, the supplementary relationship depends on the durability of the combine.

Another common supplementary enterprise is hogs following beef animals fed on corn. In this case, at the same point in time, the hogs forage in the droppings and utilize corn that would otherwise be wasted. If no hogs are produced, cattle production is unaffected. Hog production can be increased until the corn available in the droppings is fully utilized. Beyond that point, more hogs can be grown only by diverting corn fed to the cattle to the hog enterprise.

Small flocks of chickens, milk cows, and family gardens may all represent supplementary enterprises on some farms. In each case, labor or some other input is available for use on a small scale and, rather than let it go idle, a small enterprise is undertaken.

Joint Products

Products that result from the same production process are termed joint products. In the extreme case, the two are combined in fixed proportions, and production of one without the other is impossible. Production possibility curves for joint products of this type are presented in Figure 5–2D; each level of resource results in a production possibility curve that is a point. No substitution among products is possible. Conceptually, joint products produced in fixed proportions can be handled in the same manner as single-output production situations.

Most of agricultural examples of joint products fall in a second category: joint products with variable proportions. Introduction of new varieties in crops or breeds in livestock may affect the proportions in which the joint products are produced. The "meat-type" hog has relatively less lard and more meat than a "fat-type" hog. Different varieties of wheat produce varying proportions of straw and grain. Thus, the proportions may be changed by varying technologies, husbandry, or cropping practices usually associated with the fixed inputs. For such products, a narrow range of product substitution may exist, as suggested on the right diagram Figure 5–2D.

MAXIMUM REVENUE COMBINATION OF OUTPUTS

The concept of marginal rate of product substitution is similar to that defined in factor-factor relationship (Chapter 4). The marginal rate of product substitution, *MRPS*, refers to the amount by which one output changes in quantity when the other output is increased by one unit along an isoresource curve (input use remaining constant).

The Marginal Rate of Product Substitution

The marginal rate of product substitution is defined as the slope of the production possibility or opportunity curve. Thus,

$$MRPS \text{ of } Y_1 \text{ for } Y_2 = \frac{\Delta Y_2}{\Delta Y_1}$$

Marginal rates of product substitution for output combinations possible from seven units of input (Table 5–1C) are presented in Table 5–2. Computation is similar to marginal physical products and marginal rates of input substitution discussed in preceding chapters and is not repeated here. In Table 5–2, as the amount of Y_1 produced increases, the amount of Y_2 sacrificed steadily increases. This is due to

Table 5–2 Computing the Marginal Rate of Product Substitution

Production Possibilities for X = 7				
Y_2	Y_1	ΔY_2	ΔY_1	*Marginal Rate of Product Substitution of Y_1 for Y_2 = $\Delta Y_2/\Delta Y_1$*
43	0			
		−1	7	−1/7
42	7			
		−2	6	−1/3
40	13			
		−4	5	−4/5
36	18			
		−6	4	−3/2
30	22			
		−8	3	−8/3
22	25			
		−10	2	−5
12	27			
		−12	1	−12
0	28			

the decreasing marginal physical products displayed by the production functions.

As indicated, the *MRPS* represents the slope of the production possibility curve. As with the marginal physical product and marginal rate of substitution, there are two measures of the *MRPS:* the average or approximate measure and the exact measure. The approximate measure is computed from a table or between points on a graph; it averages all the slopes or exact *MRPSs* between two points on a production possibility curve. Consider the output combinations $Y_2 = 36$, $Y_1 = 18$ and $Y_2 = 30$, $Y_1 = 22$, on the production possibility curve for seven units of input. The *MRPS* of Y_1 for Y_2 between these combinations is

$$\frac{\Delta Y_2}{\Delta Y_1} = \frac{30 - 36}{22 - 18} = \frac{-6}{4} = \frac{-3}{2} = -1.5$$

This computation is presented in Table 5–2 and in Figure 5–3. From Figure 5–3, it can be seen that the *MRPS* of $-3/2$ represents the average slope between the two output combinations. It means that between these points a 1-unit increase in Y_1 will necessitate a 1.5-unit decrease in Y_2.

The exact *MRPS* is the slope at any point on the production possibility curve and can be determined by drawing a tangent at the point

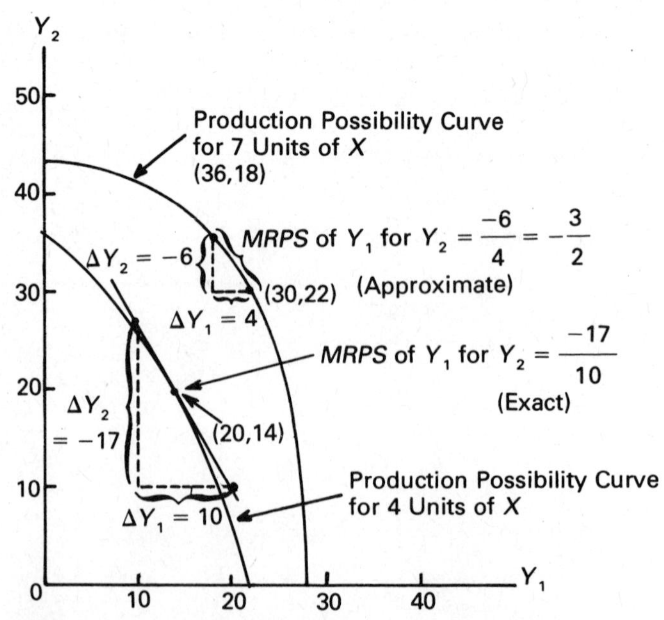

Figure 5–3. Approximate and exact measures of the marginal rate of product substitution.

in question and then measuring the slope of the tangent. This is illustrated in Figure 5–3 on the production possibility curve for four units of input. At the point $Y_2 = 20$ and $Y_1 = 14$, a tangent has been drawn. The slope of this tangent is -1.7 and therefore the exact *MRPS* at the point is -1.7. An increase in Y_1 by "one unit" could only be possible if Y_2 is decreased by "1.7 units."

The product-product relationships, discussed earlier in this chapter, can be defined in terms of *MRPS*. Whenever the *MRPS* or the slope of the opportunity curve is negative, the relationship among products is competitive, that is, an increase in one product necessitates a decrease in another product; when *MRPS* or the slope of the opportunity curve is positive, the relationship among products is complementary, that is, an increase in one product brings about an increase in the other product; when the *MRPS* is zero (or undefined), the relationship among products is supplementary, that is, an increase in one product does not change or affect the output of the other product (see Figures 5–1A, 5–1B, and 5–1C).

The Isorevenue Line

Total revenue is the value of the output produced. For example, if 36 units of Y_2 and 18 units of Y_1 were produced and if price per unit of these products were \$1 and \$2, respectively, then total revenue = \$1 · 36 + \$2 · 18 = \$72, or, in symbolic notation,

$$\text{Total revenue} = P_{Y_1} Y_1 + P_{Y_2} Y_2$$

where Y_1 and Y_2 represent the total amount of the two products.

A line representing any given total revenue can be drawn on a graph. Consider a total revenue of \$80. When $P_{Y_2} = \$1$ and $P_{Y_1} = \$2$, the \$80 revenue could be earned by selling 80 units (\$80 divided by \$1) of Y_2 and no Y_1 or 40 units (\$80 divided by \$2) of Y_1 and no Y_2. Other combinations of products will also earn \$80. For example, 20 units of Y_1 and 40 units of Y_2, 30 units of Y_1 and 20 units of Y_2, 10 units of Y_1 and 60 units of Y_2 would all earn \$80 in revenue. When graphed, these points lie on a straight line (Figure 5–4A) called an *isorevenue* or equal revenue line.

The isorevenue line in Figure 5–4A passes through all combinations of Y_1 and Y_2 that earn a revenue of \$80. The line is straight because the output prices do not change regardless of the amount of output sold. Thus, the location of an isorevenue line for any total revenue can easily be determined by computing the points on the axes and connecting these points with a straight line. The point on the Y_2 axis is always equal to TR/P_{Y_2} while the point on the Y_1 axis equals TR/P_{Y_1}. These points determine the amount of either Y_2 or Y_1 needed to earn the total revenue when the other product is not produced.

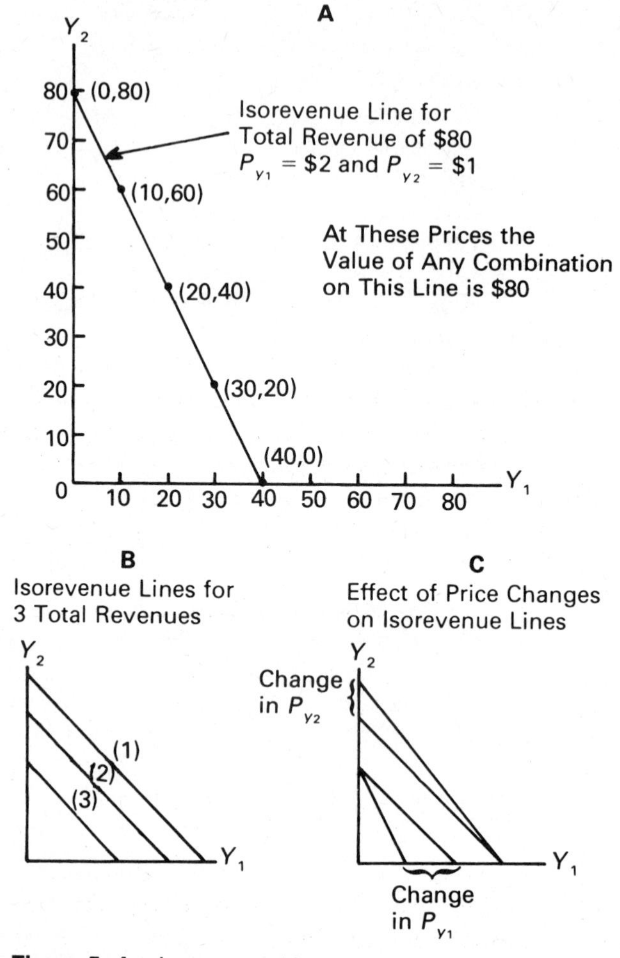

Figure 5–4. Isorevenue Lines.

The distance of the isorevenue line from the origin is determined by the magnitude of the total revenue. As total revenue increases, the isorevenue line moves away from the origin. Thus, in Figure 5–4B, the isorevenue line labeled (1) represents a higher total revenue than (2) and (3), while (2) represents a higher total revenue than (3). The isorevenue lines in Figure 5–4B are parallel because output prices have not changed.

The slope of the isorevenue line is determined by the output prices. The slope of a straight line may be measured between any two distinct points. For convenience, the slope can be measured between the points where the isorevenue line intersects the output axis. The Y_2 axis is intersected at the point $Y_1 = 0$ and $Y_2 = TR/P_{Y_2}$; the Y_1 axis is intersected at the point $Y_1 = TR/P_{Y_1}$ and $Y_2 = 0$. The slope of the isorevenue line is

$$\text{Slope of isorevenue line} = \frac{\text{rise}}{\text{run}} = \frac{0 - \dfrac{TR}{P_{Y_2}}}{\dfrac{TR}{P_{Y_1}} - 0} = -\frac{\dfrac{TR}{P_{Y_2}}}{\dfrac{TR}{P_{Y_1}}} = -\frac{P_{Y_1}}{P_{Y_2}}$$

Thus, the output price ratio is the slope of the line. The negative sign, of course, means that the isorevenue line slopes downward to the right. The slope of the isorevenue line in Figure 5–4A is $-P_{Y_1}/P_{Y_2} = -2/1 = -2$. When the output prices remain constant, isorevenue lines representing different total revenues are parallel. But a change in either price will change the slope of the isorevenue line. These effects are illustrated in Figure 5–4C. Any decrease in the price of Y_1 shifts the intersection of the isorevenue line further from the origin on the Y_1 axis. Similar changes are evidenced on the Y_2 axis by a change in the price of Y_2.

Isorevenue lines in this chapter and the isocost lines in Chapter 4 appear the same on a graph. The slopes of both are determined by the relevant price ratios. The two should not be confused, however. They are used for different purposes; isorevenue lines are used for revenue optimization, isocost lines for cost minimization.

The Revenue-Maximizing Combination of Outputs

The production possibility curve presents all possible combinations of two products that could be produced using a given amount of variable input. Only one combination of outputs will be produced in practice. The two relevant questions are: what combination should be produced and how can that combination be determined?

Total costs are constant for all output combinations on a production possibility curve. Profits from the limited, but variable, resource will be the greatest, or losses the smallest, if the output combination returning the maximum total revenue is selected. To avoid confusion with the most profitable combination of outputs, which would be determined when inputs are unlimited and therefore may be on a different production possibility curve, the revenue-maximizing combination of outputs on a given production possibility curve will be called the *maximum revenue* combination.

When the production possibilities are presented in a table, the total revenue can be computed for each output combination and the maximum revenue combination selected. This has been done in Table 5–3, using the production possibilities presented for seven units of input in Table 5–1C and the prices $P_{Y_1} = \$2$ and $P_{Y_2} = \$1$. The max-

Table 5–3 Computing the Maximum Revenue Combination of Products Given Seven Units of Input (P_{Y_1} = \$2, P_{Y_2} = \$1)

Production Possibilities for X = 7		*Revenue from Y_2*	*Revenue from Y_1*	*Total*
Y_2	Y_1	P_{Y_2} = \$1	P_{Y_1} = \$2	*Revenue*
43	0	\$43	\$ 0	\$43
42	7	42	14	56
40	13	40	26	66
36	18	36	36	72
30	22	30	44	74
22	25	22	50	72
12	27	12	54	66
0	28	0	56	56

imum revenue combination of products is 30 units of Y_2 and 22 units of Y_1. The total revenue earned is \$74. Every other combination of outputs in Table 5–3 returns less.

Calculating the total revenue from each output combination and selecting the maximum revenue combination is only feasible for a small number of combinations. An infinite number of output combinations exist along the production possibility curve; computation of the total revenue resulting from all combinations would be impossible. The maximum revenue combination of output on a production possibility curve can be determined using the criterion

$$MRPS \text{ of } Y_1 \text{ for } Y_2 = -\frac{P_{Y_1}}{P_{Y_2}}$$

or, using the expression for the *MRPS*,

$$\frac{\Delta Y_2}{\Delta Y_1} = -\frac{P_{Y_1}}{P_{Y_2}}$$

The left side of the criterion represents the slope of the production possibility curve, and the right side the slope of the isorevenue line. The maximum revenue point is that point on the graph where the isorevenue line is tangent to the production possibility curve. For the production possibility curve resulting from seven units of input, the maximum revenue combination is determined in Figure 5–5A. An isorevenue line with a slope of -2 (because $-P_{Y_1}/P_{Y_2} = -\$2/\$1$) is drawn tangent to the production possibility curve. The point of tan-

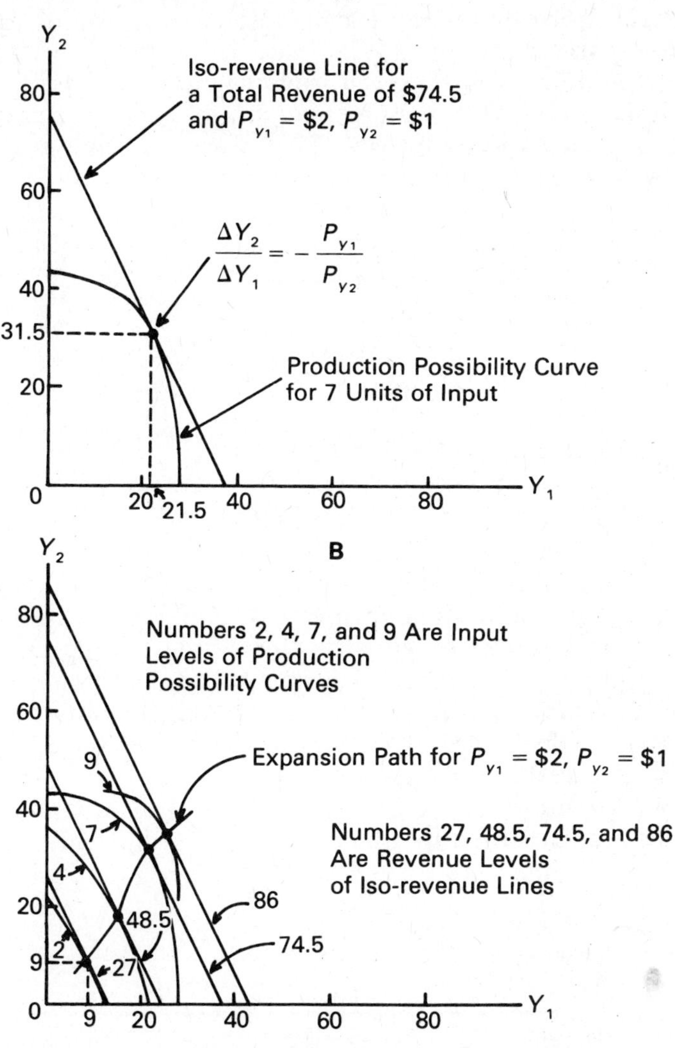

Figure 5–5. Selecting the maximum revenue combinations of products on a production possibility curve.

gency occurs at $Y_2 = 31.5$ and $Y_1 = 21.5$; the total revenue at this point is $\$1 \cdot 31.5 + \$2 \cdot 21.5$ or $74.50. When the isorevenue line is drawn, the total revenue is not known; all that is known is that the point of tangency will determine the maximum revenue for the output combinations on the production possibility curve. In general, the farther the isorevenue line is located from the origin, the higher will be the total revenue. In Figure 5–5A, any other isorevenue line drawn farther from the origin would not touch the production possibility curve and therefore would represent a total revenue unattainable with seven units of input. Any isorevenue line drawn closer to the origin would represent a lower total revenue.

The revenue earned by seven units of input is $74.50 in Figure 5–5A; the maximum revenue combination in Table 5–3 is $74. The same data were used in both cases. The answer on the graph is more accurate because more output combinations are considered.

Further consideration of the criterion for determining the maximum revenue combination of outputs will yield some insight into its meaning. Suppose that the criterion

$$\frac{\Delta Y_2}{\Delta Y_1} = -\frac{P_{Y_1}}{P_{Y_2}}$$

were rewritten as follows

$$P_{Y_2}(\Delta Y_2) = -P_{Y_1}(\Delta Y_1)$$

where ΔY_2 will be negative. The criterion as rewritten states that at the maximum revenue point the increase in revenue due to adding a minute quantity of Y_1 is exactly equal to the decrease in revenue caused by the reduction in Y_2. Thus, there is no incentive to change the output combination. When $-P_{Y_2}(\Delta Y_2) > +P_{Y_1}(\Delta Y_1)$, the amount of Y_1 should be decreased in favor of Y_2. When $-P_{Y_2}(\Delta Y_2) < +P_{Y_1}(\Delta Y_1)$, then Y_1 should be increased at the expense of Y_2.

The maximum revenue criterion can be used when production possibilities are presented in a table. Table 5–4 shows this procedure

Table 5–4 Computing the Maximum Revenue Combination of Products Using the Maximum Revenue Criterion (P_{Y_1} = \$2, P_{Y_2} = \$1)

Production Possibilities for X = 7				
Y_2	Y_1	ΔY_2	ΔY_1	*MRPS of Y_1 for Y_2*
43	0			
		−1	7	−1/7
42	7			
		−2	6	−2/6
40	13			
		−4	5	−4/5
36	18			
		−6	4	−6/4
30	22			
		−8	3	−8/3
22	25			
		−10	2	−10/2
12	27			
		−12	1	−12/1
0	28			

for the example being considered. The price ratio of -2 lies between the *MRPS* of $-6/4$ and $-8/3$; therefore, the most profitable combination of outputs is 30 of Y_2 and 22 of Y_1. Computationally, this method does not seem to offer any advantages over computing the total revenue for each output combination. However, it does serve to explain more fully the theory of enterprise combination.

Ordinarily the production possibility curve is used when the amount of variable input is constant at a particular value; thus, only one production possibility curve would be relevant. However, by assuming varying amounts of the variable input to be available, an expansion path for outputs can be derived. Such an expansion path would have an interpretation that is comparable to that of an expansion path for inputs (see Chapter 4).

Figure 5–5B contains production possibility curves for 2, 4, 7, and 9 units of input (derived from the data in Table 5–1). The production possibility curve for 9 units of input does not extend to the axes because of the assumption that input use will never be extended into Stage III; thus, the most input used on either product is 7. An isorevenue line with a slope of -2 is drawn tangent to each production possibility curve. The line connecting these maximum revenue points is called an output expansion path. For each level of input, the maximum revenue combination of outputs will fall on the expansion path. However, only one combination of outputs is the optimum, or high profit, combination resulting from the use of the optimum amount of input (where marginal cost equals marginal revenue). All other points are the maximum revenue points, given the limited amount of input. The optimum combination cannot be located geometrically on Figure 5–5B, but it will be some point on the expansion path.

Opportunity Cost and Marginal Criterion for Resource Allocation

Maximum revenue from a limited amount of input was shown to occur when

$$\frac{\Delta Y_2}{\Delta Y_1} = -\frac{P_{Y_1}}{P_{Y_2}}$$

and this was subsequently rewritten as

$$-P_{Y_1}\Delta Y_1 = P_{Y_2}\Delta Y_2$$

where ΔY_2 was negative. But the decrease in Y_2 could only be caused by shifting some amount of input, X, from enterprise Y_2 to enterprise Y_1. Denote the amount of input shifted by ΔX. Dividing both sides of the above expression by ΔX and multiplying both sides of the equality by -1 gives

$$P_{Y_1}\frac{\Delta Y_1}{\Delta X} = P_{Y_2}\frac{\Delta Y_2}{\Delta X}$$

or

$$P_{Y_1}MPP_{XY_1} = P_{Y_2}MPP_{XY_2}$$

$$VMP_{XY_1} = VMP_{XY_2}$$

Thus, revenue from the limited amount of input, X, will be a maximum when the value of the marginal product of the input is the same in each enterprise. (The notations MPP_{XY_1} and VMP_{XY_2} are used to denote the *MPP* of X used on Y_1 and the *VMP* of X used on Y_2, respectively.)

Equating the *VMP*s of the input in the two enterprises leads to the identical solution obtained from the production possibility curve. The two criteria are compared in Table 5–5, where the production functions from Table 5–1 are repeated. For two units of input, one unit would be applied to Y_1, where it would earn \$14, and the second to Y_2, for an earnings of \$12. The total revenue would be \$26. (The second unit could also go to Y_1 and earnings would be unchanged.)

Table 5–5 Comparing the Marginal Criterion for Resource Allocation and the Production Possibility Curves (P_{Y_1} = \$2, P_{Y_2} = \$1)

X	Y_1	MPP_{XY_1}	VMP_{XY_1}	X	Y_2	MPP_{XY_2}	VMP_{XY_2}
0	0	0	0	0	0		0
1	7	7	\$14	1	12	12	\$12
2	13	6	12	2	22	10	10
3	18	5	10	3	30	8	8
4	22	4	8	4	36	6	6
5	25	3	6	5	40	4	4
6	27	2	4	6	42	2	2
7	28	1	2	7	43	1	1

Units of Input Available	*Solution Equating VMP*			*Solution Using Production Possibility Curves (Figure 5–5B)*		
	Y_1	Y_2	*TR*	Y_1	Y_2	*TR*
2	7	12	\$26	9	9	\$27.00
4	13	22	48	15.5	17.5	48.50
7	22	30	74	21.5	31.5	74.50
9	25	36	86	25.5	35.0	86.00

From the production possibility curve for two units of input in Figure 5–5B, the maximum revenue combination of outputs is 9 each of Y_1 and Y_2. Reference to the production functions in Figure 5–1A shows that 0.7 units of X used on Y_1 and 1.3 units of X used on Y_2 will produce 9 units of each. Nine units of each output produce a total revenue of \$27, slightly more than the allocation using "average" marginal criteria. Thus, the geometric approach is slightly more accurate, but for any practical situation the difference is negligible. Solutions from other input amounts are similarly close.

This alloction of inputs between products can also be viewed in terms of opportunity cost. It demonstrates the cost in terms of the value of an alternative product that is given up rather than the purchase price of the variable input. As long as the *VMP* in one enterprise, which is sacrificed, equals the *VMP* in the other enterprise, which is gained, the opportunity costs for both enterprises are equal and total returns are maximum. If the value of the marginal product given up exceeds the value of the marginal product gained, then the opportunity cost exceeds the value added and returns are below the maximum; the sufficient condition of profit maximization is not met.

TWO EXAMPLES

Most of the problems involving the allocation of limited inputs among different enterprises have traditionally been solved through budgeting and, in more recent times, through linear programming techniques. The objective of these procedures is, of course, the same—to use the limited inputs in such a way that the enterpreneur's objective, specified here as profit maximization, is achieved.

Production Possibility Curve Given

As shown earlier, the product-product problems lend themselves to marginal analysis using the procedures applied in factor-product (Chapter 3) and factor-factor relationships (Chapter 4). For example, a product-product relationship could be described by the following equation:

$$Y_1 = 100 - 0.0065Y_2^2 \qquad (5.1)$$

where 100 is the maximum amount of Y_1 that can be obtained from the limited amount of input. The maximum amount of Y_2 possible, 124, can be determined by setting Y_1 equal to zero in (5.1) and solving for Y_2. Equation (5.1) assumes a specific amount of input—a change in the input amounts would change the equation. The equation suggests that Y_1 and Y_2 are competitive for the limited input. An increase in one product requires a decrease in the other.

The output combinations shown in Table 5–6 are derived by substituting different values for Y_2 in equation (5.1) and solving for Y_1. Thus, if $Y_2 = 0$, then $Y_1 = 100$; if $Y_2 = 10$, then $Y_1 = 99.35$; or if $Y_2 = 100$, $Y_1 = 35$; etc.

The opportunity curve shown in Figure 5–6 is based on the output combinations presented in Table 5–6. The opportunity curve is concave to the origin, indicating that the *MRPS* of Y_1 for Y_2 or Y_2 for Y_1 is increasing. For example, increasing Y_2 from 10 to 20 units is possible by giving up 2 units of Y_1 or 0.20 units of Y_1 for each unit increase of Y_2; when increasing Y_2 from 120 to 124 units, 1.60 units of Y_1 have to be given up for each unit increase of Y_2 (see Table 5–6).

The difference between the average and exact *MRPS* can be explained in the same way as the difference between the average and exact *MPP* or the average and exact *MRS* in the factor-factor relationship. The average *MRPS* denotes the average slope between intervals such as Y_2 between 0 and 10 units and Y_1 between 100 and 99.4 units; and Y_2 between 10 and 20 units and Y_1 between 99.4 and 97.4 units in Table 5–6. The exact *MRPS*, on the other hand, refers to the exact slope at a precise coordinate on the opportunity curve such as Y_2 at 10 and Y_1 at 99.4 or Y_2 at 30 and Y_1 at 94.2 etc. Since *MRPS* is increasing, the exact *MRPS* for any given output combination is higher than the average *MRPS* as shown in Table 5–6.

The exact *MRPS* can be derived from equation (5.1) directly

$$\frac{dY_1}{dY_2} = -0.013Y_2 \tag{5.2}$$

Table 5–6 Output Combinations and *MRPS* of Y_1 and Y_2 ($Y_1 = 100 - 0.0065Y_2^2$)

		MRPS Y_2 for Y_1	
		Average	*Exact*
Y_2 (1)	Y_1 (2)	$\frac{\Delta Y_1}{\Delta Y_2}$ (3)	$\frac{dY_1}{dY_2}$ (4)
0	100.0		—
		−0.06	
10	99.4		−0.13
		−0.20	
20	97.4		−0.26
		−0.32	
30	94.2		−0.39
		−0.46	
40	89.6		−0.52
		−0.58	
50	83.8		−0.65
		−0.72	
60	76.6		−0.78
		−0.85	
70	68.1		−0.91
		−0.97	
80	58.4		−1.04
		−1.11	
90	47.3		−1.17
		−1.23	
100	35.0		−1.30
		−1.37	
110	21.3		−1.43
		−1.49	
120	6.4		−1.56
		−1.60	
124	0		−1.61

This *MRPS* equation defines the slopes of the opportunity curve, Figure 5–6, for any combination of Y_1 and Y_2. The slopes or the exact *MRPS* are presented in column 4, Table 5–6.

If the prices of Y_1 and Y_2 were \$5 and \$6 respectively, then the optimum combination of Y_1 and Y_2 is obtained by equating *MRPS* with the price ratio or the slope of the isorevenue line.

In this case,

$$\frac{dY_1}{dY_2} = -\frac{P_{Y_2}}{P_{Y_1}} \tag{5.3}$$

or

$$-0.013Y_2 = -\frac{6}{5}$$

and

$$Y_2 = 92.3$$

and

$$Y_1 = 100 - 0.0065(92.3)^2 = 44.6$$

Thus, the combination of Y_1 and Y_2 that maximizes revenue consists

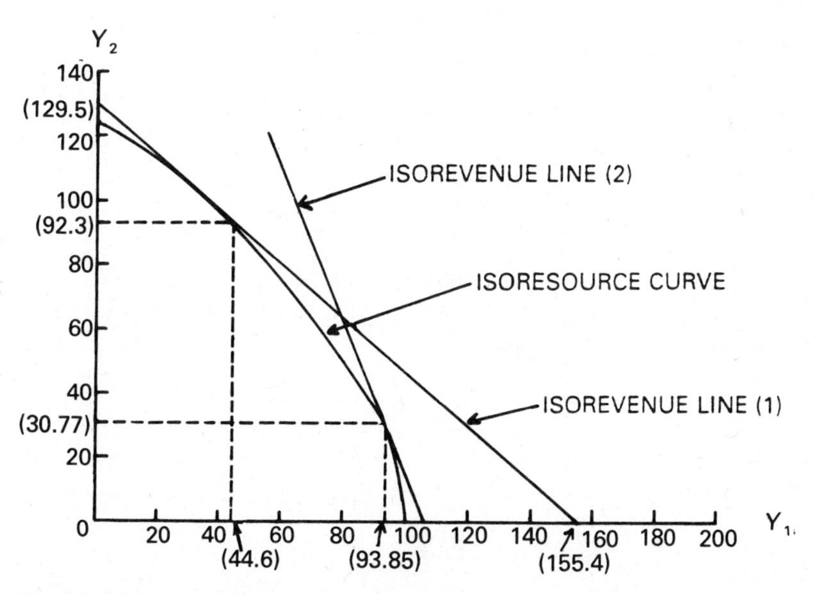

Figure 5–6. Selecting the maximum revenue combinations of products on a production possibility curve.

of 44.6 units of Y_1 and 92.3 units of Y_2. The resulting total revenue is $\$5 \cdot 44.6 + \$6 \cdot 92.3 = \$776.80$. The same solution can be obtained graphically, as suggested in Figure 5–6, by the tangency of the isorevenue line (1).

A change in relative prices also changes the slope of the isorevenue line. If the price of Y_2 were to decline from \$6 to \$2 and the price of Y_1 were to remain at \$5, the new optimum combination of Y_1 and Y_2 suggests a production of 93.8 units of Y_1 and only 30.8 units of Y_2. The combinations of Y_1 and Y_2 are calculated as above and are derived in Figure 5–6 where isorevenue line (2) is tangent to the opportunity curve.

Derivation of the Production Possibility Curve

Production possibility curves can be derived directly from production functions. The procedure used is the analytical equivalent of the graphic analysis presented in Figure 5–1. Suppose two products, Y_1 and Y_2, are produced with a single input, X, where the quantity of X is constant. The quantity of input used on each product has to be identified in some distinctive manner; let X_1 be the amount of X used to produce Y_1 and let X_2 be the amount of X used to produce Y_2. Thus, $X_1 + X_2 = X = K$.

When the production functions are algebraically simple, the procedure is straightforward. For example, given the two Cobb-Douglas functions,

$$Y_1 = 1/2X_1^{1/2}$$
$$Y_2 = X_2^{1/2}$$

then the inverse production functions are given by

$$X_1 = 4Y_1^2$$
$$X_2 = Y_2^2$$

These inverse production functions are then substituted directly into the constraint expressing the input limitation to obtain the production possibility curve. In this case,

$$X_1 + X_2 = K$$

and

$$4Y_1^2 + Y_2^2 = K$$

This is an ellipse centered at $Y_1 = Y_2 = 0$. Solving to express Y_2 as a function of Y_1 in the positive quadrant gives

$$Y_2 = (K - 4Y_1^2)^{1/2} \tag{5.4}$$

where Y_1 is zero or greater. When Y_1 is not produced, then $Y_2 = (K - 0)^{1/2} = K^{1/2} = X^{1/2}$, which is the amount of Y_2 that would result when all of X is used to produce Y_2. Alternately, when Y_2 is zero, then $(K - 4Y_1^2) = 0$ and $Y_1 = 1/2K^{1/2} = 1/2X^{1/2}$, which is the output of Y_1 resulting when all of X is used to produce Y_1. The production possibility equation (5.4) is actually a family of curves that has as a parameter the input amount, K. As K increases, the production possibility curves shift outward from the origin.

The product expansion path is found by equating

$$\frac{dY_2}{dY_1} = -\frac{P_{Y_1}}{P_{Y_2}}$$

where, by differentiating (5.4) directly,

$$\frac{dY_2}{dY_1} = -\frac{4Y_1}{(K - 4Y_1^2)^{1/2}} = -\frac{4Y_1}{Y_2}$$

The expansion path is

$$-\frac{4Y_1}{Y_2} = -\frac{P_{Y_1}}{P_{Y_2}}$$

or

$$Y_2 = \frac{4P_{Y_2}}{P_{Y_1}}Y_1 \tag{5.5}$$

which gives the ratio of Y_1 and Y_2 that will maximize total revenue from the sale of Y_1 and Y_2. To obtain any particular solution, given K and the prices, the expansion path equation must be solved simultaneously with (5.4). This is required to determine the points of intersection between the expansion path and the production possibility curve. Therefore, from (5.4) and (5.5),

$$4rY_1 = (K - 4Y_1^2)^{1/2}$$

and

$$Y_1 = \left[\frac{K}{16r^2 + 4}\right]^{1/2} \tag{5.6}$$

where $r = (P_{Y_2}/P_{Y_1})$. In the solution process, the negative solution would fall outside the first quadrant and is ignored.

To illustrate the solutions obtained, suppose $P_{Y_1} = \$2$, $P_{Y_2} = \$1$, and 36 units of input are available. The production possibility curve is

$$Y_2 = (36 - 4Y_1^2)^{1/2}$$

and the product expansion path is $Y_2 = 2Y_1$. To obtain the solution,

$$Y_1 = \left(\frac{36}{4 + 4}\right)^{1/2} = 2.12$$

and so $Y_2 = 4.24$. Total revenue is $2(2.12) + $1(4.24) = $8.48. Substituting into the inverse production functions shows that

$$X_1 = 4Y_1^2 = 18$$
$$X_2 = Y_2^2 = 18$$

and for the price ratios given, the input is divided equally between the enterprises.

As noted above, the production combinations on the product expansion path determine the maximum return from the given amount of input. For output combinations on the expansion path, the value of the marginal product of the input is equal in each enterprise but may not be equal to the input price. To determine profit, the input price or, more correctly, the opportunity cost of the input must be considered.

If the input is being purchased with a fixed sum of capital, then the opportunity cost of the input is its purchase price. (In this instance, capital is actually the limiting resource, but its scarcity is reflected by the limitation in input.) For the numerical example just presented, the *VMP* of the input is about $0.12 in each use. If the input costs more than that, then less than 36 units should be purchased; if the input costs $0.12, then the optimum is being used in both enterprises. If the input is owned and not resaleable, then its opportunity cost is simply its *VMP* in the production of Y_1 and Y_2, assuming only two products on the farm. In this case, the manager's goal of maximizing revenue is achieved by expanding along the expansion path until all of the input is utilized. (This is similar to the familiar "guns versus butter" example given in principles textbooks: A nation is viewed as possessing a total bundle of resources which it does not want to sell but will use only to produce products desired by society.)

As a final note, deriving production possibility curves from production functions often leads to rather complex algebraic forms. The curve itself, (5.4), and the expansion path, (5.5), are often not difficult to obtain, but their simultaneous solution, as given in our example by (5.6), can become quite intractable. Changes in the Cobb-Douglas functions given above, or selection of algebraic equations with more complex inverse functions, will make the example more difficult to solve. Usually, the problem can be more easily approached by considering the direct allocation of inputs to enterprises.

INTERMEDIATE AND FINAL PRODUCTS

The art of farm management centers around a knowledge of the competitive, complementary, and supplementary relationships among farm enterprises. The farm manager tries to combine enterprises to take maximum advantage of supplementary and complementary relationships. This becomes complicated because the relationships between enterprises differ depending on the input considered. Thus, wheat and soybeans may be supplementary enterprises with respect to combine use because one may be harvested in June and the other in September. With respect to land, assuming single cropping, they are competitive because if an acre of land is used to grow wheat, it cannot be used to grow soybeans the same season. Two crops may be competitive with respect to labor use during the spring planting time, but supplementary with respect to labor use during the harvesting periods.

Livestock enterprises usually compete with crop enterprises for capital, at least within a given production period. And, when pastureland may be suitable for crops, livestock and crops also compete for land. However, livestock enterprises can usually be planned to be supplementary to crops with respect to labor use. For example, farrowing can be timed so that hog labor requirements are small during harvest and high when crop needs are small. Livestock enterprises can be planned to require increased labor during off-seasons.

Livestock enterprises differ from crop enterprises in that major portions of the inputs in livestock enterprises may consist of crops grown on the farm. Products grown on the farm and used as inputs for other farm products are called intermediate products. Forage and hay are intermediate products for livestock enterprises. Corn fed to hogs is an intermediate product; corn sold is a final product.

The most appropriate combination of intermediate products can only be selected by considering the needs of the livestock enterprises. Thus, suppose that for a given farm the available supply of land, capital, labor, etc., will permit the production of grain and forage as shown in Figure 5–7A. Suppose further that the livestock enterprise selected by the farmer has an isoquant map as represented in Figure 5–7B. This isoquant map would be derived, as explained in Chapter 4, from a production function depicting meat production as a function

of the amount of grain and forage fed. Isoquants can be derived for any amount of meat production; for convenience, however, the isoquants in Figure 5–7B are labeled 1, 2, and 3 to denote increasing amounts of production.

The problem is to coordinate grain and forage production in such a way that the maximum amount of meat is produced from the limited resources. The outputs represented by the production possibility curve are the same as the inputs for the isoquants. Therefore, the two can be imposed on one graph (Figure 5–7C). Because isoquants can be derived for any possible level of meat production, there will be one isoquant tangent to the production possibility curve. This tangency determines the highest possible level of meat production. *OB* amount of forage and *OA* amount of grain will be grown. Any other combination of forage and grain would result in lower output of meat. It would be represented by an isoproduct curve closer to the origin than isoquant 2 in Figure 5–7C. The solution in this case is independent of input and product prices.

At the point of tangency, the *MRPS* of forage for grain (the intermediate products) is equal to the *MRS* of forage for grain in the production of meat (the final product); in other words, the *MRPS* in production is equal to *MRS* in consumption. If, for example, *MRPS* and *MRS* equals 2, it means that in production to increase forage by one unit, two units of grain must be sacrificed, but in consumption the unit of forage gained will substitute or replace two units of grain.

If *MRPS* and *MRS* are not equal, the final product, meat, is reduced. For example, if *MRPS* of forage for grain is greater than *MRS* of forage for grain, it means that one unit of forage gained resulted in a greater sacrifice of grain in production than the benefits obtained from substituting that unit of forage for grain in consumption. If the opposite is true—*MRPS* is less than *MRS*—there is further incentive to substitute forage for grain because one unit of forage substitutes for more grain in consumption than must be sacrificed in production.

In the above example, the intermediate products are not traded on the market. The situation represents a farm where all intermediate products, forage and grain, are sold through a secondary enterprise, livestock. Thus, it would be coincidental if the isorevenue line were tangent at the same output (forage–grain) combination as *MRPS* and *MRS*. In the above case, economic efficiency is achieved when

$$\underset{\text{Production}}{\frac{\Delta G}{\Delta F}} = \underset{\text{Consumption}}{\frac{\Delta G}{\Delta F}}$$

Prices of grain and forage do not matter, because the objective is maximization of output of the secondary enterprise. The farm is self-sufficient in production of the intermediate products. (The case is similar to a closed economy in trade theory where trade of intermediate products is not permitted.)

A more typical situation is illustrated in Figure 5–7D. The graph in Figure 5–7D is identical to the graph in Figure 5–7C except that the former includes an isorevenue line. The isorevenue line in Figure 5–7D is tangent to the opportunity curve at a different forage and grain combination than determined by the tangency of the isoquant and opportunity curve. The prices of the intermediate products, forage and grain, suggest a rotation that includes more grain and less forage than the rotation which maximizes meat production.

The isorevenue line traces the points of all combinations of products that, when sold, result in the same revenue. Thus, in Figure 5–7D, the combination OA' of grain and OB' of forage is equivalent in value to the combination OA'' of grain and OB'' of forage needed to increase meat output to a level represented by isoquant 3. On the other hand, the combination of OA' and OB' does not enable the farmer to produce on isoquant 3; it produces an excess of grain and a shortage of forage. Given the market prices, the producer would sell $A''\ A'$ of grain and purchase $B'\ B''$ of forage. This trade would increase the output of meat from isoquant 2 to isoquant 3. (This case is often referred to as an open economy where trade is possible. It demonstrates the advantages of trade.)

In Figure 5–7D, the managers' objectives for both the product-product and factor-factor relationships are met. The maximum possible returns from the intermediate products are obtained and the output of the secondary product is produced with the least cost combination. The isorevenue line in production is identical to the isocost line in consumption. This case can be restated as:

$$\underbrace{\frac{\Delta G}{\Delta F} = -\frac{P_f}{P_g}}_{\text{Production}} = \underbrace{\frac{\Delta G}{\Delta F} = -\frac{P_f}{P_g}}_{\text{Consumption}}$$

The *MRPS* of forage for the grain on the opportunity curve is equal to the slope of the isorevenue line, and it is also equal to *MRS* of forage for grain on the isoquant, which equals the slope of the isocost line. The isorevenue line for one enterprise is the isocost line for the other, and the opportunity cost of the intermediate products is their market value—not their cost of production on the farm. Trading intermediate products enables the farmer to enlarge the final enterprises without altering the bundle of fixed resources that form the basis for the production possibility curve. Thus the enlargement of the intermediate enterprise (from isoquant 2 to isoquant 3) is due solely to the trade of intermediate products.

The final equilibrium point on isoquant 3 will represent a least cost combination, but it may not be the profit-maximizing combination. If the final equilibrium amounts are not the optimum, the manager will either sell more grain, buy less forage, and produce fewer livestock, or will add livestock units until the added return equals the added cost.

GENERAL EQUIMARGINAL PRINCIPLES

The economic view of the production process has increased in complexity. The remaining chapters in this section of the text will present somewhat different aspects of production theory; therefore, a summary of some of the principles presented, along with some logical extensions, seems useful at this time. The general view of the production process can be obtained by studying the input side—the allocation of variable inputs among competing uses. This approach, based on production functions, leads to the same results as the production possibility curves discussed earlier in this chapter. As always in this section, pure competition is assumed in both the input and output markets.

The input view of the production process is based on the marginal increment or marginal product. The marginal product is measured in physical units, bushels, tons, pounds, etc. The economist and farm manager, however, are interested in the revenue earned by the enterprise, and so the marginal product is multiplied by the product price to obtain the value of the marginal product. Thus, if one pound of fertilizer increases wheat yield by one-half a bushel and wheat sells for $4.50 a bushel, the value of the marginal product of fertilizer is $2.25.

Suppose that two different enterprises on the farm use two different variable inputs, each of which may be readily purchased in the marketplace. The question the farm manager must answer when he does not have unlimited capital to purchase inputs is: How much of each input should be purchased for use in each enterprise? If some input is being used in each enterprise, the manager may decide to compare the value of the marginal product of each input. If so, he may find that X_1 (used on Y_1) has a marginal earnings or value of the marginal product of $4, and X_2 (used on Y_2) has a value of the marginal product of $16. Should he decide to purchase more X_2 to use on Y_2 because one unit of X_2 will return four times as much as X_1? Not necessarily. In fact, he can't make a wise economic choice until he knows the cost of X_2 relative to the cost of X_1. Suppose X_1 costs $1 and X_2 costs $8. Then $1 worth of X_1 will return $4 while $1 worth of X_2 will only return $2. X_1 is a better buy because it earns more relative to its cost (on the margin). It will continue to be a better buy until, as it is purchased and applied, its marginal product, and thus marginal earnings, drops to the same level as X_2.

The marginal criterion used is

$$\frac{VMP_{X_1Y_1}}{P_{X_1}} = \frac{VMP_{X_2Y_2}}{P_{X_2}}$$

For the example in the last paragraph, the ratios are \$4/\$1, which is larger than \$16/\$8. Marginal returns from the two inputs will be balanced when X_1 is purchased and used to the point where its marginal earnings are \$2, so that \$2/\$1 = \$16/\$8, or \$2 = \$2. When input use is increased until the ratio is equal to one, an added dollar spent earns exactly \$1 and the optimum amount of input is being used. The optimum amount of input cannot be used in one enterprise unless it is used in all enterprises; otherwise the marginal earnings of the inputs would be out of balance. (Notice that the ratio of the value marginal product divided by the unit price is not "average return per dollar spent.")

The general equimarginal criterion states that the ratio of the value of the marginal product of an input to the unit price of the input (VMP_X/P_X) be equal for *all* inputs in *all* enterprises. This presumes, of course, that input use in all enterprises is in the zone of economic relevance (Stage II) and that managers seek to maximize profits. When the criterion is fulfilled, a dollar spent on any enterprise will have the same marginal earnings, that is, will add a similar amount to total revenue. To fulfill the equimarginal principle, all complementary and supplementary relationships among products must be expanded until competitive relationships exist.

One Input—Several Products

To allocate a limited amount of a variable input among several enterprises, the production function and product prices must be known for each enterprise. Next, the *VMP* schedule must be computed for each enterprise. Finally, using the opportunity cost principle developed earlier, units of the input are allocated to each enterprise in such a way that the profit earned by the input is a maximum.

Decisions regarding resource use are made using *VMP* as a guide. Profits from a limited amount of variable resource are maximized when the resource is allocated among the enterprises in such a way that the marginal earnings of the input are equal in all enterprises. In symbolic notation, this can be stated as

$$VMP_{XA} = VMP_{XB} = VMP_{XC} = \ldots = VMP_{XN}$$

where VMP_{XA} is the value marginal product of X used on product A,

VMP_{XB} is the value marginal product of x used on product b, and so on; N is the number of enterprises under consideration.

The opportunity cost principle works in this way: Suppose an input is used on two enterprises, corn and soybeans. If the *VMP* of the input is $10 for corn and $6 for soybeans, removal of one unit from soybean production and adding it to corn production would reduce returns on beans by $6 and increase returns from corn by $10—a net gain of $4. If, after the unit had been transferred, the *VMP* of the input were $9 on corn and $7 on beans, transfer of another unit from beans to corn would increase returns by $2. If, after this second transfer, the *VMP* of the input were equal to $8 in each enterprise, further transfer of the input would decrease rather than increase returns from the limited stock of input.

Production functions for three enterprises, *A*, *B*, and *C*, are presented in Table 5–7. The enterprises are labeled *A*, *B*, and *C* because the data are hypothetical; they could represent any three enterprises on a farm that use or "compete for" an input. Machinery must be used for several different enterprises, fertilizer may be spread on several fields and different crops, and labor may be allocated to various crop and livestock enterprises. Much of farm management is composed of decisions relating the use of a limited amount of input on several enterprises.

The manager is in a planning situation; he is faced with the production functions in Table 5–7 and has to decide on his future actions. Suppose he has five units of *X* and can allocate them one unit at a time to any of the three enterprises. According to the opportunity cost principle, he will allocate each successive unit of input to the use where its marginal return, *VMP*, is the largest. The first unit of *X* earns $20 in *A*, $18 in *B*, and $14 in *C*. Thus, the first of the five units is applied to enterprise *A*. Having applied the first unit to *A*, the second unit can earn $16 in *A*, $18 in *B*, and $14 in *C*; the second unit is applied to *B*. After applying the first and second unit, the third unit will earn

Table 5–7 Allocating a Limited Amount of Variable Input Among Three Enterprises (P_A = $2, P_B = $1, P_C = $2)

Enterprise A			*Enterprise B*			*Enterprise C*		
X	*Y*	VMP_{XA}	*X*	*Y*	VMP_{XB}	*X*	*Y*	VMP_{XC}
0	0		0	0		0	0	
		$20			$18			$14
1	10		1	18		1	7	
		16			13			12
2	18		2	31		2	13	
		12			11			10
3	24		3	42		3	18	
		10			9			8
4	29		4	51		4	22	
		8			7			6
5	33		5	58		5	25	
		6			6			4
6	36		6	64		6	27	

\$16 in A, \$13 in B, and \$14 in C; the third will go on A. Continuing, the best use of five units of X is:

Units of X	*Enterprise Used on*	*VMP*
1st	*A*	\$20
2nd	*B*	18
3rd	*A*	16
4th	*C*	14
5th	*B*	13

Two units of input go on A, two on B, and one on C. Used on this manner, the five units of input will earn \$81. No other allocation of the five units among the three enterprises will earn as much. After the fifth unit of input is applied, *VMP* is \$12 in A, \$11 in B, and \$12 in C. The marginal earnings of X are not quite equal in each enterprise. This is due to the discontinuities, or "lumpiness," of the data in Table 5–7. More refined data, such as provided by a graph, would be necessary to equate the marginal earnings exactly. The principles used would be the same, however.

As yet, the cost of the five units of input has not been discussed. The cost of the inputs raises several interesting problems; let us review the opportunity cost principle discussed earlier. Suppose the inputs cost \$6.50 per unit and the manager has \$32.50 of capital, enough to purchase five units. He could then make \$48.50 (less fixed expenses) by buying the five units and using them as described. Before doing so, however, he would want to be sure he could not earn more than \$48.50 profit by using the \$32.50 of capital in some off-farm enterprise—for example, the stock market or a business in the next town. The first point, then, is that the capital will be used to purchase inputs for enterprises A, B, and C only if it won't earn more elsewhere.

Next, suppose that the manager already owns the inputs, having bought them in the past. The cost of the inputs may be \$1 or \$1000, but in either case it has already been paid. Will the manager now use the inputs on enterprises A, B, and C? Not necessarily. If he can sell the inputs to, say, his neighbor for more than \$81, he will do so. If the inputs are durables, such as land, he may choose to sell the services of the input for a given time period, such as a year, rather than utilize them himself. Only if he cannot dispose of the inputs for more than they will earn on his farm will he use them. To summarize, if the manager has capital to purchase the inputs, the opportunity cost is determined by the earnings of that capital in alternative uses. If he already owns the inputs, the opportunity cost is represented by the disposal value of the inputs, regardless of their original cost. Thus, when discussing input use for one or more enterprises, the manager must have considered and rejected all alternative uses for his capital or inputs.

What is the maximum amount of input needed for enterprises *A*, *B*, and *C*? To find out, the manager must determine the most profitable amount of input for each enterprise. When inputs cost $6.50 per unit, the optimum amounts are 5 for *A*, 5 for *B*, and 4 for *C*. Profit is $77. Thus, the manager would never use more than a total of 14 units of inputs on *A*, *B*, and *C*, no matter how many units he could afford to buy. To review the definitions presented at the beginning of this chapter, if the manager has sufficient capital to purchase 14 or more units of input, capital is termed "unlimited"—meaning he does have enough capital to use the optimum amount of input in each enterprise. If the manager does not have enough money to purchase 14 units of input, then capital is "limited."

The marginal allocation procedure lends itself to an algebraic solution. As an example, consider the following two production functions: The first production function, introduced earlier as (3.3), depicts corn response to nitrogen on irrigated land:

$$C = 65.54 + 1.084N_C - 0.003N_C^2 \tag{5.7}$$

The second production function describes grain sorghum response to nitrogen on irrigated land in the same area:

$$S = 68.07 + 0.830N_S - 0.002N_S^2 \tag{5.8}$$

where C is corn in bushels per acre, S is grain sorghum in bushels per acre, N_C is nitrogen in pounds used on corn land, and N_S is nitrogen in pounds used on sorghum land.

Assume the farmer has 100 pounds of nitrogen available for two acres—one acre to be used for corn and one to be used for sorghum production—and that the price of corn is $3.00 a bushel and the price of grain sorghum is $2.50 a bushel.

Using the principles illustrated earlier, the allocative equations would be

$$VMP_{NC} = VMP_{NS}$$

or

$$P_C MPP_{NC} = P_S MPP_{NS}$$

From (5.7),

$$\frac{dC}{dN_C} = 1.084 - 0.006N_C$$

and from (5.8),

$$\frac{dS}{dN_S} = 0.830 - 0.004N_S$$

Multiplying the MPPs by their respective prices gives

$$VMP_{NC} = (1.084 - 0.006N_C)(\$3.00) = \$(3.252 - 0.018N_C) \quad (5.9)$$
$$VMP_{NS} = (0.830 - 0.004N_S)(\$2.50) = \$(2.075 - 0.01N_S) \quad (5.10)$$

If $N_C + N_S = 100$ pounds of nitrogen, then $N_C = 100 - N_S$. Equating the *VMP*s and substituting $N_S = 100 - N_C$ into the equation leads to the solution

$$N_C = \frac{2.177}{0.028} = 77.8$$

and

$$N_S = 100 - 77.8 = 22.2$$

In this example, the corn acre would get 77.8 pounds of nitrogen and the sorghum acre 22.2 pounds. This allocation equates the value of the marginal products and assures the largest return from 100 pounds of nitrogen. Substituting 77.8 pounds of nitrogen into VMP_{NC}, equation (5.9), and 22.2 pounds into VMP_{NS}, equation (5.10), demonstrates that the *VMP*s are equal at $1.95.

If the nitrogen allotment were increased from 100 to 200 pounds, then the corn acre would get 113.5 pounds and the sorghum acre 86.5 pounds. The resulting *VMP*s are $1.21, 74 cents less than when 100 pounds of nitrogen are available. However, the total returns from nitrogen are higher in the second case.

The above analysis is based on the assumption that input use is in Stage II for all enterprises. If the quantity of the input is not sufficient to extend input use to the point where *APP* is a maximum in all enterprises, the input will be used to the point of maximum *APP* in some enterprises, the ones that are the most profitable, and no input will be used on the rest of the enterprises. Thus, when sufficient quantities of the variable input are not available to extend input use into Stage II for all enterprises, it will pay the manager to leave some fixed resources idle (Problem 5–8).

Two Inputs—Two Outputs

Consider the case in which two inputs, X_1 and X_2, can be used to produce two products, Y_1 and Y_2. When the inputs are used in the first enterprise, the equimarginal principles dictate the following equality:

$$\frac{P_{Y_1} MPP_{X_1Y_1}}{P_{X_1}} = \frac{P_{Y_1} MPP_{X_2Y_1}}{P_{X_2}}$$

or

$$\frac{VMP_{X_1Y_1}}{P_{X_1}} = \frac{VMP_{X_2Y_1}}{P_{X_2}}$$

Thus, the marginal earnings of each input must be the same per unit of cost, even within a specific enterprise. When both ratios equal one, the optimum has been reached. These results are identical to Chapter 4. Dividing both sides of the above equality by P_{Y_1} gives

$$\frac{MPP_{X_1Y_1}}{P_{X_1}} = \frac{MPP_{X_2Y_1}}{P_{X_2}}$$

This equality specifies the minimum cost combinations of inputs for the production of any level of output. Thus, the general equimarginal criterion automatically specifies the minimum cost combination of inputs within each enterprise considered.

The same conditions must hold for the use of the two inputs in the second enterprise. The equimarginal conditions for that enterprise can be written as

$$\frac{VMP_{X_1Y_2}}{P_{X_1}} = \frac{VMP_{X_2Y_2}}{P_{X_2}}$$

But marginal returns per dollar spent on inputs must be the same for both inputs in both enterprises. Thus, the general condition is

$$\frac{VMP_{X_1Y_1}}{P_{X_1}} = \frac{VMP_{X_2Y_1}}{P_{X_2}} = \frac{VMP_{X_1Y_2}}{P_{X_1}} = \frac{VMP_{X_2Y_2}}{P_{X_2}}$$

Minimum cost combinations of inputs are used in both enterprises, and marginal returns are equated between enterprises. Stated differently, production is taking place on the expansion paths at output levels where the value of the marginal return per dollar is the same in both enterprises.

As an example, consider the case where a fixed quantity of capital is available to purchase corn and protein supplement for hogs and steers. For each enterprise, a production function will describe output as a function of protein supplement and corn. Given the prices of corn and the supplement, each enterprise will have an isoquant map and an expansion path. The objective is to provide a ration where the *MRS* of corn for protein will be the same for both enterprises and the marginal returns per dollar are the same in both enterprises. Thus, a pound of protein should replace the same amount of corn when fed to hogs as when fed to steers.

The necessary condition for efficiency is met when the *MRS* are equal. Inequality of the *MRS* means that it would be possible to produce more of one or both with the same amounts of inputs or to produce the same amounts of the products with fewer inputs.

The graphic solution is shown in Figure 5–8, which is called a *box diagram*. The dimensions of the box are determined by the amount of the limited inputs purchased by the farmer. As more inputs are available, the box increases in size. In Figure 5–8, the length of the base (and top) of the rectangle represents the number of units of corn available, and the length of the sides represents the amount of protein supplement.

Because the box is measured in units of input, the isoquant maps and the associated expansion paths can be imposed directly upon it. The isoquant map for hogs is in the normal position: The hog isoquants are convex to the origin, *O*, and the hog expansion path, *OE*, passes through the points of tangency between the hog isoquants, H_1, H_2, H_3,

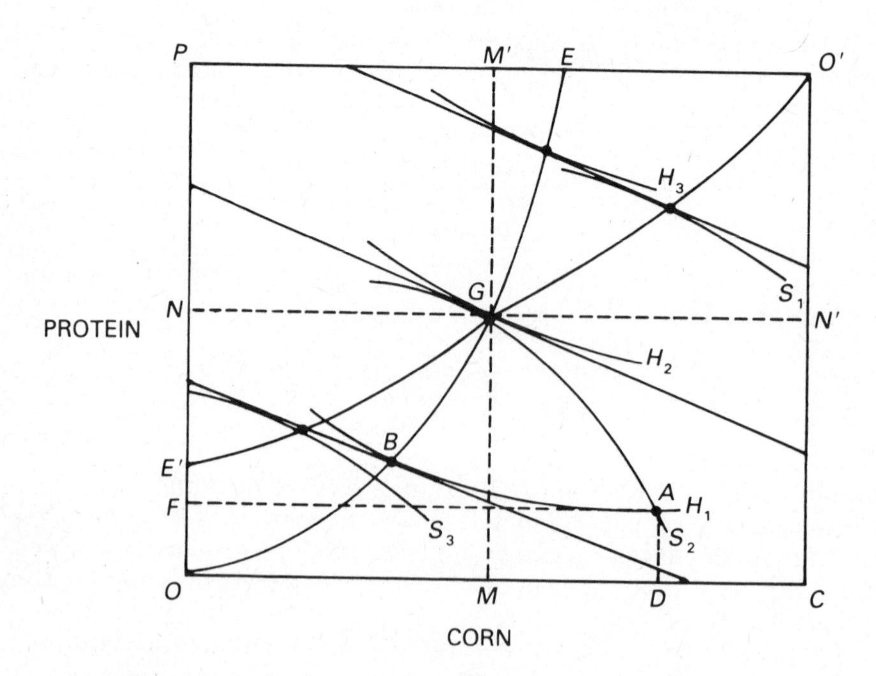

FIGURE 5–8. Allocating two inputs between two products.

and the isocost lines. The isoquant map for steers is upside down and starts at the northeast corner of the box at point O': The steer isoquants are convex to O' and the steer expansion path, $O'E'$, passes through the points of tangency between the steer isoquants, S_1, S_2, S_3, and the isocost lines. Any point within the box represents an allocation of supplement and corn between the two enterprises. Consider point A, for example. An isoquant for each product passes through A; H_1 of hogs and S_2 of steers will be produced. The hog enterprise will require OD of corn and the steer enterprise DC of corn; hogs will get OF of protein supplement while steers will get FP. The total amount of corn, OC, will be used as will the total amount of supplement, OP.

While the input allocation and resulting outputs represented by point A could be produced, A is not an efficient point. The same outputs could be produced using less resources by moving to point B on the hog expansion path and G on the steer expansion path. At these points the MRS between enterprises will be equal but some of each resource will be unused. Because G and B are presumed not to represent profit maximizing positions (input use in the enterprises is less than the optimums), the farmer will want to expand production and use all of both inputs.

Point G is the only point where the MRS are equal in both enterprises and all the input quantities are being used. At G, hogs will utilize OM of corn; steers will receive $O'M'$. Hogs will get ON of protein supplement and steers will be given $O'N'$. In general, G may not be unique. The expansion paths may cross at several points or even be coincident. Then the point that equalizes the value of the marginal return per dollar will be selected.

The above analysis presumes that the marginal earnings per dollar will be the same in both enterprises at G, but this concept cannot be shown on the graph. If the equimarginal conditions were not met at G, the manager would shift resources from one enterprise to the other—perhaps by buying a different combination of corn and protein supplement and changing the dimensions of the box diagram—until the expansion paths intersect at a point where the equimarginal conditions do hold.

More General Cases

The general equimarginal criterion requires that the ratio of the value of the marginal product to the price of an input be equal for all inputs in all uses. Of course, the common value of these ratios should never be less than one, for that would cause input use to exceed the optimum in every enterprise.

Specific cases of the equimarginal conditions can be formulated, given the production functions. For example, suppose that three inputs, X_1, X_2, X_3, are used to produce three products, Y_1, Y_2, Y_3, in such a way that

$$Y_1 = f(X_1)$$
$$Y_2 = f(X_1, X_2)$$
$$Y_3 = f(X_3)$$

Then, the equimarginal criterion specifies that input use should be such that the following ratios hold true:

$$\frac{VMP_{X_1Y_1}}{P_{X_1}} = \frac{VMP_{X_1Y_2}}{P_{X_1}} = \frac{VMP_{X_2Y_2}}{P_{X_2}} = \frac{VMP_{X_3Y_3}}{P_{X_3}}$$

Again, when all ratios equal one, the optimum amount of input is used in each enterprise. When the ratios are equal but greater than one, the marginal earnings per dollar of cost are equal for all inputs in all enterprises. Note that whether input use is at the optimum or below it, a least cost combination of inputs is always used in Y_2.

Allocation of Owned Resources

The equimarginal principles discussed above tacitly assume that all inputs are purchased. This is true for the beginning farmer or for an enterprise where inputs are completely depleted each production period. Usually, however, production is carried on with some combination of purchased and owned resources. Purchased resources can be allocated according to the equimarginal principle until all available capital is utilized. The problem is somewhat different with owned resources. These resources are often durables—inputs that last several production periods. Further, they usually fall in two groups. The highly specialized inputs—sugar beet harvesters, cotton pickers, corn pickers, fence posthole diggers and so on—can be used only for specific jobs and present no allocation problems. The sugar beet harvester has no use at all if sugar beets are not grown; under no circumstances can it be used for other purposes.

The other durable inputs, represented by tractors, plows, combines, trucks, land, buildings, etc., can be utilized in many different enterprises. A given quantity of these inputs, or, more exactly, their services, are available to the farm manager to allocate among competing enterprises. The inputs are owned, so purchase prices are not relevant; operating costs are zero or constant in all uses. The criterion for allocating such inputs is to use them in such a manner that the marginal earnings of their services are equal in all uses (and exceed operating costs, if any). In this way the value of the input services is maximized. Any other use of the inputs will result in less total earnings.

General Discussion

When several enterprises exist on one farm, the equimarginal principles specify that the optimum amounts of inputs will not be used in one enterprise unless they are used in all enterprises. Only when inputs are unlimited—available in quantities such that the optimum can be used in each enterprise—will the optimum amount of inputs be used in all enterprises. Therein lies the difference between maximizing farm profits and maximizing profits from an individual enterprise. When inputs (capital) are limited, farm profits are maximum, given the resource restrictions, when marginal earnings per added dollar cost are equalized. On the other hand, profits from any one enterprise are at a maximum, or the optimum, where $VMP_X = P_X$ or, alternately, the ratio VMP_X/P_X equals one. Thus, use of the optimum amount of input in one enterprise is consistent with earning the maximum farm profits only when inputs are unlimited. In this latter case, of course, farm management problems are greatly simplified.

Problems and Exercises

5–1. Diagram: (a) Two production functions with elasticities of production less than one and the resulting opportunity curve; (b) the range of complementarity and supplementarity on a production possibility curve. (c) Point out the economic implications of (a) and (b).

5–2. Illustrate graphically and discuss the needed resource adjustments in order to achieve a maximum secondary product when

$$\frac{\Delta X_1}{\Delta X_2} > \frac{\Delta X_1'}{\Delta X_2'}$$

where X_2 and X_1 are products produced with a given set of resources and are used as inputs (X_2' and X_1') in the secondary production. The elasticity of production for both X_2 and X_1 is less than one.

5–3. Assume a farm has 200 acres of cropland which can produce 25 bushels of wheat per acre and 50 bushels of milo per acre. (a) When the *net* prices (gross price − variable cost) of wheat are $2.00 per bushel and of milo $1.00 per bushel, what allocation of land among the two crops would bring the highest return? (b) Would there be any change in the cropping program when the net price of wheat falls to $1.50 per bushel and the net price of milo stays the same? (c) What effect would increases in the net price of wheat to $3.00 per bushel and the net price of milo to $2.00 per bushel have on the cropping plan?

5–4. Two products (Y_1 and Y_2) can be produced with a given set of inputs in the following combinations:

Y_1	Y_2
53	0
52	17
50	23
46	28
40	32
32	35
22	37
0	38

Which combination of products (Y_1 and Y_2) will maximize returns when (a) $P_{Y_1} = \$6$; $P_{Y_2} = \$2$; (b) $P_{Y_1} = \$4$; $P_{Y_2} = \$6$; (c) $P_{Y_1} = \$2$; $P_{Y_2} = \$10$?

5–5. Given production functions (5.7) and (5.8), what combination of corn and sorghum will yield maximum returns if the price of corn increases from \$3 to \$4 per bushel and the price of sorghum decreases from \$2.50 to \$2 per bushel when nitrogen is limited to: (a) 100 pounds; (b) 200 pounds; and (c) 300 pounds?

5–6. Show graphically and explain how trading primary products in a market may lead to enlargement of a secondary enterprise which uses these primary products as inputs.

5–7. Diagram: (a) Two production functions with elasticities of production less than one ($\epsilon_p < 1$) and the resulting opportunity curve; (b) two production functions with elasticities of production equal to one ($\epsilon_p = 1$) and the resulting opportunity curve. (c) What are the implications for situations (a) and (b)?

5–8. Suppose a farmer owns 10 units of a fixed factor and 20 units of a variable input, X. Assuming the classical production function presented in Table 3–1 of Chapter 3 represents the production response for each fixed factor, show that the farmer can produce the maximum output, 260, by leaving 8 units of the fixed factors idle. Explain.

5–9. Some farm management specialists often state that the farmer desires to maximize returns per dollar spent. Is this goal consistent with profit maximization when capital is unlimited? Is it consistent with the general equimarginal principle when capital is limited? In what special case would the use of the optimum also maximize returns per dollar spent?

Suggested Readings

Cramer, Gail L. and Jensen, Clarence W. *Agricultural Economics and Agribusiness*, Second Edition. New York: John Wiley and Sons, Inc., New York, 1982, Chapter 5.

Doll, John P. "The Allocation of Limited Quantities of Variable Resources Among Competing Farm Enterprises." *Journal of Farm Economics*, Volume XL, November 1959.

Heady, Earl O., *Economics of Agricultural Production and Resource Use*. New York: Prentice-Hall, Inc., 1952, Chapters 7, 8, and 9.

ECONOMIES OF SIZE AND THEIR IMPLICATIONS FOR FARMS

CHAPTER 6

All inputs are variable in the long run. Over time, durable inputs owned by the farmer wear out and can be replaced by new, more efficient inputs. Additional farmland becomes available for rent or purchase. The amount of capital available to the farmer will increase, perhaps because of accumulation through the production process or perhaps because of the increased availability of credit from lending agencies. The managerial abilities of the farmer will change because of experience, study, and maturity.

In the long run, the farmer is able to change the size of his business. He will seek to make those changes that increase the efficiency of his farming operation and enable him to more nearly achieve his goals. In fact, as discussed earlier, the manager will always change the amount of an input if it will profit him to do so. Thus, he is always in a short-run position, but striving for increased efficiency by adjusting the so-called fixed inputs. Given this perspective, the long run is often called the *planning period*. Production planning in the long run consists of enumerating and evaluating all the production possibilities the farmer could produce when he has the flexibility to consider all amounts and combinations of inputs, always utilizing the best technologies for each level of output. He may not possess the capital and other inputs necessary to actually implement all possible plans, but he needs to know these plans if he is to make efficient changes in his business.

In Chapters 2 through 5, fixed inputs were assumed to be present in the production process. That assumption is now dropped, but the assumptions of pure competition and perfect certainty are retained.

PRODUCTION IN THE LONG RUN

The presentation in this chapter will be based on the premise that the farm manager is producing in the long run with two variable inputs. Two inputs simplify the geometric presentation; the same principles would apply to three or more inputs. The collection of all durables owned by a firm is often called the "plant," and that term will be used here to signify one of the inputs—representing land, machinery, buildings, and other durables found on farms. An increase in any one of these durables would increase plant size. Defining the existence of such a productive input isn't logically necessary but makes the assumption of two variable inputs more descriptive. Production in the long run can thus be represented symbolically by the long-run production function

$$Y = f(X_1, X_2)$$

where X_1 represents the usual variable input, X_2 is plant size and may be varied, and Y is output.[1] Although all inputs are variable, the production process is defined for a particular unit of time, such as a growing season or a calendar year.

Long-run Cost Curves

All long-run cost curves are derived from long-run production functions. No new analytical concepts are needed to visualize the long-run production function. But because all inputs are considered variable, the law of diminishing returns no longer applies. The shape of the long-run production function therefore depends entirely on the technical and biological characteristics of the production process under consideration.

To minimize the cost of production in the long run, each level of output must be produced with the least cost combination of inputs. For the long run, as for the short run, the least cost combination of inputs will occur where the criterion

$$\frac{MPP_{X_1}}{P_{X_1}} = \frac{MPP_{X_2}}{P_{X_2}}$$

holds. Thus, for each plant size there will be a corresponding amount of variable input that minimizes the cost of producing a given output. This is depicted in Figure 6–1. The long-run expansion path passes through the now familiar points of tangencies between the isocost lines and the isoquants. To ensure that output is always produced using a minimum cost combination in the long run, the firm will expand (or contract) by moving along the long-run expansion path.

Long-run total costs along the expansion path can be computed as always,

$$LRTC = P_{X_1}X_1 + P_{X_2}X_2$$

and the corresponding average and marginal costs derived in the usual manner. There are no fixed costs; other than that, no new mechanics are required.

A typical long-run average cost curve, expressing cost as a function of output, is depicted in Figure 6–2. This average cost curve is drawn

[1]If X_2 is a bundle of inputs, then those inputs must be combined in least cost combinations. The same would be true of X_1.

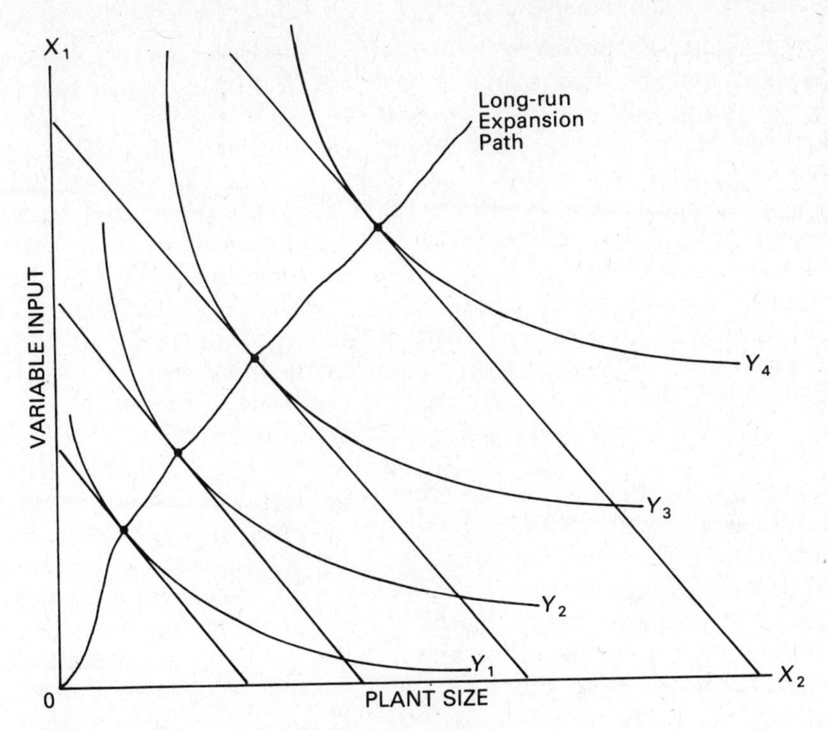

Figure 6–1. The long-run expansion path.

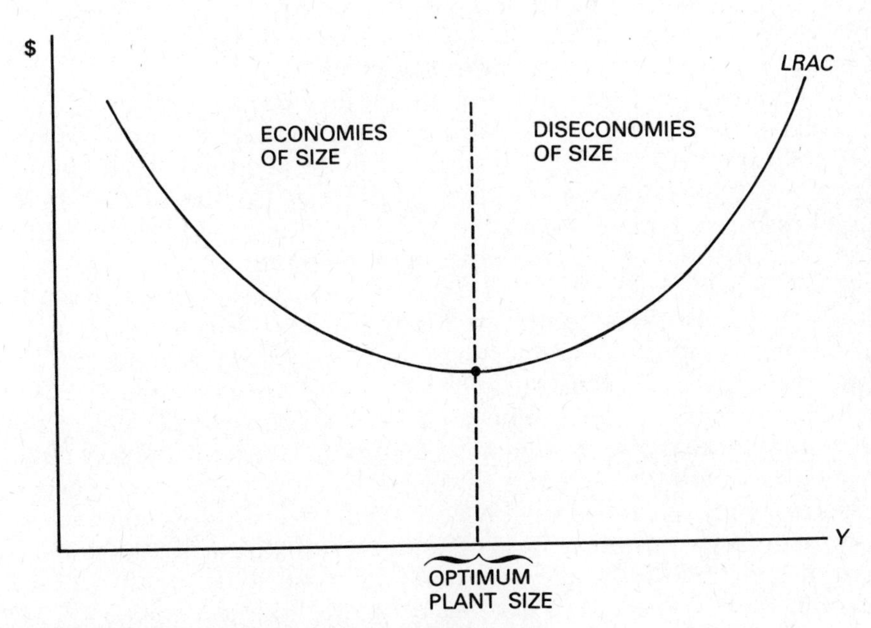

Figure 6–2. The long-run average cost curve.

with the same shape as the short-run *ATC* curve but for different reasons. When the firm is small, expansion of output usually increases efficiency, and average costs per unit of output will fall. The reasons usually cited for this decrease include specialization of labor and capital. Some efficient types of machines, such as self-propelled combines, cotton pickers, or sugar beet harvesters, can only be utilized on large acreages. Up to a point, rewards to management effort increase with size. A farmer with 100 cows has more incentive to learn proper management techniques than does a farmer with one cow. Mastery of a given technique has a larger payoff on the large enterprise. As the size of the business increases, the manager may be able to purchase inputs at a discount, thereby gaining market economies. Expansion of the firm enables workers to specialize and use more advanced or efficient technologies.

Eventually, the long-run average cost curve will turn up; costs per unit of output begin to increase as output is expanded. Reasons commonly advanced for increasing inefficiencies are managerial limitations and, in large firms, bureaucratic red tape. As firm size increases, the manager encounters increasing difficulty in maintaining control of his organization, communications and coordination become more difficult, mistakes are both more frequent and more costly. As a result, costs rise.

When *LRAC* are falling, the firm is said to be experiencing economies of size. The minimum point on the *LRAC* curve defines the optimum plant size. A plant of this size will produce the product at the lowest possible cost per unit. Diseconomies of size occur where the *LRAC* curve is rising. These cases are all shown in Figure 6–2. (*Economies of size* are sometimes called *economies of outlay*.)

In practice, the *LRAC* may not always take the form of the classic curve illustrated in Figure 6–2. Gains from specialization and losses from red tape may not exist in many real-life production situations. Figure 6–3 depicts four alternative possibilities for the *LRAC*; these suggest different types of economies or diseconomies of size. In Figure 6–3A the *LRAC* is a horizontal line, indicating that all plant sizes can produce the output at the same average cost. This case is called constant returns to size. In the second case, Figure 6–3B, *LRAC* increases with output, indicating diseconomies of size at every level of output. The *LRAC* in Figure 6–3C illustrates just the opposite—economies of size prevail at every output level. The last diagram, Figure 6–4D, illustrates a cost situation that also may be quite common. Constant average costs prevail over a rather wide range of output levels. Significant cost economies exist only when the firm is very small, and diseconomies set in only at very large outputs. Some economists might argue that the *LRAC* curve in Figure 6–3C would eventually take the shape of the one in Figure 6–3D if output were expanded further in that case.

All the *LRAC* curves described in this section would be derived

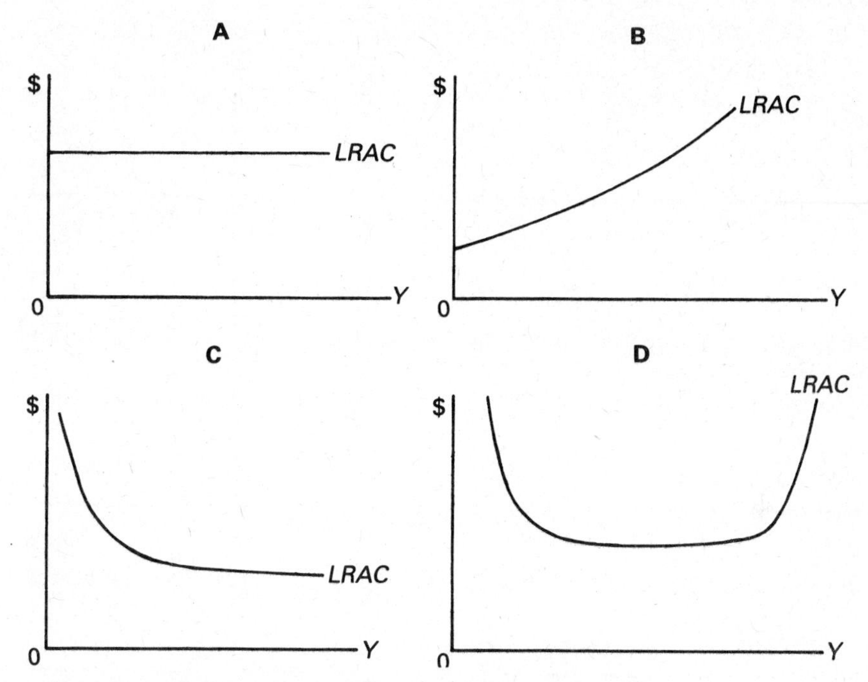

Figure 6–3. Different types of long-run average cost curves.

from corresponding long-run production functions. Similarly, long-run *APP* and *MPP* curves or equations could be presented. However, these concepts are analogous to those already presented in Chapters 2 to 4 and therefore do not need repeating. Nonetheless, it is useful to remember that a long-run production function always underlies the long-run cost curves.

Relationship Between Long-run And Short-run Cost Curves

In the short run, the farmer has a fixed plant—the number of acres, the buildings, and the size and type of equipment are all fixed in amount. He can expand output in the short run only by changing the amount of variable input. This situation can be represented as shown in Figure 6–4A. Plant size is fixed in the short run at $\bar{X}_2$. That particular plant size, $\bar{X}_2$, will produce output Y_2 at the least average total cost when combined with *OF* amount of X_1, the variable input. To produce outputs other than Y_2 in the short run, the manager must vary the amount of X_1 and, by so doing, restrict input use to the combinations represented by the vertical line above $\bar{X}_2$. For example, to produce output Y_1 in the short run, the manager would use *OG* of X_1 with the fixed plant, $\bar{X}_2$, and to produce output Y_3 in the short run, the manager would use *OH* of X_1 with $\bar{X}_2$. Movements along the line *DAB* represent short-run adjustments described in Chapters 2 to 4.

The combination of inputs at point *A* represents the least cost

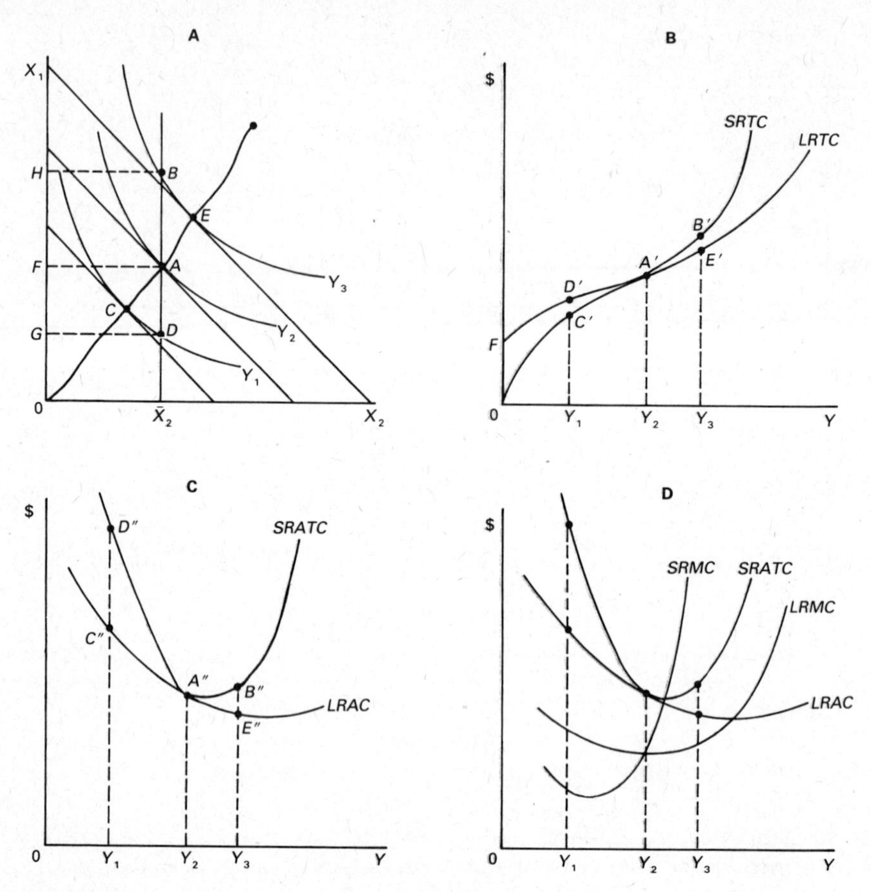

Figure 6–4. Relationship between short-run and long-run cost curves.

combination for the production of Y_2 in the long run, as do the combinations represented by C for Y_1 and E for Y_3. Thus, the combination of inputs at D, OG of X_1 and $\bar{X}_2$, necessarily costs more than the combination of inputs at point C on the long-run expansion path. Similarly, the combination at point B costs more than the combination at E. As a result, total costs in the short run along the line DAB will be higher than total costs in the long run along the segment CAE on the long-run expansion path. The exception will occur at point A, where short-run and long-run costs will be equal.

This argument is translated into total costs in Figure 6–4B. Fixed costs of amount OF are associated with $\bar{X}_2$ amount of X_2. There are no fixed costs in the long run. Short-run total costs, *SRTC*, increase with output but remain above long-run total costs, *LRTC*, until output Y_2 is reached. Point A' on *LRTC* and *SRTC* represents the cost of the input combination at A in Figure 6–4A. At A' the two cost curves are tangent. At output levels past Y_2, *SRTC* increases more rapidly than *LRTC*. The costs of the input combinations *C*, *D*, *E*, and *B* in Figure 6–4A are represented by *C'*, *D'*, *E'*, *B'* in Figure 6–4B.

The same concepts are shown for average cost curves in Figure 6–4C. Average costs are the same for output Y_2 but *SRATC* lies above *LRAC* for higher and lower outputs. Average costs for each of the points *A*, *B*, *C*, *D*, and *E* in Figure 6–4A are represented by the points *A"*, *B"*, *C"*, *D"*, and *E"* in Figure 6–4C. Note that the tangency between *LRAC* and *SRATC* does not necessarily occur at the lowest point on the *SRATC* curve. In fact, that tangency, representing equality between long-run and short-run average costs, may occur where *SRATC* is falling, at a minimum or rising. Returning to Figure 6–4A, if long-run average costs are reduced by expanding output from *A* to *E*, then short-run average costs will also drop from *A* to *B*. (As long as all functions are continuous, *A* and *B* can always be moved as close to *E* as necessary to make this statement true.) But costs along the long-run expansion path are always less than costs incurred in the short run when plant size is fixed. A similar type argument holds when costs are increasing from *A* to *E*.

Figure 6–4D includes the short-run marginal cost, *SRMC*, and the long-run marginal cost, *LRMC*. The short-run curves are derived as always. *LRMC* and *LRAC* show the usual correspondence that exists between average and marginal costs. *LRMC* is derived from the slope of *LRTC*. *LRMC* lies below *LRAC* when *LRAC* is falling, *LRMC* equals *LRAC* when *LRAC* is at a minimum, and *LRMC* lies above *LRAC* when the latter is increasing. At output Y_2, *SRTC* is tangent to *LRTC* (Figure 6–4B). The slopes are equal; therefore, *LRMC* must equal *SRMC* at output Y_2 for the plant size represented by $\bar{X}_2$. *LRMC* will not equal *SRMC* at any other output for that particular plant size. In Figure 6–4D, this unique relationship is shown by the intersection of the *SRMC* and *LRMC* at output Y_2.

In Figure 6–5, the relationship among the long-run average cost curve and several plant sizes is shown. $SRATC_1$ and $SRMC_1$ are the average and marginal costs for plant size 1. $SRMC_2$ and $SRATC_2$ are the marginal and average costs for a larger plant size, plant size 2. Each plant size represents a set of durable inputs fixed at a certain level. Many plant sizes exist between 1 and 2, but to avoid clutter their cost curves are not shown. Plant size 2 produces all outputs larger than (to the right of) amount *OM* more efficiently than does plant size 1. As output increases further, plant size 3 becomes more efficient than 2, and 4 becomes more efficient than 3. For plant sizes larger than 4, expansion of output is obtained only at increased cost per unit.

As explained, the *LRAC* will be tangent to each of the *SRAC* curves. Because of this, the *LRAC* is often called an *envelope* curve. To the left of *D* (Figure 6–5), the *LRAC* curve is tangent to short-run curves to the left of the latter's minimum. At *D*, both long-run and short-run

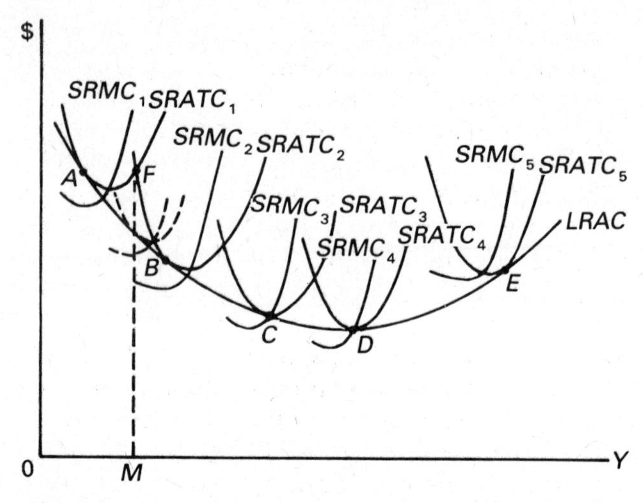

Figure 6–5. Long-run average cost for several plant sizes.

costs attain a minimum. Therefore, plant 4 represents the optimum plant size. To the right of *D*, *LRAC* is tangent to the short-run curves to the right of the latter's minimum.

The *LRAC* curve depicts the minimum average cost for each output level and thereby determines the most efficient plant size for each output level. Plant size 1 is most efficient in the production of the output corresponding to *A*, plant size 2 for *B*, plant size 3 for *C*, etc. While plant size 2 is more efficient than size 1 for all amounts of output to the right of the line *FM*, other plants will produce the amounts between *A* and *B* more efficiently than either 1 or 2. At point *F*, a plant could be built which has a *SRAC* depicted by the dashed line. This plant would produce amount *OM* at less cost per unit than either plant size 1 or plant size 2. To repeat, the *LRAC* is the envelope curve that is tangent to each *SRATC* curve at the output for which that plant size is most efficient in the long run. With the exception of plant 4 (Point *D*), this is not the most efficient output of the plant in the short run.

It is important to remember that the *LRAC* is a long-run or planning curve. Once a plant is built and production is undertaken, *the firm always operates on one of the short-run curves.*

RETURNS TO SCALE

Economies or diseconomies of size refer to the impact of output expansion upon average costs. The inputs are combined along the long-run expansion path in any ratio that minimizes the cost at each level of output. Returns to scale, either economies or diseconomies, refer to the effect of increased output on average costs when all inputs are increased in the same proportions. The definition of returns to scale is also often related directly to output. Thus, returns to scale can also be said to measure the change in output resulting from a proportionate change in all inputs. If the proportionate change in output is less than the proportionate change in inputs, diseconomies of scale result. If the change in output is equal to (greater than) the proportionate change in inputs, constant returns to scale (economies of scale) exist.

This concept can be expressed algebraically using the long-run production function

$$Y = f(X_1, X_2, \ldots, X_n)$$

where Y is output and $X_1, X_2, \ldots, X_n$ are inputs used in the production process. Let k denote the amount by which each input will be changed $(1 < k)$; returns to scale will be defined by λ where

$$Yk^{\lambda} = f(kX_1, kX_2, \ldots, kX_n)$$

The factor k^{λ} represents the change in output when all inputs are changed by the factor k.

For example, if the exponent λ equals one, the change in output is equal to the changes in the inputs. Returns to scale are constant. If the exponent λ is greater than one, the change in output exceeds the proportionate change in all the inputs, and returns to scale are increasing. Conversely, if λ is less than one, the returns to scale are decreasing.

This can be illustrated using the Cobb-Douglas function defined previously. Given two variable inputs in the long run, the function would be

$$Y = AX_1^b X_2^c$$

If each input is increased by the factor k, the result would be

$$A(kX_1)^b(kX_2)^c = k^{b+c}(AX_1^b X_2^c) = k^{b+c}Y$$

Therefore, whatever the original output was for any given amounts of inputs, the new output will be k^{b+c} times the original output. The sum of the coefficients $(b + c)$ measures returns to scale for the Cobb-Douglas production function.

Another representation of returns to scale is possible. For a long-run production function, let

$$Y^o = f(X_1^o, X_2^o)$$

be the original input and output amounts. Then , let

$$Y' = f(X_1', X_2') = f(kX_1^o, kX_2^o)$$

It will always be true that

$$\frac{X_1'}{X_1^o} = \frac{X_2'}{X_2^o} = \frac{kX_1^o}{X_1^o} = \frac{kX_2^o}{X_2^o} = k$$

but, if

$$\frac{Y'}{Y^o} < k \text{ decreasing returns to scale exist}$$

$$\frac{Y'}{Y^o} = k \text{ constant returns to scale exist}$$

$$\frac{Y'}{Y^o} > k \text{ increasing returns to scale exist}$$

Returns to scale must be measured along a *scale line* that is a straight line passing through the origin. Proportionate input changes are possible only on such a line or *ray*. Thus, economies (diseconomies) of size are the same as economies (diseconomies) of scale only when the long-run expansion path is a straight line passing through the origin. In most agricultural production situations, input proportions representing least cost combinations vary with the level of output—thus, strict interpretations of scale concepts are probably not of great value in agriculture.

When an isoquant map is made up of isoquants representing yield levels that are equally spaced, such as 0, 1, 2, 3, 4, . . ., or 0, 0.5, 1.0, 1.5, 2.0, 2.5, . . ., or 0, W, $2W$, $3W$, $4W$, . . ., where W is any number greater than zero, returns to scale can be determined by measuring the distance between the isoquants on any scale line. When these isoquants become successively farther apart, decreasing returns to scale are present; when they are equidistant, constant returns hold; when they are successively closer, increasing returns are evident.

Homogeneous Functions And Euler's Theorem

The emphasis on scale rather than size (or outlay) may have resulted from the belief by early economists that all long-run production functions were homogeneous to degree one. The definition of a homogeneous function is as follows: A function is homogeneous to degree λ if multiplication of each of its independent variables by k will increase the value of the dependent variable by k^λ. In symbols this would again be expressed

$$Yk^\lambda = k^\lambda f(X_1, X_2) = f(kX_1, kX_2)$$

This definition appears similar to the definition of returns to scale. The difference is that, for homogeneity, λ will be constant regardless of the choice for the original values of X_1 and X_2 and the size of k. For returns to scale, λ may vary with the value selected for k and the starting point on the production surface; that is, the original amounts of X_1 and X_2. The Cobb-Douglas production function is thus seen to be homogeneous to degree $(b + c)$; the quadratic function, such as (4.1), is not homogeneous.

A long-run production function that is homogeneous to degree one has the property of constant returns to scale. The expansion path will be a straight line passing through the origin (as it will for all homogeneous functions), and output will increase proportionately with inputs. Such a function is also called linear and homogeneous. Its three-dimensional production surface is often called a *ruled surface* because output increases linearly above any isocline; hence, a ruler placed with one end at the origin would touch the surface evenly above any isocline. The *LRAC* curve for a linear-homogeneous production function would be horizontal line as in Figure 6–3A.

Homogeneous functions have interesting mathematical properties.[2] One, already mentioned, is that all isoclines are linear and pass through the origin. A second property, known as Euler's theorem, can be expressed as

$$X_1\frac{\partial Y}{\partial X_1} + X_2\frac{\partial Y}{\partial X_2} = \lambda Y$$

or

$$X_1MPP_{X_1} + X_2MPP_{X_2} = \lambda Y$$

And, for homogeneity of degree one,

$$X_1MPP_{X_1} + X_2MPP_{X_2} = Y$$

where X_1 and X_2 represent amounts of the two inputs, the marginal products are measured at those input amounts, and Y is the output resulting from those input amounts. For example, consider the Cobb-Douglas function

$$Y = 10X_1^{1/2}X_2^{1/2}$$

[2]Allen, R. G. D., *Mathematical Analysis for Economists*, New York, Macmillan and Co., 1956, pp. 315–322.

when $X_1 = 4$ and $X_2 = 9$, then

$$Y = 10 \cdot 2 \cdot 3 = 60$$

But applying Euler's theorem gives

$$4\frac{\partial Y}{\partial X_1} + 9\frac{\partial Y}{\partial X_2} = 4(5X_1^{-1/2}X_2^{1/2}) + 9(5X_1^{1/2}X_2^{-1/2}) = 30 + 30 = 60$$

when the derivatives are evaluated at $X_1 = 4$ and $X_2 = 9$.

By assuming that all inputs were infinitely divisible, some classical economists arrived at the conclusion that all long-run production functions are homogeneous to degree one. Their defense of the divisibility assumption was apparently based on the belief that in the long run any machine (or other input) could be devised to be equally efficient at all sizes. Doubling inputs should therefore double output—two acres produce twice as much as one. This belief led to two rather interesting results.

Homogeneous Functions and the Three Stages of Production

Euler's theorem is often used to impute returns to fixed and variable factors in the short run. To do so, the long-run production function is assumed to be homogeneous to degree one and to display negative marginal returns to each input for some input combinations. But inputs are assumed limiting, that is, $f(0,0) = f(X_1,0) = f(0,X_2) = 0$. The second assumption is important because many homogeneous functions, such as the Cobb-Douglas function, have average and marginal products that are positive for all positive input values. When these three assumptions hold, Euler's theorem can be used to determine the symmetry of the stages of production for the input that is assumed variable in the short run and the input that is assumed fixed in the short run.

Let X_1 be variable in the short run and let X_2 represent "plant size," which is fixed in the short run but variable in the long run. Because the long-run production function is homogeneous to degree one (constant returns to scale), Euler's theorem applies:

$$X_1 MPP_{X_1} + X_2 MPP_{X_2} = Y$$

At the beginning of Stage II for the variable input, $APP_{X_1} = MPP_{X_1}$. But, from Euler's theorem,

$$MPP_{X_1} + \frac{X_2}{X_1} MPP_{X_2} = \frac{Y}{X_1} = APP_{X_1}$$

and, because X_1 and X_2 are both utilized in positive quantities,

$APP_{X_1} = MPP_{X_1}$ implies $MPP_{X_2} = 0$. At the beginning of Stage II for the variable factor, the marginal product of the "fixed" factor is zero.

At the right boundary of Stage II for the variable input, its marginal product is zero. In this case, Euler's theorem implies

$$X_1 \cdot 0 + X_2 \cdot MPP_{X_2} = Y$$

or

$$MPP_{X_2} = \frac{Y}{X_2}$$

and the marginal product of the "fixed" factor is equal to the average product of the "fixed" factor. Thus, the stages of production for the two factors are symmetrical but reversed. This is depicted in Figure 6–6.

Study of Figure 6–6 shows that Stage II is identical for both inputs. Marginal product is decreasing and less than average product for both inputs. Put more succinctly, the partial elasticity of production for each input falls between zero and one in Stage II, where these elasticities are

$$\epsilon_{Y \cdot X_i} = \frac{MPP_{X_1}}{APP_{X_1}} \quad \text{and} \quad \epsilon_{Y \cdot X_2} = \frac{MPP_{X_2}}{APP_{X_2}}$$

Thus, using the variable input in Stage II ensures that the fixed input is also utilized in Stage II.

Why is the marginal product of X_2 negative when the variable input is used in Stage I? By definition, it is always true that $Y = APP_{X_1} \cdot X_1$; but in Stage I for the variable input, $MPP_{X_1} > APP_{X_1}$ and therefore $X_1 \cdot MPP_{X_1} > Y$. But Euler's theorem is true for all input values, so that $X_1 \cdot MPP_{X_1} > Y$ implies that MPP_{X_2} is negative. When the variable input is used in Stage III and $MPP_{X_1} < 0$, then a similar type of argument shows that $APP_{X_2} < MPP_{X_2}$.

The stages of production can be identified on an isoquant map for the linear homogeneous production function. In Figure 6–7, the ridge lines are represented by the rays OD and OE; the region of substitution, Stage II for both inputs, is the cone EOD. Outside the cone, one marginal product is negative so that input substitution is not rational. That is, when the inputs prices are zero or positive, any output level can always be produced at least cost by an input combination within the cone.

When X_2 is fixed at OA in Figure 6–7, a slice through the surface along line ABC would reveal average and marginal products for the two inputs as shown in Figure 6–6. Stage I for X_1 and Stage III for

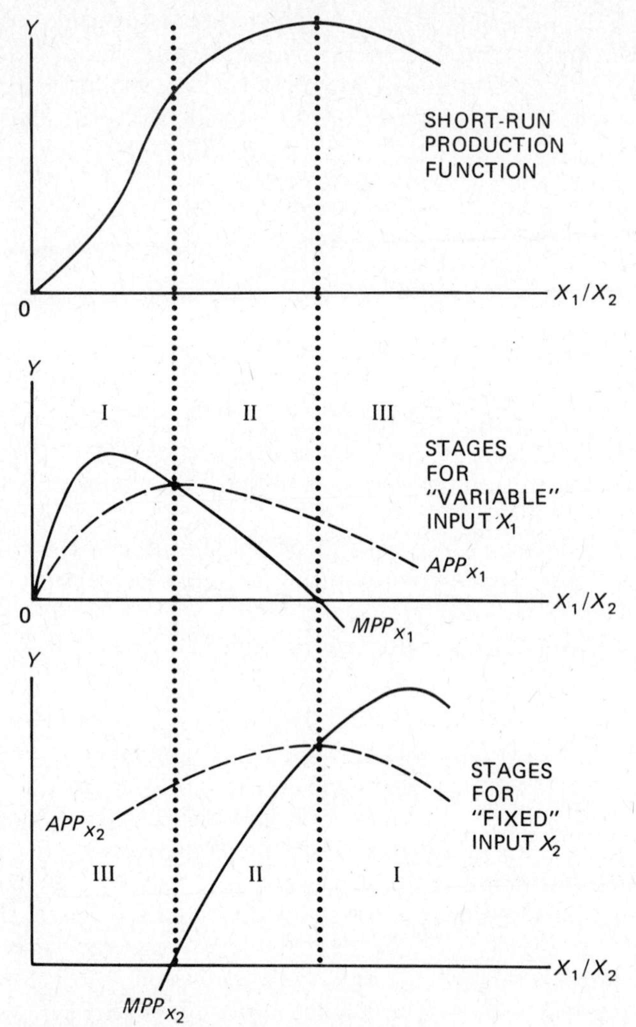

Figure 6–6. Symmetrical stages of production for the linear homogeneous production function.

X_2 fall below ridge line *OD* and the reverse is true above ridge line *OE*. As use of X_1 is expanded along line *ABC*, output eventually becomes zero.

The symmetry of the three stages of production depends on a specific set of assumptions. In addition to those listed above, the analysis requires perfect divisibility of both inputs and, as always in this textbook, constant input and output prices. While these assumptions comprise a portion of the classical analysis, they may not always hold true in empirical analyses. Mundlak has extended the analysis to homogeneous functions in general. A discussion of his analysis and re-

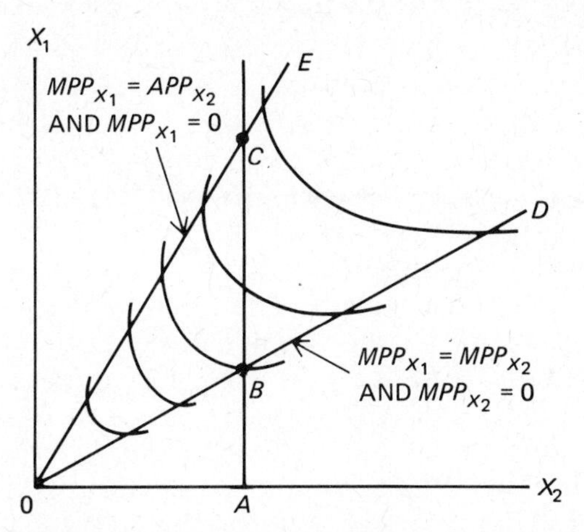

Figure 6–7. Substitution region for linear homogeneous production function permitting negative marginal products.

lated issues on the three stages of production has been presented by Seagraves and Pasour, Gates, and Johnston and Nelson. Ferguson provides an in-depth but rather more advanced discussion in his monograph on neoclassical theory.

As a digression, we wish to caution the student about the hazards of indiscriminate application of the symmetry of the three stages of production to short-run analyses as presented in Chapter 4. That chapter considered production processes for two variable inputs with one or more fixed inputs. If the two variable inputs are limitational, such as moisture and plant population, then the ridge lines will emanate from the origin, as in Figure 6–7, but will be curves and eventually intersect at the point of maximum yield. The maximum will exist in the short run due to the presence of fixed inputs and the Law of Diminishing Returns. In such a case, the yield response to X_1 and X_2 could not be homogeneous to degree one because such a function would not attain an unconstrained maximum.

The existence of Stage II can be determined by the presence of negative marginal products, but the simultaneous existence of Stage I in the short run cannot be based on the arguments of this section. Rather, each case would have to be justified by the technical or biological properties of the production process being studied.

Homogeneous Functions and Imputing Returns

By writing Euler's theorem for a production function that is homogeneous to degree one and multiplying both sides of that equation by P_Y, it is found that for any input and output level

$$X_1 \cdot VMP_{X_1} + X_2 \cdot VMP_{X_2} = TR$$

Thus, if each input is paid a unit price equal to its *VMP*, total revenue will be exhausted.

If an input, say labor, is paid less than its *VMP*, then the classical argument was that an *unearned* increment accrued to the owner of the business and the input is *exploited*. The concept that the *VMP* of an input represents a *fair* price is thus due at least in part to Euler's theorem.

The division of output between the two inputs in the short run can be demonstrated on a graph. Figure 6–8 presents a short-run production function with X_1, the variable input, and X_2, the plant size, fixed at some amount. If *OB* of X_1 is used, output *BA* will result. The problem is to impute this total product to the two inputs. Assuming the long-run production function is homogeneous to degree one, Euler's theorem will hold. When *OB* amount of input is used,

$$MPP_{X_1} = \frac{AD}{CD} = \frac{AD}{OB}$$

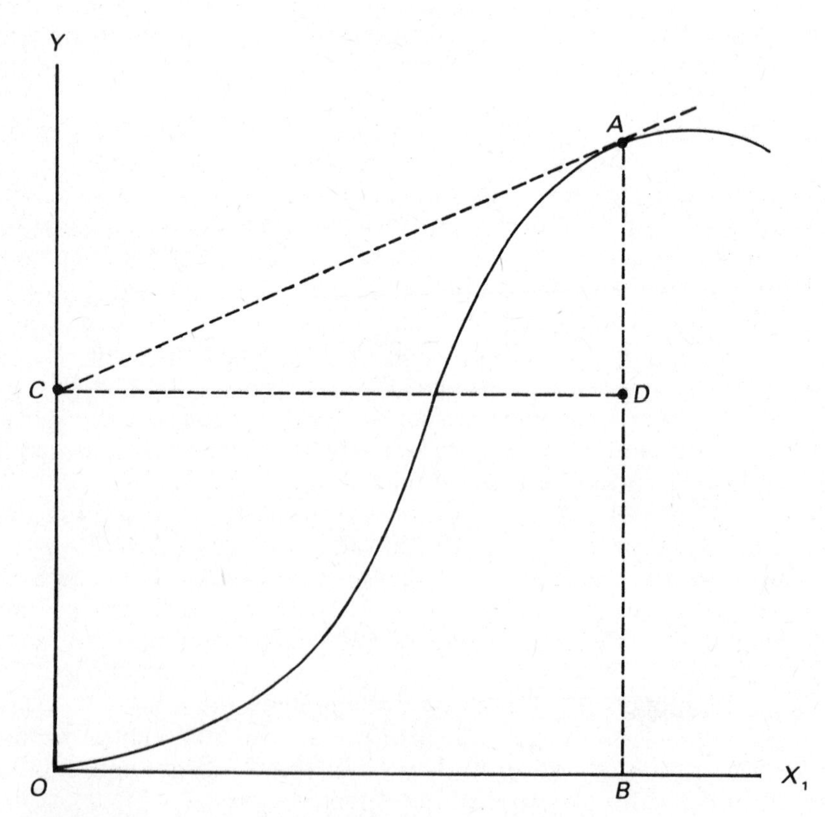

Figure 6–8. Imputing returns in the short run using Euler's theorem.

but then the product attributed to X_1 must be

$$X_1 MPP_{X_1} = (OB)\frac{AD}{OB} = AD$$

Therefore, from Euler's theorem,

$$X_2 MPP_{X_2} = Y - X_1 MPP_{X_1} = AB - AD = DB$$

and the total output is divided.

The imputation of output shares to inputs has long fascinated production economists. The problem is that inputs used in the production process are ultimately complements, although some substitution is possible. When one person owns all the productive services used in production, the question of relative shares, while of theoretical interest, is of little practical concern. When input services are owned by different people, the question becomes one of practical relevance because those services must be rewarded. By establishing a stringent set of assumptions, the classical economists attempted to resolve the issue within the production process. In general, when the production function is not homogeneous to degree one or even homogeneous, the imputation process described above is not possible; more usually, the values of inputs are determined by the market forces of supply and demand—outside the firm. Nonetheless, the problem of imputing values to inputs remains an important concept in the theory of the firm. It will be approached in a somewhat different manner in Chapter 9 when linear programming is discussed.

EQUILIBRIUM IN THE LONG RUN

A discussion of long-run costs leads in a natural way to an explanation of long-run equilibrium for the firm. The long-run equilibrium for a firm in a purely competitive industry is shown in Figure 6–9A. The diagram labeled "Industry" represents the aggregate market for an agricultural product. The demand curve is given while the supply curve, labeled *SRS* for short-run supply, represents the aggregate of all short-run supply functions, that is, the summation of all *SRMC* curves for all firms in the industry. The equilibrium price for the aggregate market is P_Y, and the industry will supply quantity Y_I.

The cost curves for a representative firm in the industry are shown in the right-hand diagram of Figure 6–9A that is labeled "One Firm." In equilibrium that firm will build the optimum-size plant, the one that permits production at the minimum point on the *LRAC* curve, and will produce output Y_F. When each firm produces Y_F, the number of firms must be sufficient to produce the industry output Y_I. In long-

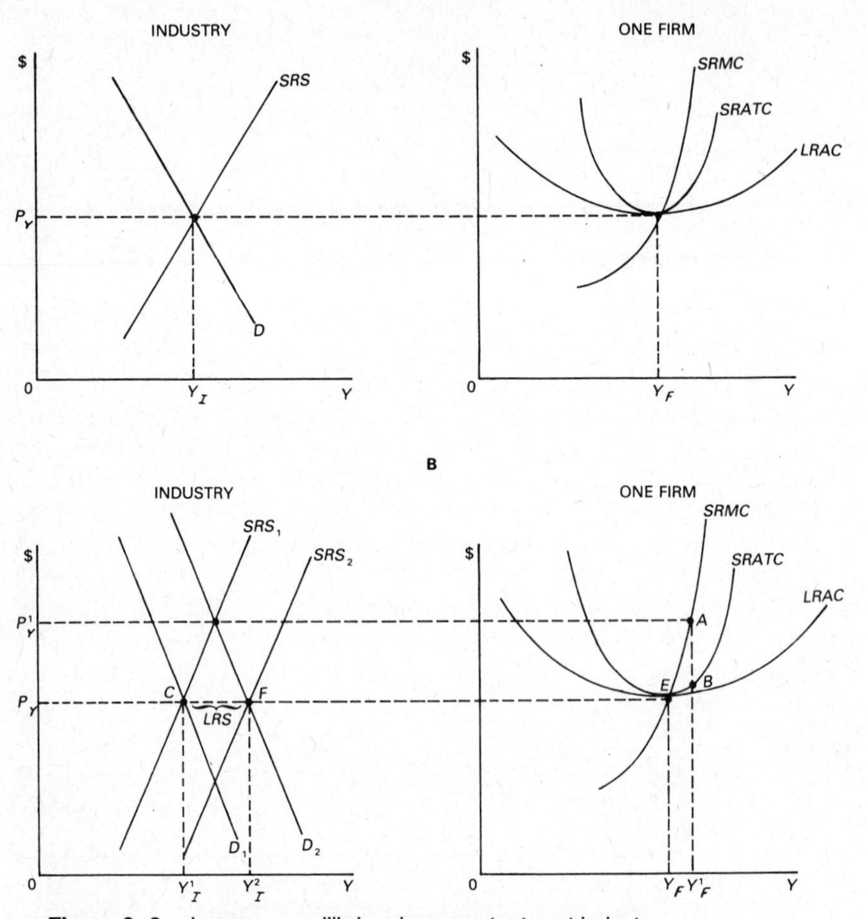

Figure 6–9. Long-run equilibrium in a constant cost industry.

run equilibrium, $P_Y = SRMC = SRATC = LRMC = LRAC$. Product price P_Y is sufficiently high to cover all variable and fixed costs, including the opportunity costs discussed in Chapter 2.

How is long-run equilibrium attained? This is best illustrated by introducing a change that will upset the current equilibrium. Suppose that aggregate demand increases, as shown in Figure 6–9B. The demand curve shifts from D_1 to D_2 and in the short run firms cannot change plant size so they expand output along SRS_1. The new higher price of output will be P^1_Y. The individual firm will respond by expanding output to Y^1_F where $P^1_Y = SRMC$. The firm is in short-run equilibrium and is earning a pure profit of AB per unit of output. It is not in long-run equilibrium. In fact, $P^1_Y = SRMC > SRATC > LRAC$. The existence of pure profits will attract new firms to the industry, industry output will increase, and price will begin to fall. Assuming there are no restrictions on entry, additional firms will enter the industry until a total aggregate output of SRS_2 is attained, price will again fall to P_Y, and industry output will be Y_I^2. Each firm will again produce Y_F, but now there will be more firms in the industry.

During the adjustment process, the firm may expand plant size past the optimum size—that needed to produce Y_F at the minimum average cost. That is, he might expand the plant to produce Y_F^1 at a lower average cost per unit. (If *SRATC* is tangent to *LRAC* at Y_F^1, then $LRMC = SRMC = P_Y^1$ at Y_F^1, but the firm is not in long-run equilibrium because $LRAC = SRATC < P_Y^1$.) But a pure profit would still exist; and as additional producers enter production and build the optimum-size plant, all producers will eventually adjust output to Y_F.

The firm's long-run supply curve is simply the point E in Figure 6–9B. Industry supply is increased by new firms entering the market and producing output Y_F. The industry supply curve is the horizontal line *CF* on the left diagram of Figure 6–9B.

In agriculture, the long-run trend in number of farms is downward. Therefore, the concept of "new firms" entering the industry requires some interpretation. Increases in output in the long run occurs because farms currently producing the product expand output while other farms, not producing the product, undertake production. In the first instance, the response can be of two types: First, higher product prices cause the farmer to expand production to equate marginal earnings to marginal cost in an enterprise already existing on the farm; second, when the farmers produce more than one product, the response can be created by shifting along the production possibility curve (or surface) and increasing output of the (now) relatively higher-priced product. In the second instance, when the product is not being produced, the response is essentially the same except the amount produced increases from zero.

Long-run Externalities

The analysis of adjustment to a change in demand is based on producer responses to a change in output price. In fact, when industry output changes, the resulting change in the aggregate demand for inputs may cause input prices to change (Chapter 4). For example, the price of feeder pigs would be expected to increase when farmers in the aggregate decide to increase hog production. Feed and corn prices may begin to increase a short time later. Thus, the impact of industry expansion on input prices should be considered.

Changes in input prices that cause shifts in long-run cost curves are called *externalities*. This term is used because these changes in costs are not the result of decisions, good or bad, on the part of the manager. The three cases corresponding to increasing input prices, constant input prices, and decreasing input prices are called *increasing cost industries, constant cost industries*, and *decreasing cost industries*, respectively.

Consider an increasing cost industry; when output is expanded, firms that enter the industry must bid for inputs, and therefore input prices rise for all producers. Figure 6–10 shows the adjustment process that results from an increase in demand. The original output price is P_Y, industry output is Y_I, and each firm, with an optimum-size plant, produces Y_F. When demand increases to D_2, firms respond along their *SRMC* curves and the industry responds along SRS_1. Output price increases to P_Y^1 and each firm produces Y_F^1. New firms respond to the existence of pure profit as explained previously but, by so doing, shift costs upward. The *LRAC* curve of a representative firm may shift upward to the left, upward vertically, or upward to the right. The result in a particular case would depend on the relative increases in the prices of fixed and variable inputs. In the case shown in Figure 6–10, the *LRAC* shifts upward to the right; the new least cost combination of inputs evidently requires more of the so-called fixed inputs relative to the variable inputs. $LRAC_2$ is the new cost curve for the representative firm. The optimum plant size now occurs at a higher output.

All firms now find themselves on the $LRAC_2$ curve—regardless of any action they might take. As explained above, they must eventually build the optimum-size plant and produce output Y_F^2. Long-run price will be P_Y^2 and industry output will be Y_I^2. Industry output has expanded but so has output of the representative firm. Thus, the possible change in numbers of firms is indeterminate.

The long-run supply curve for the firm is traced out by a line such as *NV* in the right-hand diagram, which traces the positions of the minimum point on the firm's *LRAC* as input prices change. The long-run aggregate supply curve is traced out by the line *WZ* in the left-hand diagram of Figure 6–10. Both short- and long-run supply curves have positive slopes, but a case does exist wherein they could exhibit slopes of opposite sign. In general, the number of possible combinations is large. The firm could have long-run supply functions that are positive, negative, or perfectly inelastic. Industry long-run supply curves

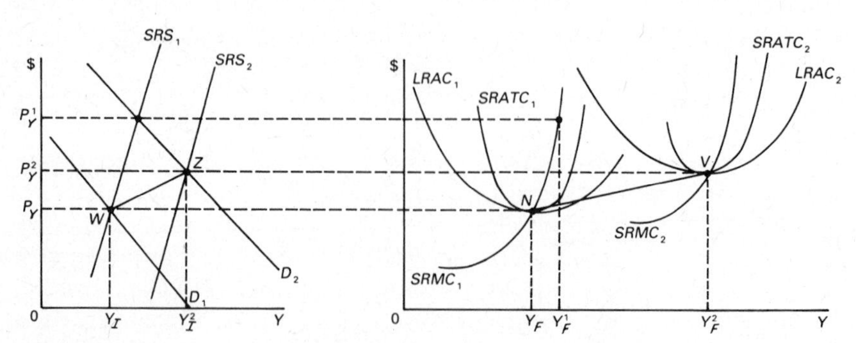

Figure 6–10. Long-run adjustments in an increasing cost industry.

could be perfectly elastic, have positive slopes or negative slopes. All cases cannot be considered in detail here (but they are excellent test questions).

A number of additional issues should be considered before exhausting the subject of long-run equilibrium. First, it is true that in the long run in pure competition the price of output is determined by the average cost of production—represented by the minimum point on the *LRAC*. Thus, if input prices are forced upward by increased demand for inputs elsewhere in the economy or the world, *LRAC* will rise and long-run output price must also rise even when the industry is doing nothing internally to cause such an increase. Or, if demand is so low that price does not cover the minimum possible average cost, the firm and industry will cease to exist in the long run.

The second point of interest is that when the firm is producing at the optimum plant size—for example, output Y_F in Figure 6–9A—then $SRMC = LRMC = LRAC = SRAC$ and the long-run production function will be homogeneous to degree one. This will be true for that one point regardless of the general properties of the production function. Thus, returns to inputs can be imputed using Euler's theorem as described previously. The classic view of long-run equilibrium is complete: Price of output equals average cost so the consumer is not being "exploited." Average costs include all real and imputed opportunity costs so the producer is earning a fair return and will neither expand nor leave the business. If all inputs are paid the value of their marginal products, the owners of all input services are rewarded fairly and the owner of the business will not retain an "unearned" increment.

Finally, this analysis has dealt with a representative firm. In the short run, all firms will not have identical fixed plants and therefore identical *SRATC* curves. In practice, most firms are somewhere on the *LRAC* curve attempting to adjust to the optimum plant size. If time permitted them all to adjust and remain stable, would they have the same *LRAC* curve? The classical economists argued in the affirmative. This would be easily justified if all inputs of a given type were homogeneous. But when inputs are not homogeneous and a firm possesses inputs of above average productivity—land, for example—the value of those inputs will be bid up in the long run by prospective producers. The opportunity costs of these inputs will rise, causing an upward shift in costs, until all firms have the same minimum average cost. The proposition that all firms would have the same optimum plant size is harder to justify, but also much less necessary to the analysis.

Expansion Pressure on Farms When *LRAC* Decreases

When economies of size exist, the firm will tend to expand. In Figure 6–11, the firm operating on $SRATC_1$ will produce output OA when the output price is P_Y^1. But OA can be produced at less cost by expanding the plant and operating on $SRATC_2$. Once the plant is expanded, OC

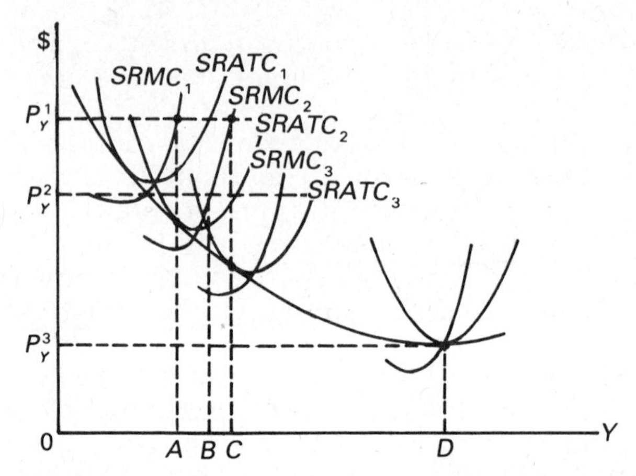

Figure 6–11. Pressure to expand when decreasing *LRAC* exists.

becomes the optimum output and further efficiencies can be gained by expanding plant size to $SRATC_3$ and so on.

If a firm is operating on $SRATC_1$ (which is irrational for long-run profit maximizing) and output price falls to P_Y^2, perhaps because of expansion of output by other firms in the industry, then the firm must expand (say to $SRATC_2$) if it is to survive. Once on $SRATC_2$, the firm's output will be *OB*, and if the manager wishes to maximize profits, he will again have pressure to expand.

Long-run equilibrium will occur at output price P_Y^3 when the firm's output is *OD*. At this output, pure profits are zero (imputed costs are paid) and the firm is producing at its most efficient level. The firm has no reason to expand or contract, and new firms will not enter the industry because profits are zero.

In primitive agriculture the long-run average cost curve may be horizontal (or approximately so) as shown by Figure 6–3A. In this situation the size of enterprise does not affect the cost per unit of output. The average cost is the same regardless of output. Thus, a large farm or enterprise would not have competitive advantage over the small farm or enterprise.

In actual practice, new technologies usually create, apart from the managerial factor, differences in returns to resources and costs per unit of output for varying sizes of farms and farm enterprises. The primary reason for changes in size is that new technologies increase profits from the production process. In addition, increased size results in more efficient use of individual inputs and thus lowers the unit cost of production as shown in Figure 6–11. Because output price is constant to the individual farmer, the larger enterprises are more profitable than the smaller enterprises.

In the competitive economy, the smaller-size enterprises are under pressure to expand to remain in business. In times of declining farm prices, the cost-price squeeze will affect small enterprises more severely than enterprises operating on or near the lowest point on the long-run average cost curve. The smaller enterprises, in this case, may not survive. The disappearance of many small farm units is generally attributed to economies resulting from the use of technology and equipment which require large capital investments. Many of these investments cannot be justified on small farms because the cost advantages can be realized only by expanding output.

Because of increases in the size and capacity of tractors and other farm machines, one-man crop farms have increased in acreage by about 50 percent in the last 15 to 20 years. Prior to the 1960s, most planting and cultivating machines for row crops—corn, soybeans, cotton, potatoes, and vegetables—were four-row. With the technology of that time, four-row equipment was the largest an operator could handle with precision.

Since then, such equipment has become technically more advanced, more automated, so the machine operator can monitor more rows. Farmers have shifted to six-row planters and cultivators for cotton and six-to-eight-row equipment for corn and soybeans. Planting and cultivating vegetable crops with four-bed, eight-row machines is common. Wheat, barley, and rice producers are using wider tillage machines and wider drills. The cutting width of the combine-harvester has been increased from 12–14 feet to 18–24 feet.

Some of these machines represent large investments and are developed to perform specialized tasks; thus, they contribute to the trend toward greater specialization and less diversification. Farms that in the past had five or six enterprises now have but two or three.

The increase in the size of farm operations may lead to efficiencies in the use of labor and to buying and selling advantages. However, purchasing economies generally do not extend beyond carload or truckload lots, that is, quantity discounts are often granted for full truckloads delivered to one stop. Additional discounts may be given to the very large purchaser who is willing to take delivery at the convenience of the vendor. An example is tractor diesel fuel delivered in the winter.

In general, quantity discounts of this type range from 5 to 10 percent. Bailey, from a survey of Montana wheat farms, reports that total discounts below list price on purchased inputs varied from $1.05 per crop acre on 3000-acre farms to $1.84 per crop acre on 9000-acre farms, a difference of only 79 cents per crop acre in an area where the total variable cost is about $30 per acre. This price advantage may not be sufficient to cause farms to become larger or to force smaller farmers out of business. In fact, they do not exceed discounts rebated to members by local farm cooperatives.

Economies of size in product sales are possible but not large. Some of the larger corn growers, for example, can command a premium of up to 5 or 6 cents per bushel on grain because they contract in large lots for delivery on a regular schedule. The purchaser will pay a price premium because the service reduces his procurement costs.

Farm Size and Cost Economies

The concept of economies of size has received much recognition since early in the industrial revolution. In agriculture, size and cost economies once were considered largely a theoretical phenomenon—they still are in primitive agricultural economies. Cost economies generally occur because of substitution relationships among inputs. Enterprises in less developed agriculture are labor intensive, and the costs per unit of output do not vary appreciably as the size of the enterprise increases or decreases.

In primitive agriculture the number of alternative methods of producing an output is limited, because the substitution possibilities, due to a lack of available inputs, are limited. The inability to substitute inputs also makes farms more homogenous. Given the technology, there may be only one way to produce wheat, make hay, or milk cows. The inputs used can be considered technical complements. On a one-man, one-hoe farm, with other inputs variable, output can be doubled by adding another man and another hoe. Before the advent of mechanization, the size of farms and farm enterprises in the United States was largely limited by the amount of family labor available.

The ascendancy of human and physical capital over land and water has transformed agriculture, revolutionized living standards, and changed customs in the countryside. This technical transformation of agriculture has aided everyone, including farmers, in numerous ways. It has removed the drudgery from farm work, reduced the uncertainties of production, created a greater variety of higher-quality foods, lowered food prices relative to nonfood prices, and supplied large quantities of products to foreign lands. But it has also left a large number of small farmers in the backwaters of American economic life.

Due to the rapid changes in technology in the United States, modern agriculture is composed of farms that are more and more heterogeneous. This heterogeneity is reflected in the composition of inputs used on different farms and resulting cost structures. While the average farm has grown in size as a result of changes in the use of capital inputs and improvements in the know-how of the operator, actual farms change only in degrees. Some change rapidly, and some change very little or not at all. Examples of changes in the cost of output caused by changes in the composition of inputs and expansion in the size of enterprises will be presented in the next section.

Returns to Scale Versus Economies of Size

Most changes in agriculture are due to changes in input proportions when some inputs are fixed. Proportionality relationships refer to short-run production processes. The nature of agricultural enterprises is not conducive to changing all inputs in the same proportions because of difficulties in controlling and measuring inputs as well as the "lumpiness" or indivisibilities of durable inputs (Chapter 4). Even in the long run, the scale line—a straight line emanating from the origin—may not be identical to the expansion path.

Studies of many different agricultural enterprises support the general contention that the least cost mix of inputs will change as the size of the enterprise changes. A different mix of capital and labor is found on small wheat or cotton farms as compared to large farms. Least cost rations, too, are examples of changing input mix for animals of different weights. In other words, expansion paths in agriculture are usually curvilinear rather than linear.

The farmer strives to combine labor, land, and capital in proportions that will yield the best output results, not only in quantity but also in quality. Within the limits set by the requirements of the technology employed, the amounts of the factors can be varied to secure the least cost input combination. For example, when labor is plentiful and capital scarce, it is usual to use a relatively larger amount of labor and a relatively smaller amount of capital in the input mix.

Improvements in the technological process, or changes in the quantity, quality, or price of the factors of production, will inevitably lead to new least cost combinations when factors are technical substitutes. If the price of a factor rises without a change in its quality, this factor will be used more sparingly and other factors more freely, again leading to new least cost combinations.

Quantitative and/or qualitative changes in inputs ordinarily affect the size of enterprises and the cost of output. A decrease in the average cost per unit of output can result from changes in technology that change the marginal productivity of factors, the factor prices and the input mix. An increase in the average cost per unit of output resulting from a change in the input mix is a cost diseconomy. Changes in inputs that affect the size of enterprises and bring about changes in costs are common in today's agriculture.

COST AND RETURNS STUDIES—EXAMPLES

Studies documenting decreasing average costs and increasing returns to size of farms are becoming more numerous. Madden has reviewed many studies on the economies of size in farming and compiled an excellent bibliography on the subject. More recently, Miller and his colleagues studied farms producing wheat, feed grains, and cotton in seven field crop regions of the United States to determine the importance of economies of size and to examine the effects of resulting cost economies on farm structure. While empirical studies are usually applicable only to specific areas and enterprises, they do furnish some general guidelines for anticipated outcomes for the same enterprises in other areas. More important, such studies are time-dated; they are based on specific production practices and technologies used during a given period of time. Time is particularly important because technologies change rapidly.

Most such studies are relatively recent, probably because the technologies favoring larger farms are of recent origin. While the number of studies measuring cost economies on farms increased during the 1950s, such studies were still largely directed toward measuring cost efficiencies of small or medium-size farms. Research was not directed toward larger farms or larger-size farm enterprises at that time because large farms were few in number, farm equipment was primarily adapted to small or medium-size farms, and research techniques needed refinement.

During the last decade numerous studies have attempted to measure cost economies associated with different-sized farm enterprises in many parts of the country. (See the Suggested Readings at the end of this chapter.) Efficiency in most studies is measured using average total cost curves or by a curve of cost/revenue ratios. While studies of this type may or may not include all possible factors that affect the ultimate viability of the farm enterprises, they suggest the pressure to adjust that has been exerted on producers of hogs, beef, wheat, dairy, and other products.

Technology has pushed the minimum-size production units beyond the capital and managerial capabilities of many managers. The gap is widest in livestock and in some specialty crops, narrower in feed grains, wheat, and cotton areas. The magnitude of the changes that have already occurred can be seen by examining trends in the size of farm enterprises over time. The following examples, though subject to restrictions as mentioned previously, indicate trends and changes taking place in agriculture.

Cost and Returns for Crop and Livestock Enterprises

Cattle. A trend toward fewer and bigger livestock enterprises has been apparent for several years. For example, McCoy and Olson reported that in Kansas in 1940 slightly more than three-fourths of all cattle feeders handled 25 or fewer head. The cattle they fed amounted to 23 percent of total fed cattle marketings. These small feeders were the dominant group in both number of operators and in cattle fed. By 1967, only 4 percent of the cattle were fed in groups of 25 or fewer. Since the early 1960s, most of the net increase in cattle feeding in major producing states occurred in herds of more than 1000 head (Figure 6–12).

A trend toward fewer and larger units also is apparent in grass-fed cattle and calf operations, but the degree of change is not as extensive as for grain-fed cattle. Cow herd operations are more difficult to mechanize and automate than feed lots.

Analyses of farm record data have suggested that many adjustment

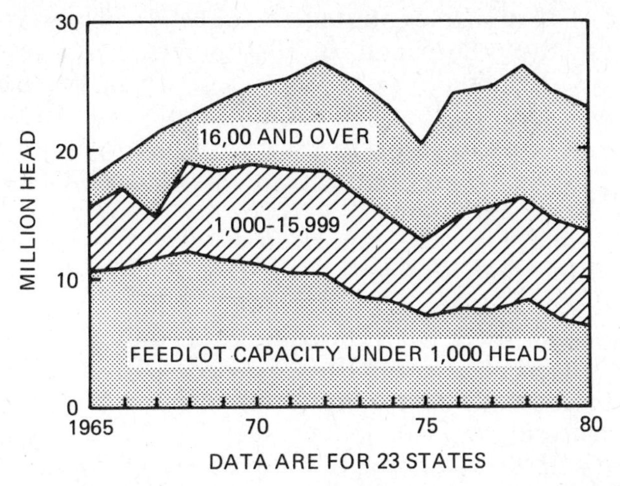

Figure 6–12. Fed cattle marketed by feedlot capacity. (*Source*: USDA, *1981 Handbook of Agricultural Charts*, Washington, D.C.)

problems in livestock enterprises are due to fixed nonfeed costs. One of the major problems facing the feeder seems to be insufficient volume to effectively use livestock equipment and buildings. Farms handling two lots of cattle per year instead of one can reduce nonfeed costs per pound of gain from 2.2 to 0.7 cents, depending on the size of the enterprise. The nonfeed costs on farm feed lots have varied from 8.6 cents per pound for one lot of 40 head to 4.4 cents per pound gain for a lot of 925 head and from 6.4 cents per pound gain by two lots of 40 head to 3.7 cents by two lots of 925 head. Major cost economies appear to be achieved on farm feed lots of 450-head capacity.

Similar conclusions are suggested by studies using synthetic-farm budgets. Heady and Gibbons, for example, compared the effects of different cattle feeding methods and systems (varying in degree of labor and capital intensiveness) on costs per steer fed, profit maximization, and stability of returns. They reported no important cost advantages for cattle feeding operations above the size of 400–500 steers. When the cost of labor is considered ($2.50 per hour), large cattle feeding enterprises may achieve cost advantages by adapting more highly mechanized systems rather than intensive labor systems. Because the cost of farm labor rises as off-farm employment opportunities expand, larger and more specialized cattle feeding operations can be expected. In addition, large feeding enterprises have an advantage in cattle marketing.

Hogs. Research results show a pronounced trend toward fewer and bigger hog operations. In 1940, slightly more than two-thirds of the hogs produced were from enterprises marketing 50 or fewer head. By

1967, units of that size accounted for only 7.5 percent of the hogs marketed. On the other hand, in 1940 enterprises marketing more than 150 head accounted for only about 7 percent of the production—by 1967 they accounted for 69 percent. As with cattle, most producers handled small herds in 1940; 91 percent of the operators had 50 or fewer hogs while less than 1 percent had more than 150 head. By 1967, the percentages were 36 and 29 percent, respectively.

Bauman and Eisgruber, more than 20 years ago, reported on farms producing hogs with different sizes of sow herds. The 50- to 60-sow herd size reduced costs (compared with 10-sow herds) in all major categories—feed, labor, building use, equipment use, and miscellaneous. Differences in total costs between 10- and 60-sow herds amounted to about $2.20 per hundred pounds gain or about $5.00 per hog marketed. Net returns per man hour increased from about $1 for 10-sow herds to about $4.50 for 50- to 60-sow herds.

In 1969, Van Arsdall and Elder used linear programming models to determine potential cost minimizing farm plans for Illinois cash grain and hog farms under varied resource situations, including different combinations of field machinery. On a one-man farm with four-row equipment (1M4R), a gross income of about $20,000 was needed to break even (Figure 6–13) at that time. Average costs dropped rapidly as annual production reached $55,000. Major enterprises on the optimum 574-acre Illinois farm included corn, soybeans, and hogs. Production per man on hog farms of optimal size ranged from 130 to 150 litters annually, plus sufficient corn for feed. Gross output averaged $55,000 to $60,000 per man. Costs per dollar of gross income was about $0.70 and did not vary appreciably among different-size farms (Figure 6–14).

Dairy. Buxton and Jensen's study shows substantial economies of size in dairy farming. Using economic-engineering, a synthetic approach, and considering available dairy technology, they demonstrated that the average cost per unit of output decreases sharply as the size of dairy enterprises increases to 80 cows or more (Figure 6–15). The same figure shows the relationship between gross income and dollar cost of gross income. For one-man dairy farms, costs per dollar of gross income are much higher than on larger farms. Total investment per dollar of net returns varies from $32.05 on a one-man, 40-cow farm to $17.21 on a four-man, 141-cow dairy farm.

Similar findings were noted in a Michigan study by Shapley. A dairy farmer using modern technology and operating near optimum levels could offer wages to hired help that are competitive with industry wages in Michigan. The results in both studies indicate that large dairy farms have substantial cost advantages; therefore, if operators are to seek higher net returns, they must expand dairy operations.

The size of herd suggested by the studies cited above is consid-

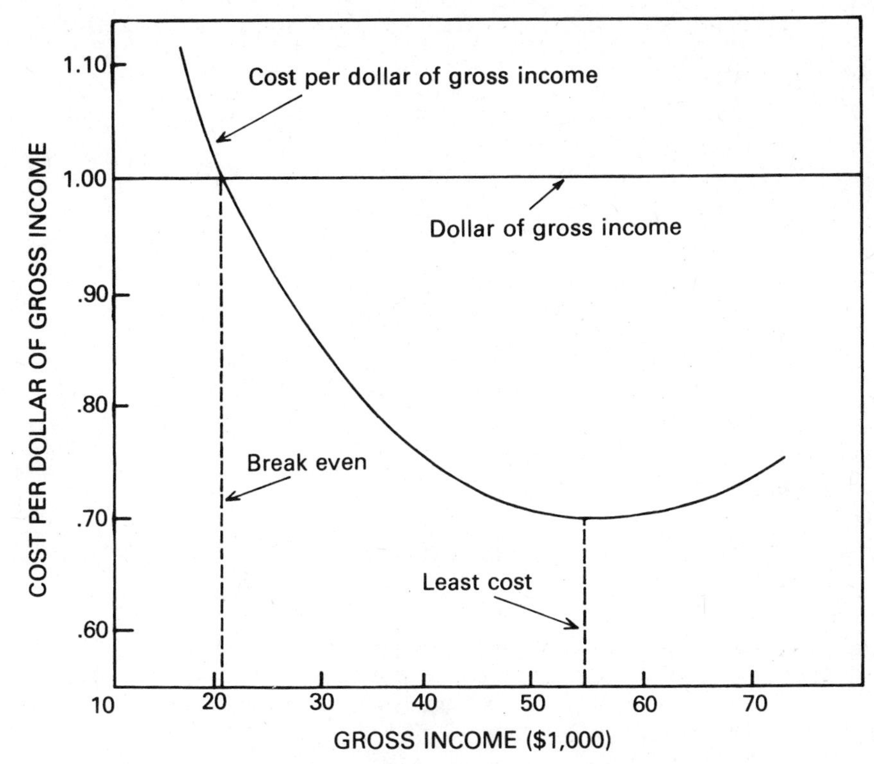

Figure 6–13. Least cost per dollar of gross income for a one-man hog farm. (*Source*: Van Arsdall, R. N. and Elder, W. A., *Economies of Size of Illinois Cash Grain and Hog Farms*, Illinois Agricultural Experiment Station Bulletin 733, February, 1969.)

erably larger than those presently found on the average dairy farms. This suggests that current trends toward fewer and larger dairy farms will continue. The speed of the adjustment will depend on the condition of existing dairy facilities, the availability of investment capital and land, the supply of labor and such institutional factors as farm organizations, government programs, and taxation.

Cotton And Wheat. Studies of cotton and wheat farms reviewed by Madden indicate that economies of size in crop production are not as pronounced as those associated with livestock production. Large farms are not necessarily more efficient than highly mechanized one-man farms. Units studied were one-man, 440-acre, irrigated cotton farms in Texas and a 1600-acre wheat-summer-fallow farm in Oregon.

The incentive for increasing farm size beyond the technically optimum one-man farm may not be to reduce costs per unit of output but to increase the volume of output and total income. The long-run average cost does not change appreciably as the size of an enterprise

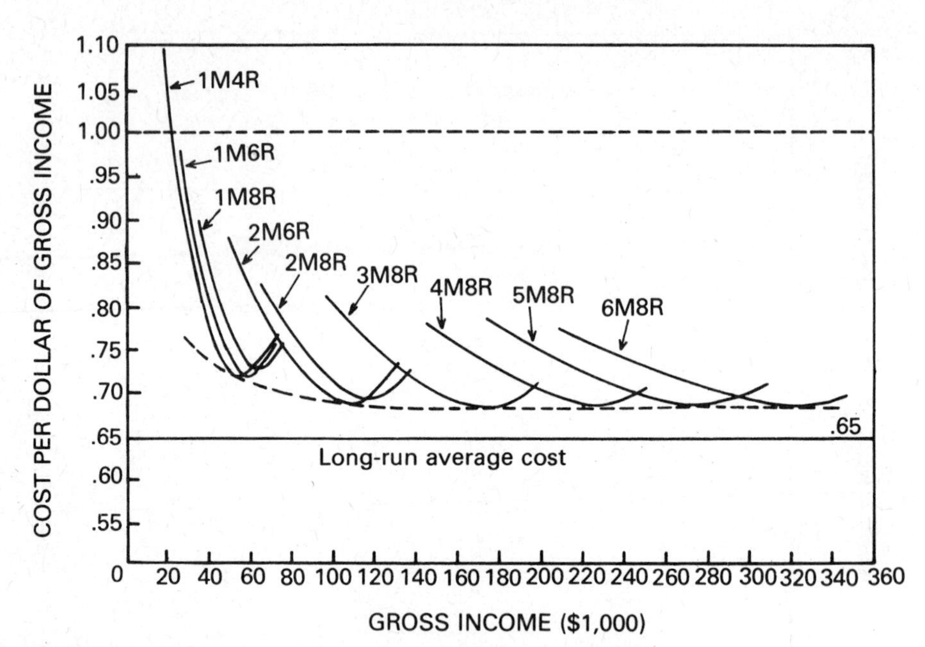

Figure 6–14. Short-run and long-run average cost curves for hog farms in Illinois. (*Source*: Van Arsdall, R. N. and Elder, W. A., *Economies of Size of Illinois Cash Grain and Hog Farms*, Illinois Agricultural Experiment Station Bulletin 733, February, 1969.)

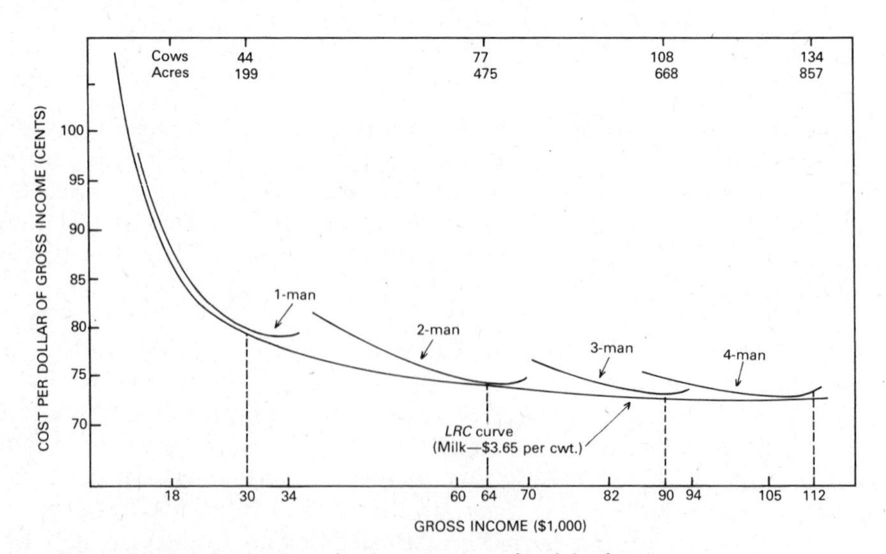

Figure 6–15. Derivation of unit cost curves for dairy farms. (*Source*: Buxton, M. B. and Jensen, H. R., *Economies of Size in Minnesota Dairy Farming*, Minnesota Agricultural Experiment Station Bulletin 488, 1968.)

expands from a one-man to a two-man, three-man, or larger operation. Most of the internal economies, such as overcoming the cost of indivisible factors by increasing size, enterprise specialization, or adoption of new technology, can be realized, as illustrated in the previous pages, on one-man farms.

Costs and Returns by Economic Class of Farms

Tweeten has developed data that give further evidence of decreasing average costs and economies of size in American agriculture. He defines size in terms of gross sales receipts, dollars, as shown in Figure 6–16. His emphasis was on explaining the persistence of low resource returns to the majority of U.S. farms. Whether the so-called break-even points occur at the economic size levels indicated in his 1960 and 1965 data depends in part on the acceptance of the measure he used for the opportunity cost of labor.

Using 1960 data, Tweeten computed what he called a long-run unit cost of production by economic class of farms. The cost per unit of output averaged $2.67 on Class VI farms and $0.91 on Class I farms.[3] The illustrated relationship suggests that for that particular time period most of the economies of size appeared to be achieved by Class II farms—farms with an annual output of $25,000 per farm or more.

He concluded that the average farm with gross sales under $25,000 did not recover all production costs in 1960. Farms with sales over $25,000 were able to earn a profit. By 1965, using a similar analysis, he concluded that the break-even point increased to $30,000 gross sales. His data provide additional evidence that the break-even point is shifting to higher production levels as a result of technological development and inflation. The dividing line is never very sharp and undoubtedly varies from one type of farming to another. However, the income advantage of the large farm is strongly indicated, as it has been by the trend toward farm consolidation and increasing size of farm enterprises.

External Economies in Agriculture

In addition to internal economies, external economies also exist in agriculture. The difference between internal and external economies is that the former are the direct result of the actions taken by the operator—adopting a new variety, overcoming input lumpiness, changing the resource unit, specialization, marketing—while the latter occur as a result of forces outside of the farm. For example, improved transportation facilities, stability in government programs, access to the banking and credit systems, availability of computers, improvements in chemicals and machinery, and publicly supported research and ed-

[3]Farms are classified as Class I through Class VI based on the value of gross sales. As shown in Figure 6–16, Class I farms are the largest and Class VI are the smallest.

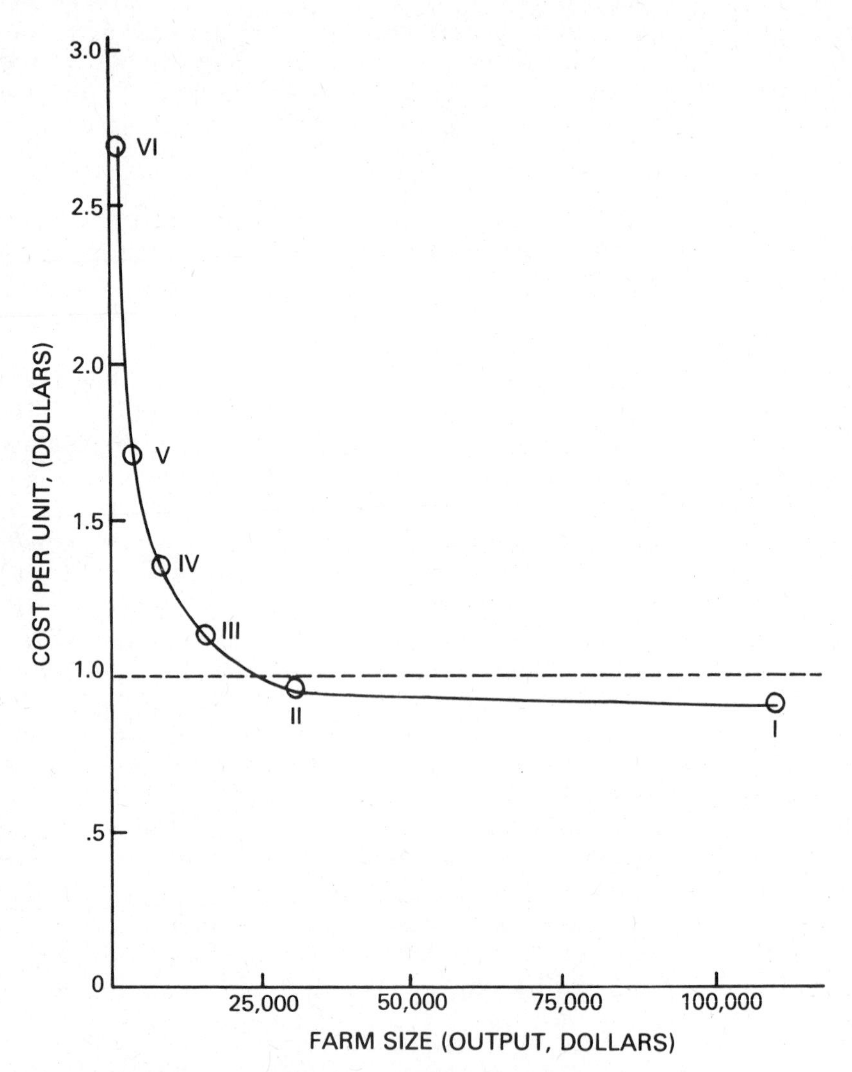

Figure 6–16. Long-run unit cost of farm production by economic class of farms in 1960. (*Source*: Tweeten, L. G., "Low Returns in a Growing Farm Economy," *American Journal of Agricultural Economics*, Volume 51, November, 1969.)

ucation all affect individual farms, but are external to farm enterprises. In most instances, they are not controlled by individual operators.

Agricultural research is an externality that has state as well as regional implications. Research affects an area's comparative advantage and thus has an important effect on the relative strength or weakness of an area's agricultural enterprises. Research findings and their application can cause a change in the profit from crops produced in an area. Some crops will increase in an area while others may decrease in importance.

Hybrid corn and hybrid sorghums represent an example of how technological development changes an area's competitive position and, thus, has an important effect on farm enterprises and incomes. In the early 1930s, Kansas produced about 3 million hogs and 3.5 million cattle a year. Then hybrid corn became available to feed hogs and cattle. Compared with the Corn Belt states, Kansas had relatively little corn acreage, so hybrid corn gave the Corn Belt states a competitive edge until the late 1950s when hybrid sorghum became available. By then, Kansas was producing less than one-third as many hogs and about the same number of cattle as in the early 1930s. More people in the United States were eating more meat per person and meat exports had risen, but Kansas producers were not supplying even their "depression" share of the market. Hybrid corn had given the Corn Belt states a competitive advantage in finishing cattle and feeding hogs.

All of that changed when hybrid sorghums became available. The first hybrid sorghum seed in Kansas was distributed commercially in 1957. By 1966, five Kansas counties produced more sorghum than all 105 Kansas counties produced in 1956, the year before hybrid seed became available. Within a few years, the sorghum-producing areas of the Great Plains, called the "Milo Belt" by some, became a feed-surplus rather than a feed-deficit area.

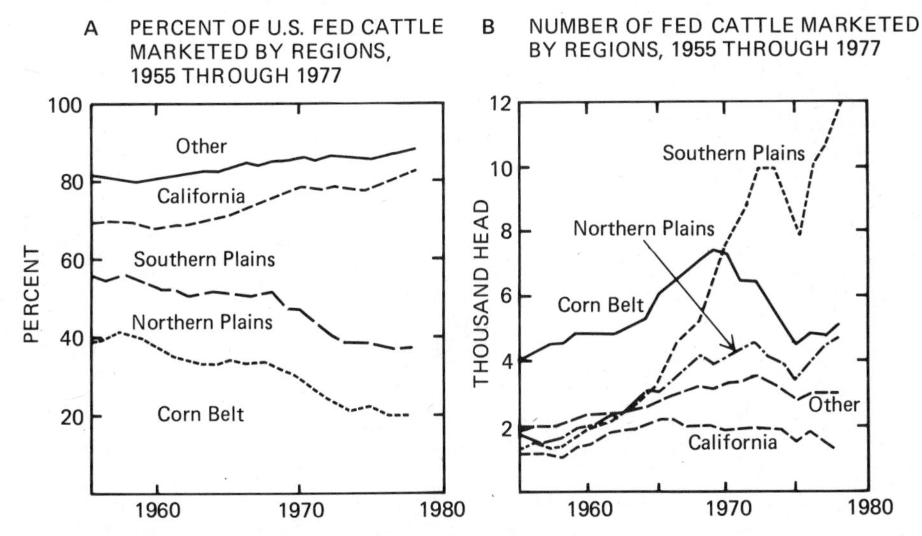

Figure 6–17A. Percent of U.S. fed cattle marketed by regions, 1955 through 1977.

Figure 6–17B. Number of fed cattle marketed by regions, 1955 through 1977 (Source for 6–17A and B: Boyken, Calvin C., "Structural Characteristics of Beef Cattle Raising in the United States," in Lyle P. Shertz et al., *Another Revolution in U.S. Farming*, Agricultural Economics Report No. 441, USDA-ERS, December 1979.)

By January 1, 1970, Kansas had nearly six times more cattle on feed than January 1, 1957. The increase in cattle feeding in Texas has been even more dramatic. Thus, the Milo Belt became a *bona fide* competitor with the Corn Belt. In 1956, the Corn Belt fed nearly 50 percent of U.S. cattle and the Milo Belt about 20 percent; in 1978, the percentages were about 20 and 43 for the Corn Belt and the Southern Plains, respectively (Figure 6–17). The number of cattle fed in the Corn Belt has not decreased, but nearly all the gains in beef feeding in recent years have been in the Great Plains area—Texas, Oklahoma, Kansas, Colorado, New Mexico and Nebraska.

Suggested Readings

Armstrong, David L. *Can Family Farms Compete?—An Economic Analysis, Corporation Farming*. Department of Agricultural Economics, Nebraska Experiment Station Report 53, 1969.

Bailey, R. W. *The One-Man Farm*. ERS-519, USDA, August 1973.

Bauman, R. H. and Eisgruber, L. "Moderately Large Hog Enterprises." *Economic and Marketing Information*, Purdue University, April 1960.

Bilas, Richard A. *Microeconomic Theory*, Second Edition. New York: McGraw-Hill Book Company, 1971, Chapters 6, 7, and 8.

Butcher, Walter R. and Whittlesey, Norman K. "Trends and Problems in Growth of Farm Size." *Journal of Farm Economics*, Volume 48, December 1966, pp. 1513–1519.

Buxton, M. Boyd and Jensen, Harald R. *Economies of Size in Minnesota Dairy Farming*. Minnesota Agricultural Experiment Station Bulletin 488, 1968.

Carter, H. O. and Dean, G. W. "Cost Size Relationships for Cash Crops in a Highly Commercialized Agriculture." *Journal of Farm Economics*, May 1961, pp. 264–277.

Cohen, Kalman J. and Cyert, Richard M. *Theory of the Firm*. Englewood Cliffs, N.J.: Prentice-Hall, Inc., 1965, Chapter 6.

Donaldson, G. F. and McInerne, J. P. "Changing Machinery Technology and Agricultural Adjustment." *American Journal of Agricultural Economics*, Volume 55, December 1973, pp. 829–839.

Ferguson, C. E. *The Neoclassical Theory of Production and Distribution*. Cambridge: Cambridge University Press, 1969.

Gates, John M. "On Defining Uneconomic Regions of the Production Function: Comment." *American Journal of Agricultural Economics*, Volume 52, February 1970, pp. 156–158.

Heady, Earl O. and Gibbons, James R. *Cost Economies in Cattle Feeding and Combinations for Maximization of Profit and Stability*. Iowa Experimental Station Research Bulletin 562, July 1968.

Henderson, James M. and Quant, Richard E. *Microeconomics Theory, A Mathematical Approach*, Second Edition. New York: McGraw Hill Book Company, 1971, Chapters 3 and 4.

Johnston, R. S. and Nelson, A. G. "On Definition of The Economic Region of the Production Function." *American Journal of Agricultural Economics*, Volume 53, February 1971, pp. 109–111.

Madden, Patrick J. *Economies of Size in Farming*. Agricultural Economic Report No. 107, ERS, USDA, February 1967.

Marshall, Alfred. *Principles of Economics*, Eighth Edition. New York: MacMillan and Co., Ltd., 1920, bk. V, Chapters IV and V.

McCoy, John H. and Olson, Ross. *Structural Changes in the Livestock Economy of Kansas—Changes in the Number and Size of Livestock Production and Marketing Units*. Kansas Agricultural Experiment Station Circular, January 1970.

Miller, Thomas, A., Rodewald, Gordon E., and McElroy, R. G. *Economics of Size in U.S. Field Crop Farming*. Agricultural Economics Report No. 472, ESS, USDA, July 1981.

Mundlak, Yair. "A Note on the Symmetry of Homogeneous Production Function and the Three States of Production." *Journal of Farm Economics*, Volume 40, August 1958, pp. 756–761.

Seagraves, J. A. and Pasour, E. J., Jr. "On Defining Uneconomic Regions of the Production Function." *American Journal of Agricultural Economics*, Volume 51, February 1969, pp. 195–202.

Shapley, Allen E. *Alternatives in Dairy Farm Technology to Meet High Labor Costs*. Research Report 80, Michigan Agricultural Experiment Station, January 1969.

Shertz, Lyle, P. et al. *Another Revolution in U.S. Farming?* Agricultural Economics Report No. 441, USDA, December 1979.

Tweeten, Luther G. "Low Returns in a Growing Farm Economy." *American Journal of Agricultural Economics*, Volume 51, November 1969, pp. 810–811.

Tweeten, Luther G. "Economic Factors Affecting Farm Policy in the 1970's." Paper presented at the Agricultural Policy Review Conference, Raleigh, North Carolina, December 4, 1969.

Upchurch, M. L. "Implication of Economies of Scale to National Agricultural Adjustment." *Journal of Farm Economics*, Volume 43, December 1961, p. 1246.

Van Arsdall, R. N. and Elder, William. *Economies of Size of Illinois Cash Grain and Hog Farms*. Illinois Agricultural Experiment Station Bulletin 733, February 1969.

Viner, Jacob. "Cost Curves and Supply Curves." In K. E. Boulding and G. J. Stigler, eds. *American Economics Association Readings in Price Theory*, Homewood, Ill.: Richard D. Irwin, Inc., 1952.

THE PRODUCTION PROCESS THROUGH TIME

CHAPTER 7

In the study of a subject, the most basic and fundamental concepts must be introduced first. These concepts then serve as a skeleton around which firmer fabric can be draped. The most fundamental concept in the economic theory of the firm is the production function. In the last five chapters, the characteristics of the production function and the marginal concepts of resource allocation based on the production function were discussed in great detail—always under the assumptions of perfect certainty and timelessness. Future prices, yields, and other events relevant to the production process were assumed to be known, and problems unique to the passage of time were not considered. In this chapter, the economic picture of the production process will be broadened by a consideration of the effects of time on the production process. Background material and analytical techniques will be developed to make time analyses amenable to the marginal principles previously developed.

The type of theory presented in previous chapters is called static because it ignores the fact that production can only occur with the passage of time. Static analyses regard production as a network at a point in time rather than a flow through time. Although the analysis presented in this chapter does include time as a variable, perfect certainty is assumed. In this chapter, inputs, outputs, costs, and revenues are time-dated, but the analyses are not dynamic, as the term is usually applied in economic theory. Nevertheless, time does introduce a number of unique aspects into our consideration of the production process and, by considering these phenomena, we can begin the first, albeit small, step away from the concept of instantaneous production.

The chapter is divided into four major sections, each discussing a different aspect of the impact of time on the production process. First, the most general effects of time will be discussed. Time is seen to have an effect on the manager's objectives, which might vary with the planning horizon. Second, the impact of time within a year is considered. The nature of the time of occurrence of input services or output of product are explored in an elementary fashion. The concept of a variable-length production period is introduced. Cash flows are discussed—a cash flow problem could occur even when yields and prices are known. The third section introduces intertemporal (inter-year) analyses. Compounding costs and discounting revenues are discussed; these techniques permit costs in one period to be compared to revenues in another. Net present values and the elements of investment are presented. In the fourth section, the implications of the theory of investment in durable resources are explored using examples from agriculture. Finally, to provide an application of the methods presented in this chapter, the valuation of farmland is considered.

A GENERAL PICTURE

The analysis of the effects of time on production must include the impact of time within a production period as well as over a sequence of production periods. Within a period, inputs and outputs are considered as functions of time; thus, a variety of problems associated with time flows must be considered. Timeliness of input application, the optimal length of the production period, and cost and revenue flows are among the important problems. When a sequence of periods is considered, then the time preference for money, the flow of profit over time, and the economic logic determining investment in durables must be developed.

The manager purchases inputs to obtain the productive services embodied within them. In previous chapters, the distinction between input quantities and services were not emphasized, the reason being that the production services of the so-called variable inputs were assumed to be completely utilized in one production period. Quantities and services were thus synonymous. Durable resources, however, are defined as those resources which provide productive services for more than one production period. In large part, the fixed costs discussed in previous chapters were costs associated with durable inputs. While these fixed costs were acknowledged, their exact nature was ignored.[1] The introduction of time into the analysis permits a more complete discussion of these costs.

In the purest sense, a farmer or other businessman can be pictured as a manager with a sum of capital. At any point in time, he is interested in diverting this capital to its highest use, that is, to that use which will earn the most profit per unit of time. He undertakes a particular type of production because his capital earns more in that use than in any other. In this state the manager is perfectly flexible; he can, and will, divert his capital to its best use. He is the ultimate maximizer.

Once a particular type of production such as farming is selected, the next purest form is constructed by assuming the manager has all of his capital in liquid form at the beginning of the production period and receives all his capital and profit in liquid form at the completion of the production process, where "liquid" means currency or some

[1]The makeup of fixed costs were not omitted through any weakness in the analysis but rather because of a search for simplicity. The principles of production functions, cost curves, and marginal allocation were introduced without all the clutter of time and durable inputs. Because the analysis was assumed to be timeless, the length of the production period could be selected to be that period of time in which all services of all inputs are completely utilized.

other perfectly negotiable form. In this type of production process, fixed costs arise from the purchase of input quantities at the beginning of the production period which are completely utilized by the end of the period but which cannot (or will not) be varied during the period. At the end of the period no services remain in the input, and production decisions relevant to the next production period can be made independently of past decisions. Labor hired on an annual basis or land rented on a yearly contract are examples of this type of fixed cost. The manager incurs these types of fixed costs only when he decides to produce. In essence, this describes the type of fixed cost assumed present in previous chapters.

When the manager receives all his capital and profit in liquid form at the end of the production process, he again is free to invest his capital in its highest earning use. Once he decides to produce, he is committed for the production period, but if at the end he realized his capital earned less in farming than it would in, say, a savings and loan account, he is able to place all his capital in a savings and loan account.

A more realistic production process arises when the manager must invest in some durable resources if he is to undertake production. These durable inputs embody a lump of services too large to be used in one production period. At the end of the production period, the manager has some profit, some liquid capital, and some durable inputs possessing unused services.[2] A decision to produce or not to produce during the next production period cannot be made without consideration of these unused services. Present decisions are thus influenced by past decisions. A manager with a stock of durable inputs may rationally react quite differently to output or input price changes than the manager with liquid capital.

The abstract picture of a business, then, is that of a manager subtly guiding his capital through time in search of maximum profit. Profit in this sense is not a once-over lump sum at a point in time but rather a stream of receipts into the future—in actuality, a rate of capital growth. The production process can be visualized as a capital "pool" with input costs flowing in at various dates and output revenues flowing out at other dates. Usually, part of the capital invested in the process is in the form of durable services.

TIME WITHIN A YEAR

Time Within the Production Period

In this section we will outline briefly the general concepts that arise when time is considered within the production period. Because of the

[2]One of these durables with unused services is management skill acquired from previous production. Thus, as a manager ages, his commitment to "durable" management services adaptable only to his business increases.

potential complexity, a framework to analyze time-related problems will not be developed. Rather, our purpose will be to provide an introduction to the problem.

The static model regards production essentially as instantaneous; input quantities disappear and output quantities appear as if by magic. If time does pass, it does so in a very benign fashion—production response is never distorted. In a more general sense, all input and output quantities can be expressed as functions of the amount of time that has elapsed since the beginning of the production period. Viewed in this context, the timeliness of application, or occurrence, of the input can be as significant in determining production response as the total quantity of input applied. Similarly, for a host of reasons, total output might vary as a function of time.

Examples are numerous in agriculture. Corn yields are determined not only by the quantity of rainfall but also by its distribution throughout the growing period, and especially during the stressful silking and tasseling period. Livestock cannot be fed beyond their stomach capacities or ability to utilize certain nutrients. If a combine is too small or unreliable, a wheat farmer in the Great Plains may suffer reduced yields from wind damage (shattering) or hail. A season's hay yield may be increased or decreased by three cuttings rather than two, dairy cows may be milked more often, etc.

When considered as a function of time, inputs and outputs may be classified as point, sequential, and continuous. Point inputs are applied in their entirety at one point in time. Seeding, one-time fertilization, spraying, and harvesting of grains represent point inputs. Rainfall in crop production and feed in livestock production are examples of sequential inputs. Temperature, as it affects crop yields and milk production, is perhaps the most obvious continuous input in agriculture. Pasture for grazing is continuously available but perhaps not continuously utilized. For each of the three types of time-dependent inputs, timeliness of application can be important; for inputs applied in sequence, response to one application may depend on the magnitude of the previous application.

Most agricultural outputs occur either at a point in time, such as grain harvest or livestock slaughter, or are sequential, such as cutting alfalfa or milking cows. Continuous outputs are perhaps most common in industrial processes.

Although the examples presented above are biological in nature and for single enterprises, the same concepts can be applied to other inputs on the farm. The availability of labor throughout the year can be an important determinant of the enterprises found on a farm; as explained in Chapter 5, supplementary, complementary, and competitive relationships among enterprises can be created by surpluses or deficits in the supply of labor, machinery, and other inputs throughout the year. Costs and revenues can occur at single points or sequentially, but the farmer must maintain funding for family expenses, production costs, and mortgage payments. Cash flows within the production period or year therefore become important. Even when yields and prices are known with certainty, revenues and expenses could be poorly timed, resulting in unpaid bills. Capacities of durables, such as machinery, storage bins, or milking parlors, become important for planning the timeliness of production and marketing.

Although further examples are possible, it seems clear that the production process with all physical and financial inputs time-dated becomes an extremely complex system. Although some aspects have been analyzed in theoretical and applied studies of the farm management—time is too important to be neglected—many of the implications of time on the farm business are undoubtedly not well understood. Frisch has noted that the guiding consideration should always be to determine how the time-related services affect production response, that is, the shape of the production function. He warns that many of the effects may be more complex than need be considered in practice.

Further analysis of time within the production process will not be pursued in this text; in the next section we set aside the issues raised here and consider the effects of the profit-maximizing objective on a variable-length production period. In a more advanced presentation, Dillon has developed a classification for inputs based on their potential effects on output through time; he develops a number of interesting research models. His chapter on time provides a summary of literature for students interested in further study of the subject.

Profit Maximization for a Variable-length Production Period

When time is introduced into the economic analysis, the goal of the manager will, in some instances, need to be revised. While other time periods may sometimes be relevant, the usual accounting period for most firms is a year. The manager wishes to maximize profits for each accounting period. He is no longer interested in maximizing profits from one production process but rather seeks to maximize profits per unit of time and will organize his business to attain that end. The time period itself is not important because of the fixed relationship among units of time. A year has 365 days. Therefore, profit per day is maximized simultaneously with profit per year.

If the production process is such that the period of production lasts exactly one year, assumed here to be the relevant time period, then equating marginal revenue with marginal cost maximizes profits to the production process and for the year. Maximization of profits for the production process is consistent with maximizing annual profits for the firm.

Only in exceptional cases will the period of production correspond to the calendar year. When the two are different, two important situations exist: 1) the length of the production period is fixed and 2) the length of the production period is variable.

The production period is fixed in length for many important agricultural production processes. Crops are notable examples. Corn must be planted in the spring and harvested in the fall. Winter wheat must be planted in the fall and harvested the next summer. The time required to bring crops to fruition is determined by nature, not by the farm manager. Some latitude exists as to timeliness of operations, but production cycles cannot be changed to the extent that more than one cycle can be squeezed into one year. Hence, maximization of profits to the production process is the same as maximization of profits for the year, and the criterion equating marginal revenue to marginal cost for the production process remains valid.

When the production process is variable in length, the criterion requiring maximization of profit for an enterprise does not give the same solution as the criterion requiring maximum profits per unit of time. In this case, the appropriate objective is maximum profit per unit of time. An example will suffice to demonstrate this.

Livestock feeding operations are typical examples of an agricultural production process that can be varied in time. Among the important time decisions facing the livestock feeder are:

1. What type of ration and feeding system should be used? The choice of diet will affect the rate of gain and time when the animal is marketed.
2. When should the animals be marketed to obtain the best possible price?
3. Given the ration and market price, how long should the production process be to maximize profits per unit of time?

An example will be given to answer the third question. The first two, although important, are beyond the scope of this chapter.

The production process to be considered is a drylot cattle feeding operation.[3] A 600-pound animal is to be purchased and fed a specified ration for an unspecified length of time. Upon sale of the animal, another will be purchased and the production process repeated. Costs to be considered are the purchase price of the animal and feed costs. For simplicity, other costs of production are assumed to be constant. The animal can be bought and sold for $0.20 a pound. Differences in quality due to length of feeding period are ignored in this example. They could be included by varying the selling price with quality, which would be a function of time. Costs and returns for this example are presented in Table 7–1. As demonstrated in previous chapters, these

[3]This example was simplified from data developed by Faris, J. Edwin, "Analytical Techniques Used in Determining the Optimum Replacement Pattern," *Journal of Farm Economics,* Volume 42, November 1960, pp. 755–66.

Table 7–1 Costs and Returns of a Feeder Cattle Enterprise Through Time

Units of Time (10 Days)	*Total Costs*	*Total Revenue*	*Profit*	*Marginal Profit*	*Average Profit*
0	\$120.00	\$120.00	0		. . .
2	124.05	122.24	−\$ 1.81	−\$ 0.91	−\$ 0.91
4	128.43	128.32	−0.11	0.85	−0.03
6	133.45	137.28	3.83	1.97	0.64
8	139.43	148.16	8.73	2.45	1.09
10	147.70	160.00	13.30	2.29	1.33
12	155.58	171.84	16.26	1.48	1.36
14	166.39	182.72	16.33	0.04	1.17
16	179.44	191.68	12.24	−2.04	0.76

Source: Simplified from example presented by Faris, J. Edwin, "Analytical Techniques Used in Determining the Optimum Replacement Pattern," *Journal of Farm Economics,* Volume 42, November 1960, pp. 755–66.

are derived from a production function relating output, weight of the feeder, to the variable input, time. A graph of the data in Table 7–1 is presented later in Figure 7–1.

The data in Table 7–1 are derived from the estimated total time cost and total time revenue functions:

$$TTC = 120 + 2N + 0.0067N^3 \quad (7.1)$$

$$TTR = 120 + 0.6N^2 - 0.02N^3 \quad (7.2)$$

where N is time in ten-day feeding units.

In the total time cost function, \$120 represents the purchase price of the animal and the remaining two terms, $2.0N$ and $0.0067N^3$, represent the feeding cost related to time. The total time revenue function also includes \$120, 600 pounds of the original weight of the animal sold at \$0.20 per pound plus the weight gain, $0.6N^2 - 0.02N^3$, over time.

From examination of Table 7–1, profit from the feeder animal is at a maximum at 14 units of time. The Marginal Profit column indicates that the marginal profit of time becomes very small between 12 and 14 units. Therefore, the exact profit-maximizing amount falls somewhere between these limits.

The profit-maximizing units of time can be calculated exactly using the procedures outlined in Chapter 3. The profit equation can be derived by subtracting TTC, equation (7.1), from TTR equation (7.2). This gives

$$\text{Profit} = TTR - TTC \quad (7.3)$$

or

$$\text{Profit} = -0.0267N^3 + 0.6N^2 - 2N$$

and the marginal profit equation from (7.3) is

$$\frac{d(\text{Profit})}{dN} = -0.08N^2 + 1.2N - 2 \qquad (7.4)$$

At the point where total profit is a maximum, the marginal profit is

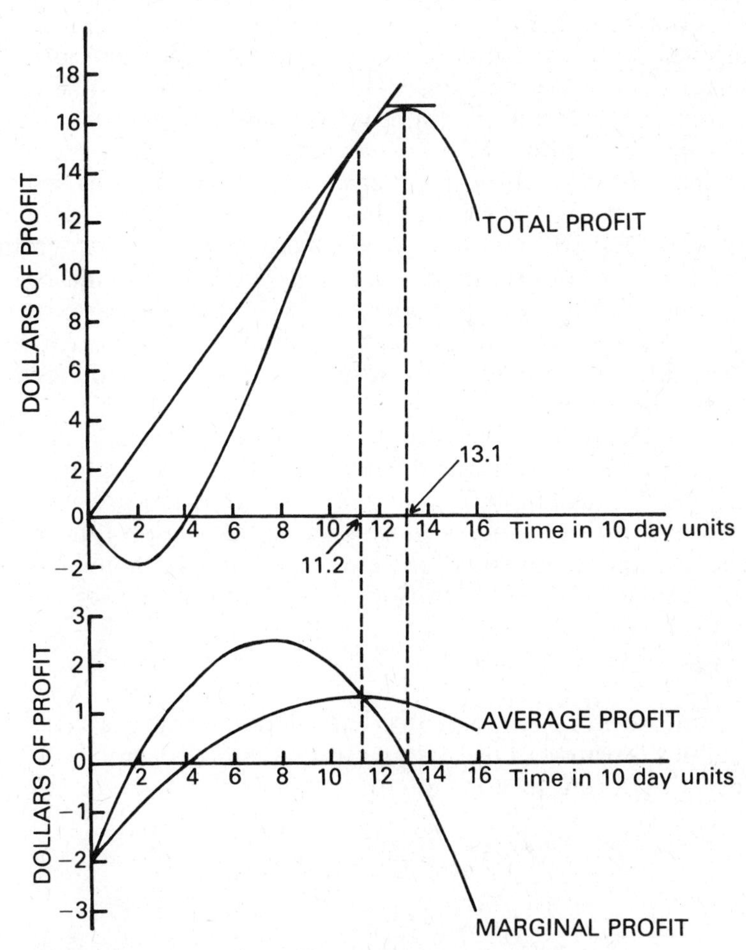

Figure 7–1. Total, marginal, and average profit of time resulting from feeder cattle enterprise.

zero. Thus, equating equation (7.4) to zero,

$$0.08N^2 - 1.2N + 2 = 0$$

$$N = \frac{1.2 + \sqrt{1.44 - 0.64}}{0.16} = 13.1$$

Reference to Figure 7–1 also shows that the maximum profit point—the units of time at which the profit curve is horizontal—is 13.1 units. At this point, of course, the marginal profit is zero and the marginal profit curve intersects the time input axis. (Note that all curves are profit curves rather than physical input-output curves. When the input has a positive price, the marginal physical product is not zero when profit is a maximum. In this case the marginal curve is a profit curve, not a physical curve, and thus is zero when profit is a maximum.) Using the marginal criterion presented in earlier chapters, profit from a single animal or any multiple thereof fed simultaneously is a maximum at 13.1 units or 131 days.

While Figure 7–1 indicates that marginal profit is also zero when $N = 2$, that solution is inferior to the solution $N = 13.1$. When $N = 2$, marginal profit is still increasing, as shown in Figure 7–1, indicating that total profit is not yet maximum. The sign of the second derivative can be used to test for the appropriate solution. In this case the second derivative of equation (7.4) is positive when $N = 2$ and negative when $N = 13.1$. Thus, profit is a minimum when $N = 2$, and profit is a maximum when $N = 13.1$.

The cattle feeder's interest is in maximizing yearly income; he wants to earn the maximum profit per unit of time rather than per animal. The data in Table 7–1 suggests that average profit is a maximum near 12 units of time. A more precise calculation can be derived using the average-profit equation. This equation is obtained by dividing the profit equation (7.3) by units of time, N. Thus, average profit is

$$\text{Average Profit} = -0.0267N^2 + 0.6N - 2$$

At the point where average profit is at the maximum, its slope is zero. Setting first derivative of the average-profit equation equal to zero, the answer to the optimum units of time, N, can be obtained:

$$\frac{dAP}{dN} = -0.0534N + 0.6$$

$$N = 11.2$$

This solution and Figure 7–1 show that maximum average profit occurs at 11.2 units, where marginal and average profits are equal. To maximize annual income from cattle feeding in this example, the feeder should sell each lot after 112 days of feeding.

The profit per animal in the first case, when total profit is maximized, 13.1 units of time, is $16.74; when average profit is maximum, 11.2 units of time, the profit per animal is $15.35. However, the turnover of animals within a year varies. In the first case the turnover, 365/131, is 2.8 and in the second case the turnover is 365/112 or 3.3. The difference in turnover of animals on feed reflects differences in yearly net income or profits from the two different maximizing objectives. The longer feeding period and thus the lower yearly turnover results in yearly profit of $46.87, and the shorter feeding period yields a yearly profit of $50.66. Of course, a higher selling price would bring about larger differences between the two practices. Calculations can be made using different price and input-output situations following the principles outlined above.

As an explanation of the solutions, the solution of 13.1 units assumes feed to be the variable input, the animal the fixed input, and time to be available in infinite quantities. The solution of 11.2 is determined assuming feed and animals to be variable and the amount of time to be limited. In this solution, the feeder is "applying" feed and animals to a fixed amount of time. If the feeder does not intend to repeat the feeding operation—that is, if he feeds just one lot a year—then he should maximize returns to the animal rather than to time.

Cash Flow Analysis

The optimal allocation of resources within a year depends on the availability of sufficient capital at the required points in time. As noted above, when time is introduced into the production process, the distribution of costs and revenues must be considered. The general technique for evaluating a current or future financial program is called a cash flow analysis. Because of the financial interdependence of enterprises, cash flow analyses are usually developed for the whole farm business.

Cash flow analysis indicates the amount of cash flowing into and out of the farm business over a specific period of time. Cash flow statements and income statements both show inflows and outflows of money, but differ in their treatment of several important accounting entries. A cash flow statement includes nonfarm items such as income taxes, nonfarm income, and living expenses and gives a complete accounting of debt transactions by showing principal payments and proceeds of new loans, whereas the income statement shows only interest payments. The cash flow statement more fully reflects purchases and sales of capital items such as breeding livestock, machinery, and real estate. Expenses associated with capital items are ordinarily shown on the income statement as a relatively constant annual depreciation allowance. The full amount of any capital sales or purchases is shown in the cash flow statement covering the period in which they occur.

The accounting period for a cash flow statement should fit the particular situation and the goal for which the cash flow analysis is intended. A year is the usual accounting period, and cash expenses and income usually should be grouped into monthly or quarterly summaries.

When used alone, annual cash flow statements can be hazardous. A yearly cash flow statement neglects the seasonal variation in sources and uses of funds. A farm business may show a favorable cash position for an entire year but may have extended periods during the year when cash expenses exceed cash revenues, necessitating the use of short-term credit and reserves to smooth out the fluctuations.

For farm enterprises having relatively constant monthly cash flows—a continuous farrowing swine operation or a Grade A dairy, for example—quarterly or semiannual summaries may be sufficient. (Additionally, annual cash flow projections over a period of five years or even longer may be necessary to determine the payout of a particular investment and to set up an acceptable loan repayment schedule.) Table 7–2 shows an example of a cash flow projection that points out the hazard of running a projection based on an annual summary. During the year shown, the farmer's operation would generate $2100 net cash

Table 7–2 An Example of a Cash Flow Projection

Name: Everett Everson Address: Rt. 2, USA Date Prepared: 198X

	Annual Estimate	*First Quarter*	*Second Quarter*	*Third Quarter*	*Fourth Quarter*
CASH INFLOW ITEMS	ESTIMATED CASH FLOW				
1. Livestock: Hogs	$10,200	$ 5700			$ 4500
2. Cattle	43,000		21,000		22,000
3.					
4. Crops: Milo	4300	2800			1500
5. Wheat	5800			5800	
6.					
7. Custom (machine) work					
8. Patronage dividends					
9. Agri. program payments					
10. Other farm income:					

Table 7–2 (*Continued*)

	Annual Estimate	*First Quarter*	*Second Quarter*	*Third Quarter*	*Fourth Quarter*
CASH INFLOW ITEMS	ESTIMATED CASH FLOW				
11.					
12.					
13.					
14. TOTAL FARM CASH INFLOW (Add lines 1–13)	$63,300	$ 8500	$21,000	$ 5800	$28,000
15. Nonfarm business income and wages					
16. Nonfarm dividends and interest					
17. Gifts, inheritance, & other nonfarm income					
18. TOTAL CASH INFLOW (Add lines 14–17) (except loans)	$63,300	$ 8500	$21,000	$ 5800	$28,000
CASH OUTFLOW ITEMS	ESTIMATED TOTAL CASH OUTFLOW				
19. Labor hired	2100		1050	1050	
20. Repairs on machinery	1300	200	400	300	400
21. Interest[a]	900	400		500	
22. Rent of farm, pasture	3000			1500	
23. Feed purchased	2400	900		1500	
24. Seed, plants purchased					
25. Fertilizer, lime	2500	1500	1350		
26. Machine hire					
27. Livestock expense	300				300
28. Gasoline, fuel, oil					
29. Taxes	1050	100	450	200	300
30. Insurance	3800	2000	900		900
31. Utilities	750	250		500	
32. Conservation expenses					
33. Other: Trucking	200		200		
34. Building repairs	1100		1100		

Table 7–2 (*Continued*)

	Annual Estimate	*First Quarter*	*Second Quarter*	*Third Quarter*	*Fourth Quarter*
CASH OUTFLOW ITEMS	ESTIMATED TOTAL CASH OUTFLOW				
35.					
36.					
37. TOTAL FARM CASH OF OPERATING EXPENSES (Add lines 19–36)	$19,400	$ 5000	$ 4150	$ 5550	$ 4700
38. Livestock purchases[a]	27,500		15,000	12,500	
39. Machinery, equipment (cash payments, principal)[a]	5500	5500			
40. Buildings (cash payments, principal)[a]	4000				4000
41. Land purchases (cash payments, principal)[a]					
42. TOTAL FARM CASH OUTFLOW (Add lines 37–41)	$56,400	$10,500	$19,150	$ 18,050	$ 8700
43. Family living expenses	4800	1200	1200	1200	1200
44. State income tax					
45. Federal income tax & social security					
46. Nonfarm business expenses					
47. Other nonfarm; family cash outflow					
48. TOTAL CASH OUTFLOW (lines 42–47) (Except operating loan payments)	$61,200	$11,700	$20,350	$ 19,250	$ 9900
49. NET CASH FLOW (+ or −) (Line 18 minus line 48) (Except loan receipts and operating loan payments)[b]	$ 2100	$−3200	$ 650	$−13,450	$18,100
50. PROJECTED OPERATING LOAN BALANCE (Oper. loan carried over from last period $7,000)[c]		$10,200	$ 9550	$ 23,000	$ 4900

[a]Principal payments on all loans not a part of this operating budget go on lines 38–41. All interest goes on line 21.
[b]Add negative net cash flow figures of each period to projected operating loan balance of previous period to arrive at projected operating loan balance for each period.
[c]The purpose of line 50 is to provide information for estimating the amount of operating borrowings needed in each period. The cash inflow and outflow items above do not include receipt or payment of operating loans.

flow (line 49). For the year, the farmer would appear to have no cash flow problems; however, a breakdown by quarters shows that in the first and third quarter he would need to borrow an additional $3200 and $13,450, respectively. At the end of the year he would owe $2100 less than at the beginning, but this annual total does not indicate his borrowing needs during the year.

Cash Flow as a Tool in Future Planning. The primary concern in financial management is future cash flows. To be adequately informed of future financial expectations, the manager should be aware of 1) the timing and amount of projected cash flows, 2) their sources and applications, and 3) the impact of these flows on the profitability and financial progress of the farm business. A cash flow projection *plus* projected balance sheets and income statements should be prepared.

A cash flow projection forecasts the future cash needs of the farm business. The manager can utilize this forecast to plan financing and to exercise close control over the liquidity position of the firm. The concept of cash flow is very simple—"a recorded projection of the amount and timing of cash expected to flow into and out of the farm business during the accounting period under consideration."

Preparing a cash flow projection for an annual farming operation begins with the crop and livestock programs. When developing a crop program, the manager will determine the number of acres to be planted to each crop and specify the amount of fertilizer, seed, and chemicals to be applied to each crop as well as the amount of labor needed to produce and harvest that crop. This physical information can be converted to costs which are expected to occur at particular times during the year. Revenue from the crops can be projected from the amount and timing of sales.

When developing the livestock program, the amount and timing of livestock purchases and sales must also be projected. Based on the total number of livestock, the manager can estimate the total feed requirements (including both feed grown and feed to be purchased), breeding expenses, and other anticipated livestock expenses.

Information on machinery and equipment purchases and sales and unusual repairs can be obtained from a planned machinery program. Operating expenses can be estimated from the previous year's records, modified to reflect anticipated changes for the planning period.

The Cash Flow Projection Form used for Table 7–2 has been developed to study the annual operating loan needs of a farm business. It may also be used to study the feasibility of short-run farm business changes. In the example, the cash flow projection is made on a quarterly basis. In general, the analysis may be monthly, bimonthly, quarterly, semiannual, or annual.

Lines 1 through 18 are for cash inflow items. Farm cash inflow items are recorded on lines 1 through 13 and nonfarm income items involved in the total family business operation are recorded on lines 15 through 17.

Lines 19 through 48 are for cash outflow items. Farm cash operating expenses are recorded on lines 19 through 36. Cash payments on livestock, machinery, equipment, buildings, and land are recorded on lines 38 through 41. Nonfarm business and family expenses including income taxes are recorded on lines 43 through 47.

After totaling the cash inflow and cash outflow items, subtract the cash outflow items (line 48) from the cash inflow items (line 18) to arrive at the net cash flow on line 49. The net cash flow for each period may be positive or negative.

The projected operating loan balance is determined on line 50 and is an estimate of the operating debt the farmer must carry in each period. Operating loans carried over from the period just prior to the projection period are recorded on line 50 under the heading "Projected Operating Loan Balance." If no operating loan existed at the end of the last period and if a surplus of cash was available, this amount should be recorded in the same space and designated "Surplus." The January projected operating loan balance ($10,200) in line 50 is determined by combining the operating loan carried over from last period ($7000) with the first quarter net cash flow from line 49 ($3200). Each period thereafter, the projected operating loan balance is determined by combining the previous period projected operating loan balance (line 50) with the net cash flow (line 49) for that period.

Solving Cash Flow Problems. Cash flow analyses will indicate problems that may exist when cash receipts do not cover cash expenditures. When this happens, the entire business should be analyzed to determine ways of increasing cash inflows and/or reducing cash outflows.

Substantial increases in cash inflows are usually not possible. Operating revenues can sometimes be increased through higher crop yields and improved livestock production. In smaller farming operations, surplus family labor can be used to generate nonfarm income.

Cash outflows can be reduced by eliminating or reducing some operating expenses and by reducing or delaying capital expenditures, especially machinery and equipment. Family living expenses can sometimes be reduced, particularly in the short run.

Budget data such as contained in Table 7–2 often provide sufficient information for solving cash flow problems. Other methods of analysis, such as linear programming, can also be used to solve cash flow and capital investment problems in more detail.

TIME OVER A PERIOD OF YEARS

Compounding Costs and Discounting Revenues

The farm manager is often required to purchase inputs that will be used for several production periods and over a period of years. Also, some investments may not earn a return for several years. As a result, the manager must compare costs incurred in one or more years to revenues that might accrue in one or more, but possibly different, years. Compounding and discounting are the primary tools for comparing the value of money received at different points in time.

The basic premise of compounding or discounting is that a dollar received now does not necessarily have the same value to a person as a dollar to be received a year from now. An individual's choice of a particular allocation of money over time is called his or her *time preference*. As a simplifying assumption, the manager's time preference is assumed to be measureable at some rate of interest, i. That is, if the manager is indifferent between \$100 now and \$110 after one year, then his or her time preference is reflected by a 10 percent rate of interest. In this section we will first develop the methods of compounding and discounting and then return again to consider the interest rate as a measure of time preference.

The unit of time for our study of compounding and discounting will be a year. Compounding and discounting can be applied to shorter time periods, such as quarters, months, weeks, or even days. The principles are the same in each case. The year is perhaps the most common time unit considered for such studies; interest charges within the year are accounted for by the opportunity cost of capital.

Compounding Present Costs. Suppose that you are a farmer who has the opportunity to buy a tract of forest land for \$100. For simplicity, this tract has no annual upkeep, and you can sign a contract now to sell the tract for \$150 at the end of five years. Should you buy the tract of land?

First, assuming that you want to invest the \$100 in some type of production process rather than spend it on a consumption item, you should evaluate alternative production possibilities. Suppose the best alternative investment is a savings and loan in your town which pays 5 percent per year on savings accounts. If you were to invest your money in the savings and loan, you would earn money as follows: At the end of the first year you would receive 5 percent of your investment, or \$5. Thus, if you invest P(for present) dollars at i interest rate per year, you would own at the end of one year exactly $P + Pi$ dollars. In the example, this would be \$100 + (\$100) (0.05), or \$105. This is not a mathematical derivation; it is an equation representing your agreement with the savings and loan association. Notice that the $P + Pi$ could be written $P(1 + i)$. Thus \$105 = \$100(1.05).

If you reinvest your money, at the end of the second year you will receive as interest 5 percent of the reinvested amount or (\$105)·(0.05) = \$5.25 and you will have in total \$105 + \$5.25 = \$110.25. In symbols, you now have the amount earned at the end of the first year, $P + Pi$, plus the interest that amount earned during the second year, $(P + Pi)i$, which when summed is $(P + Pi) +$

$(P + Pi)i$ and can be written

$$P(1 + i) + Pi(1 + i)$$

which after factoring $(1 + i)$ results in

$$(P + Pi)(1 + i)$$

Factoring a P from the left term gives

$$P(1 + i)(1 + i) = P(1 + i)^2$$

Thus, \$110.25 = \$100(1.05)^2. One hundred dollars invested at 5 percent compounded annually—meaning that the interest earned in the first year earns interest in the second year—is equal to \$110.25 at the end of two years.

If you invest \$110.25 at 5 percent for a year, it will earn \$5.51 and you will have \$115.76. Thus, \$115.76 = \$100(1.05)^3. The \$115.76 earns \$5.79 the fourth year, and you have a total of \$121.55, or \$100(1.05)^4. For the fifth and final year, the \$121.55 earns \$6.08, and your total amount of money is \$127.63, or \$100(1.05)^5.

The easy way to determine the value of \$100 compounded annually at 5 percent is to apply the compounding factor directly. The compounded value is equal to \$100(1.05)^5, which becomes \$100(1.2763) = \$127.63, because $(1.05)^5 = 1.2763$. In general, the compounded value, F (for future value), of a present sum P invested at an annual interest rate i for n years is

$$F = P(1 + i)^n$$

This procedure is called *compounding*, and in economic applications the amount being compounded is often an outlay or expenditure, hence the term "compounding costs." For convenient study, the quantities derived above in the text are summarized in Table 7–3.[4]

Investing the \$100 in the savings and loan, assumed here to be the next best investment opportunity, results in an earning of \$27.63 and a total amount of \$127.63. Your basic dilemma was comparing \$100 now with \$150 at the end of five years. You now find that \$100 invested now will increase to \$127.63 five years from now. Your cost and return figures are now dated at the same period in time (five years hence) and are comparable. A rational person would spend \$127.63 at one point in time to receive \$150 at the same point in time. Thus, you should buy the forest tract. The \$50 you receive by so doing is

[4]The compounding expression for one year, $F = P(1 + i)$, and for two years, $F = P(1 + i)^2$, were derived in the text. To save space, the expression was not derived for three, four, or five years. The student should try it. If that's too easy, prove that the general expression is true.

Table 7–3 The Procedure for Compounding Costs (Original Amount *P* Is \$100, Annual Interest Rate Is 5 Percent)

Year	*Beginning Amount*	*Interest Earned by End of Year*	*Ending Amount = Beginning Amount + Interest*	*Compounding Formula*
1	\$100.00	\$100.00(0.05) = \$5.00	\$105.00	$\$105.00 = \$100(1.05)^1$
2	105.00	105.00(0.05) = 5.25	110.25	$110.25 = 100(1.05)^2$
3	110.25	110.25(0.05) = 5.51	115.76	$115.76 = 100(1.05)^3$
4	115.76	115.76(0.05) = 5.79	121.55	$121.55 = 100(1.05)^4$
5	121.55	121.55(0.05) = 6.08	127.63	$127.63 = 100(1.05)^5$

divided into two parts: \$27.63, which represents what you could have earned anyway (opportunity cost), and \$22.37 profit from the forest tract.

Earlier the businessman was described as a profit maximizer who undertakes a production process (such as purchasing forest tracts) only if it will earn more than the usual types of investments, such as savings and loan accounts or other capital investments. The rate of capital growth within the business is called the internal rate of return (or interest). The interest rate paid by common investments such as accounts in banks and other financial agencies is called the market or external rate of return (interest). Production is undertaken when the internal rate of return is greater than the external rate of return.

The internal rate of return can be found using the compounding formula. For the forest tract, the internal rate of return is that value for i such that $\$150 = \$100(1 + i)^5$. Solving this equation yields a value for i of 8.5 percent.[5]

Although policy implications will not be developed in detail, a hypothetical example will demonstrate the uses of the concepts discussed above. Suppose government policy makers decided for some reason that capital investment in the private sector (farming, forestry, manufacturing) was increasing too rapidly and that market interest rates should be increased. The reasoning behind this would be that as market interest rates increased, businessmen would withdraw funds from present investments and/or would hesitate to make additional investments. The effectiveness of the increase in interest rates would depend on the relative magnitudes of the internal and external rates

[5]To solve, let $(1 + i) = X$. Then, $\$150 = \$100X^5$, or $1.5 = X^5$. Converting to logarithms gives

$$\log 1.5 = 5 \log X$$

$$\log X = \frac{\log 1.5}{5} = \frac{0.17609}{5} = 0.03522.$$

Therefore, $X = 1.085$ and $i = 0.085$, the internal rate of interest.

of interest. If the internal rate of interest in the forest tract (or other business) is equal to or only slightly above the external rate, a slight increase in the external rate would cause businessmen to disinvest (withdraw capital) with all due haste. However, if, as in the example, the internal rate is much higher than the external rate, a slight increase in market rates would have no effect on capital use.

If the capital used to undertake production is borrowed at the market rate of interest, the same principles still hold true, except that interest rates now represent a cash, rather than an imputed, cost. In this case, if the internal rate is much higher than the external, the only effect of a small interest rate increase is to divert profits from the businessman to the owners of the capital.

Discounting Future Revenues and Net Present Value. Costs incurred at one point in time cannot be compared with validity to revenues forthcoming at a later date. The solution presented above enables costs to be compounded to the date the revenues are forthcoming. The appropriate equation, where P is the sum at the present time, F its future value, i the interest rate, and n the number of years, was shown to be

$$F = P(1 + i)^n$$

Dividing both sides of this equation by $(1 + i)^n$, the following equation is obtained:

$$P = \frac{F}{(1 + i)^n}$$

Thus, if a payoff, F, is due n years in the future, its present value, P can be determined using the above expression where i is the interest rate. This procedure is known as discounting future returns.

What is the present value of \$127.63 received at the end of five years if the appropriate discount rate is 5 percent? In this case, the discounting expression is

$$P = \frac{\$127.63}{(1.05)^5} = \frac{\$127.63}{1.2763} = \$100$$

But, as seen in the compounding example above, the future value of \$100 compounded five years at 5 percent is \$127.63. Discounting, then, is the opposite of compounding.

Suppose that in the forestry example presented above, rather than compounding the costs forward, you decided to discount the \$150 received at the end of five years back to the present at a 5 percent interest rate. Then,

$$P = \frac{\$150}{(1.05)^5} = \frac{\$150}{1.2763} = \$117.53$$

Thus, $150 received at the end of five years has the same value to you as $117.53 received now. By investing $100 today, you receive the equivalent of $117.53 today. Therefore, you buy the tract of land and make a profit of $17.53. But $17.53 compounded five years at 5 percent is equal to $22.37. The profit earned is the same whether you decide to discount or compound.

Discounting can be used to determine the present value of the future income stream earned by a durable input. Suppose a farmer has a chance to buy a small, second-hand tractor for $850. For simplicity, assume that the tractor has a remaining serviceable life of four years and will be sold for $150. Should he invest in this durable input?

The farmer must first evaluate the worth of the tractor to him. Suppose he determines that after paying all costs of other inputs used within each annual production period, the tractor will add to his yearly revenues the amounts of $300, $250, $200, and $50 at the end of years one through four, respectively. Assuming the appropriate discount rate is 6 percent, the discounting calculations are shown in Table 7–4.

The total added revenues attributed to the tractor over the period are $800; this, added to the salvage value, gives a total of $950 the farmer can realize from purchasing the tractor.[6] This $950 is not a lump sum accruing at any point in time; rather, it is the sum of an income stream occurring over a four-year-span. This stream must be discounted back to the present and compared to the present cost. When

Table 7–4 Discounting Future Earnings of a Durable Input

Year	*Added Revenue at Year's End*	*Added Revenue Discounted at 6 Percent*
1	$300	$\$300/1.06 = \$283.02$
2	250	$250/(1.06)^2 = 222.50$
3	200	$200/(1.06)^3 = 167.92$
4	200[a]	$200/(1.06)^4 = 158.42$
Total	$950	$831.86

[a]$50 earnings plus $150 salvage value.

[6]Durable inputs provide the added problem of depreciation. Depreciation is considered in detail below. For this example, added revenue due to the durable input must include payments for depreciation as well as profits per se.

this is done as in Table 7–4, the total discounted present value of the tractor's future earnings is \$831.86. The farmer would thus lose the equivalent of \$18.14 by purchasing the tractor. He could invest \$831.86 at 6 percent per year and withdraw \$300, \$250, \$200, and \$200 in that order at the end of each year; in other words, he could have the same amounts of money at the same times by investing \$18.14 less. Therefore, the present cost of the tractor is greater than the discounted value of its future earnings, and the farmer should not buy. In general, investment in a durable input is considered profitable only if the sum of the discounted future revenues is greater than the present cost of the input.

The discounting analysis can be expressed more succintly in mathematical terms. Because the purpose of the analysis is to determine whether an investment that earns an income stream over time is profitable, the *net present value* (*NPV*) of the investment can be determined by subtracting the current cost from the sum of the discounted earnings. For the tractor example, *net present value* is

$$NPV = -\$850 + \$831.86 = -\$18.14$$

This can be expanded to

$$NPV = -\$850 + \frac{\$300}{(1.06)} + \frac{\$250}{(1.06)^2} + \frac{\$200}{(1.06^3} + \frac{\$200}{(1.06)^4}$$

To obtain a general formula for *net present value*, let C_0 equal the current cost or expenditure and R_t the added revenue in year t. Then,

$$NPV = -C_0 + \sum_{t=1}^{N} \frac{R_t}{(1 + i)^t}$$

where N is the life span of the investment. This formula can be applied to any investment problem. R_t can vary by year and can be zero in any year. Our forestry example is a special case where $R_1 = R_2 = R_3 = R_4 = 0$ and $F = R_5$. The *NPV* for the forestry example is therefore

$$NPV = -C_0 + \frac{R_5}{(1 + i)^5} = -\$100 + \$117.53 = \$17.53$$

The net present value of an investment is used to determine if a current expenditure should be made to earn a future income flow. To be considered feasible, an investment should have a positive *NPV*; if a manager has to select from several investment alternatives, he should choose the one with the highest *NPV*.

A special case of the *NPV* formula has been widely used in agriculture to determine the value of farmland. To derive this formula, the assumption required is that farmland will produce a constant return, R, each year into perpetuity. Thus, the discounted income stream is

$$\sum_{t=1}^{\infty} \frac{R}{(1+i)^t} = \frac{R}{(1+i)} + \frac{R}{(1+i)^2} + \frac{R}{(1+i)^3} + \frac{R}{(1+i)^4} + \frac{R}{(1+i)^5} + \cdots$$

where the terms continue to expand without an end. But this infinite sum can be shown to have a value of R/i (see Van Horne). The net present value for farmland is therefore

$$NPV = -C_0 + \frac{R}{i}$$

where C_0 is the price per acre, R is the perpetual earnings per acre, and i is the investor's discount rate. The rational land buyer must require the NPV of the purchase to be zero or positive; thus, R/i is the maximum price he can pay per acre for land. But buyers must compete for land and the buyers willing to pay R/i per acre will earn an annual return of i on their investment. Because i is their time-preferred discount rate, the maximum price of land should be R/i. Thus, the price of a durable asset becomes equal to the value of the discounted earnings.

What Is the Interest Rate? The methods of compounding and discounting presented above are standard types of mathematical computations used in computing interest. The only unique aspect of these formulas is the interest rate i. The meaning of this interest rate is of primary importance in economic theory of the firm.

When the manager is pictured as an individual with a bundle of capital at a certain point in time, investment in a business is not his only opportunity. He may choose to spend it all immediately or to put it in a checking account (earning no interest) to spend for consumption goods over some future period. What he will do depends on the satisfaction, or utility, if you like, he will derive from spending the money now versus spending it in the future. Viewed in this context, there is no reason to assume the existence of a general rule stating that a dollar now is worth either more or less than a dollar in the future. For consumption, each person's time preferences can be different. In some cases, such as retirement or education of children, a dollar in the future might be worth more than a dollar now. Those who hold large amounts of cash during inflationary periods apparently are willing to sacrifice future buying power for the security gained from certain possession of the cash.

Just why the individual decides to invest a portion of his money in a productive process is an issue much broader than can be discussed adequately here. Presumably, the opportunity to forego consumption of a sum of capital now to create a future flow of income could be based on a combination of goals including a steady future income, a chance to increase income, a chance for wealth of great magnitude, personal satisfaction, creation of an estate, or power. But once the individual decides to invest rather than spend, to become an entrepreneur rather than a consumer, he should never accept less than the market rate of interest. As a businessman, to accept less would be irrational.

The interest rate used to discount or compound sums of money should be at least as large as the current or market rate of interest. How much higher it might be depends on the manager's opportunity costs. Ideally, the manager should be aware of all production possibilities open to him and select the most profitable. If the oil business is more profitable than the grocery business, the grocer will make the switch. Once he has selected the most profitable business, the appropriate rate of interest to use in compounding or discounting is the internal rate of interest earned in the next most profitable business. In actual fact, because of fixed investments in management abilities, managers usually are not able to skip from business to business. Grocers do not know how to run oil businesses. Because of this and other institutional limitations, the market rate of interest is commonly used in budgeting, discounting, and compounding. The interest rate, then, represents an opportunity cost. But from here on, for simplicity, the discount rate will be referred to as the market rate of interest. The reader should remember that one of the "markets" could be an alternative production investment.

Usual farm budgeting procedures include a charge for capital owned by the farmer. Owned capital must earn a return comparable to usual market interest rates. Profits earned by the farm are defined as over and above what the capital, labor, and management abilities of the farmer could earn elsewhere. When budgeting for discounting, however, an opportunity cost in the form of interest must not be charged for the durable item being evaluated—for example, the tractor in the preceding example, because the discounting procedure accounts for the opportunity cost of the capital investment in the durable.

Discounting, Compounding, and Investment Decisions

Discounting and compounding provide a method through which economic logic can be extended to consider intertemporal resource allocation, where intertemporal means over a period of years. The manager receives a flow of income over time, but that income does not have to be spent in the same time period it is received. Through borrowing or saving, the manager can make investment choices that will increase his intertemporal income stream. Because modern farming requires large capital investments, the theory of investment is important in agricultural economics.

We will consider an elementary example consisting of two periods, which we will call years but which could be any arbitrary length of time. The manager receives M_1 income in period one and expects to receive M_2 in period two. With an interest rate of i, the present value of next year's income is $M_2/(1 + i)$, and it is assumed that this amount can be borrowed against next year's income. If the manager chooses to save all of this year's income until next year, it will have a value of $M_1(1 + i)$ at that time. By considering these discounting and compounding techniques, an intertemporal budget line can be derived, as shown in Figure 7–2.

In Figure 7–2A, dollars available in the first year are shown on the horizontal axis while dollars available in the second year are shown on the vertical axis. Initially, M_1 is available in year one and M_2 is available in year two. The intertemporal budget line must pass through the point (M_1, M_2). But M_2 can be borrowed and spent in year one; therefore, distance OA represents the total amount of money available for investment in year one, an amount given by

$$M_1 + \frac{M_2}{1 + i}$$

Or, all of M_1 can be saved for investment in year two. If so, the total amount available in year two will be given by OB, which is the amount

$$M_1(1 + i) + M_2$$

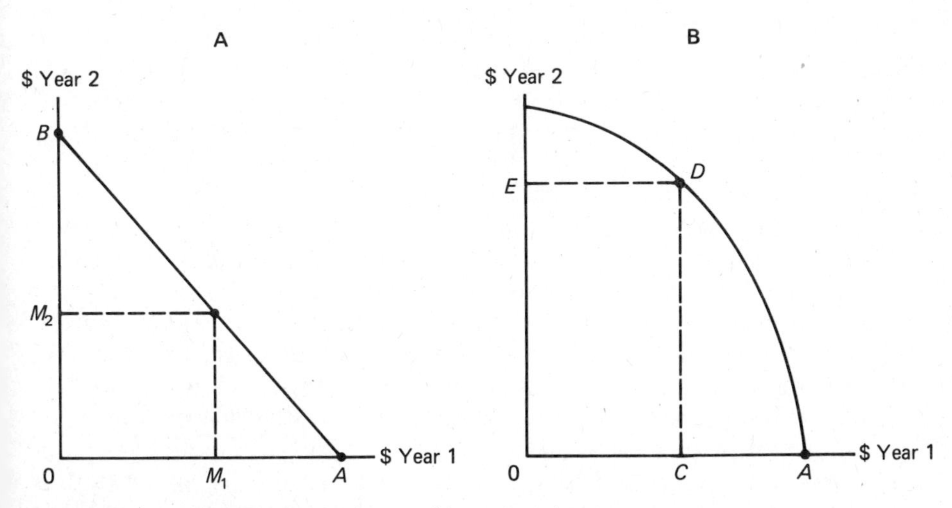

Figure 7–2. Intertemporal budget and investment curves.

The intertemporal budget line is therefore given by the straight line *BA* that passes through (M_1, M_2).

The slope of the intertemporal budget line is determined by the rate of interest. At (M_1, M_2), the manager is spending his income flows in the year they occur, neither saving nor borrowing. Suppose he decides to save one dollar from this year's consumption for an expenditure next year. This decision is represented by a one-unit movement to the left (decrease in year one) and a $(1 + i)$ movement upward (increase in year two). Thus, the slope of the intertemporal budget line is $-(1 + i)$. Increases in the interest rate will cause the budget line to become more steeply sloped; an increase in M_1 or M_2 will shift it to the right, parallel to *BA*.

The goal of the manager is to maximize his intertemporal income flow. To do so, he must consider an intertemporal investment opportunity curve as shown in Figure 7–2B. This curve represents the return in year two from investments in year one. If all income potentially available in year one, *OA* in amount, is consumed, the return in year two is zero. As consumption in year one is reduced, additional funds become available for investment. When consumption is reduced to *OC*, then *CA* is invested and earns a return of *CD* (= *OE*) in year two.

The investment opportunity curve is visualized as follows: For each amount of capital available for investment, all feasible projects are studied and the one with the maximum return is selected; where *CA* is available for investment, no other investment will return more than amount *CD*. As consumption is reduced, investment increases and the return in year two increases. But, as the investment projects increase in size, additional managerial skills are required, and the investment opportunity curve is commonly assumed to be concave for that reason. The result is that investment returns do not continually increase with investment size.

The manager will invest until returns from his investments equal the market rate of interest. The equilibrium solution is shown in Figure 7–3, where the intertemporal budget line and the investment opportunity curve are tangent. An amount of money, *PA*, is invested and earns a return of *PQ* (= *OR*). This is more than could be earned by savings alone, *PN*. Investment stops at *PA* because further increases would earn less than the market rate of interest, i.

Investment in the case depicted shifts the intertemporal budget line to the right, representing an increase in the manager's intertemporal income flow. If the investment opportunity curve is located below the budget line, perhaps because of the manager's investment abilities, then he will be able to increase his income flow more by saving at market rates. Or, when interest rates increase to very high levels, the budget line rotates clockwise about point *A* and investment is discouraged.

The example just presented is elementary but does help to explain the formation of capital. Investment in the general sense is defined as

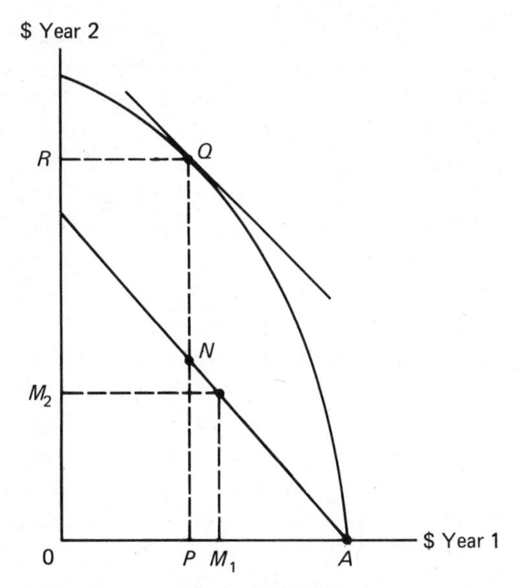

Figure 7–3. Equilibrium solution for intertemporal investment.

present consumption foregone to create capital which in turn will increase future income flows. Capital is defined to be machinery, structures, and other durables that enhance productive capacity. It should be of interest to students of agricultural economics that farmland per se does not fall under this definition. Purchasing land does not create capital but only represents a financial transfer. An investment in land that enhances its productivity represents capital formation.

Variable-length Production Process with Discounting

The cattle feeding process used as an example earlier in this chapter was carried out within the span of a year. For that reason, discounting and compounding were not needed. The increase in cost during each period was due only to the cost of feed needed for the animal; no interest was charged for the money invested. Many production processes last longer than a year. In such cases, compounding and discounting procedures must be used. An example will now be presented which assembles all the considerations of compounding and discounting presented above.

Consider a production process that requires an initial outlay of \$100. The product of this process, like that of the feeder cattle operation, comes only at one time—the time of sale. For simplicity, assume that the sale can be made only at the year's end, and annual costs are negligible. At the end of each year the manager is able to determine the sale value (total revenue) of his product.

Table 7–5 summarizes the essential revenue data for this example. At the beginning, the product can be sold for \$100, which is the purchase price. At the end of the first year, the product could be sold for \$120, etc. (These amounts are not annual earnings but represent the sale value of the product at the end of each year provided the product is not sold earlier.) As usual, these total revenues would have to be based on a production function and a product price, but these details, developed in earlier chapters, are omitted here. No annual costs are present. Therefore, a simple method of determining the most profitable length of time to grow the product is to discount future revenues. Assuming an interest rate of 8 percent, the appropriate discount factors and the discounted total revenue are shown in Table 7–5.

Discounted total revenue is at a maximum and equal to \$153.48 at the ends of years 4 and 5. Usually the maximum will be attained only in a single year. This particular feature was built in the data to illustrate a later point. To determine a specific optima, assume that the manager will produce as long as his business earns at a rate at least equivalent to the discount rate. Once he invests, he won't disinvest as long as he is doing as well as the market. Year 5 is then the year he should sell, and his discounted profit would be \$53.48. By investing in the process now, he will receive at the end of five years an amount of profit equivalent in value to \$53.48 now.

The above method is simple to apply and yields a valid solution under a specific assumption—that the production process is not repeatable. A more detailed approach to the problem will reveal the nature of this solution and suggest an alternative criterion when the process is repeatable. A more complete analysis of the example is presented in Table 7–6.

Column 1 in Table 7–6, Total Revenue, is taken from Table 7–5. Column 2, Total Cost, is the original \$100 investment compounded at 8 percent and represents the cost incurred to the end of any particular year. In this example, total costs are the \$100 investment plus the interest it could have earned elsewhere. For example, the cost at the

Table 7–5 Discounting Total Revenue for a Production Process

Year	*Total Revenue*	*Discount Factor*[a]	*Discounted Total Revenue*
0	\$100.00		
1	120.00	$(1.08)^1 = 1.0800$	\$111.11
2	150.00	$(1.08)^2 = 1.1664$	128.60
3	180.00	$(1.08)^3 = 1.2597$	142.89
4	208.80	$(1.08)^4 = 1.3604$	153.48
5	225.50	$(1.08)^5 = 1.4693$	153.48
6	239.03	$(1.08)^6 = 1.5869$	150.63
7	250.98	$(1.08)^7 = 1.7138$	146.45

[a]Using a discount rate of 8 percent, \$120 ÷ 1.0800 = \$111.11, etc.

Table 7–6 Hypothetical Example of a Production Process Through Several Years

Year	*Total Revenue* (1)	*Total Cost* (2)	*Profit*[a] (3)	*Annual Addition* *Revenue* (4)	*Annual Addition* *Costs* (5)	*Annual Percent Change*[b] *Revenue* (6)	*Annual Percent Change*[b] *Costs* (7)	*Discounted Profit*[c] (8)	*Internal Rate of Return*[d] (9)
0	$100.00	$100.00	$ 0.00					. . .	. . .
1	120.00	108.00	12.00	$20.00	$ 8.00	20%	8%	$11.11	20.0%
2	150.00	116.64	33.36	30.00	8.64	25	8	28.60	22.5
3	180.00	125.97	54.03	30.00	9.33	20	8	42.89	21.6
4	208.80	136.04	72.76	28.80	10.07	16	8	53.48	20.2
5	225.50	146.93	78.57	16.70	10.89	8	8	53.48	17.6
6	239.03	158.69	80.34	13.53	11.76	6	8	50.63	15.6
7	250.98	171.38	79.60	11.95	12.69	5	8	46.45	14.1

[a]For each year, the column 1 entry minus the column 2 entry.
[b]Computed by dividing the addition to total revenue (total costs) by the amount of total revenue (total costs) at the beginning of the year. For year 1, the percentage increase in total revenue is (20/100) = 0.20, or 20 percent.
[c]For each year, the entry in column 3 discounted by the appropriate discount factor from Table 6–4. For year 6, discounted profit is $80.34/1.5869 = $50.63.
[d]Computed by substituting the original cost, $100, in the compounding formula and solving for the interest rate that would yield a compounded amount equal to total revenue for the year in question. For year 4, the equation to be solved is $208.80 = 100(1 + i)^4$ and $(1 + i) = 1.202$.

end of year 2 is $116.64, or the $100 original investment plus $8 interest the first year plus $8.64 interest the second.[7]

Column 3 is profit determined by subtracting total costs from total revenue for each year. This undiscounted figure attains a maximum in year 6, suggesting an optimum time of six years rather than five as previously derived. Further examination will show why five is appropriate.

Columns 4 and 5 show annual additions to total revenue and total costs. When divided by the appropriate total revenue and total cost figures, the percentage changes in total revenue, column 6, and total costs, column 7, are determined. Percentage change in total revenue increases and then decreases. Percentage change in total costs, 8 percent, is constant; this need not be true when annual costs are incurred.

The percentage increase in revenue exceeds 8 percent in years 1 through 4. In year 5 the earnings of the production process are 8 percent, exactly equal to the interest rate used. (This explains why years 4 and 5 have identical discounted total revenues; the firm is earning the same interest rate as the market.) In year 6 the revenue increases only 6 percent. The manager could sell at the end of year 5, invest the money received at 8 percent and realize more revenue than by producing another year. Thus, five years, rather than six, is the optimum time span. The computation of discounted profit, column 8, verifies this. Profit from the enterprise is not increased in year 5 over year 4; thus the firm is not earning at a higher rate than the market in year 5. But discounted profit from year 6 decreases—suggesting that the market earns more than the firm in year 6. The rate of discount has an impact on the economic horizon, or the length of the planning period. As the discount rate decreases the optimum time span increases, and as the discount rate increases, the optimum time span decreases.

The general criterion for the optimum solution is that the percentage increase in total revenue must equal the percentage increase in total costs. When this is true, discounted profits are at a maximum. In the simple case in which the only cost incurred is at the start of the period, the solution can be obtained by discounting total revenue for each year and selecting the maximum (Table 7–5). When costs are incurred each year, the general criterion can be invoked and the method presented in Table 7–6 used.

The above optimum solution, five years, assumes that time is not limited or that the process is not repeatable within the time horizon of the manager. This may be true for some types of slow-growing crops such as forests, vineyards, orchards, or other production processes that

[7]An annual cost would be compounded in the same manner. Thus, if $20 annual cost were incurred in year 1, 8 percent interest would be paid on it thereafter. At the end of year 1, interest would be due on $128. Annual costs remain invested in the process until time of sale and therefore must be compounded.

require the major part of a man's lifetime. If the production process is immediately repeatable, however, a different profit-maximizing criterion applies. Time becomes valuable, and the manager should maximize the profit per unit of time, commonly called the internal rate of return.

The internal rate of return is presented in column 9 of Table 7–6. These rates were computed by solving the compounding formula for the interest rate. The internal rate of return is a maximum 22.5 percent at the end of the second year. Thus, because the process is repeatable, the manager will stop the process and reinvest the \$150 in a similar production process. If he does, at the end of two more years, or four years in all, the \$150 would grow to $\$150(1.225)^2 = \$225$. (But if he left the \$150 invested in the original production process, it would only grow to a value of \$208.80, undiscounted, after four years.) The \$225 could then be reinvested for a period of two years at an interest rate of 22.5 percent, etc.

If the manager wished to maximize his income stream rather than his investment in the business, he would withdraw all earnings above \$100 each time he repeated the production process. An investment (production) period of two years would still be the optimum, for he could withdraw \$50 for the good life every two years and still have \$100 invested. No other period of investment would yield as large an income stream over time.

Discounted Cash Flow Analysis

Cash flow analysis was presented above for the farm business within a one-year period. But investment in major capital projects requires planning over a period of years. In such cases, cash flow analysis can again be used to study the timing of expenditures and revenues.

Cash flow statements for projects of several years' duration should be presented in discounted form. Capital investment projects for some farms involve thousands and even millions of dollars, and discounting can dramatically affect the apparent returns of such projects. This is because the profitability of a new project is in part determined by the amount of cash outflow that it requires, the cash inflow that it returns, and the timing of these flows. As discussed above, a dollar received now is worth more than a dollar received a year from now. Therefore, cash outflows for a project are cheaper to the firm when they can be deferred. And the sooner cash inflows start for a new project, the more they are worth.

Cash flow analyses extending over a period of years should use the principles of discounting described above in this chapter. Discounted cash flow analysis considers the time value of money by discounting future cash inflows back to their present worth. Cash flow analyses can be used to compute an interest rate that equates the present value of the cash inflows to the present value of the investment or cash outflows. This discount rate is the true economic yield or internal rate of return of the project and can be used as shown above to determine the feasibility of the investment.

ECONOMIC ASPECTS OF DURABLE INPUTS

Purchasing Durables

A durable input should be purchased only if the sum of the discounted revenues from its use is greater than the sum of the discounted costs incurred in its purchase. The revenues from its use are actually marginal in the sense that all other durables are held constant and revenue considered results only from additions of the durable input. These revenues must be net of all actual and imputed costs except investment costs associated with the durable input being considered. Nondurable inputs costs associated with the durable, such as gasoline and grease required by a tractor, are paid out of annual revenues and are not treated as an investment cost. Similarly, interest on investment in the durable is included in the discount rate and thus should not be subtracted from added revenue.

When durable inputs can be varied, a production function can be visualized. A production function for tractors, expressed in value terms, is presented in Table 7–7, in which the data represent profits added by additional tractors. For simplicity, the underlying technical production function is omitted. The first tractor earns \$1500 the first year, \$1300 the second year, \$1100 the third year, and \$900 the fourth year. The fourth year's earnings also include the tractor's salvage value. (Four years are used to keep the problem small.) Two tractors earn \$2500, \$2200, \$1900, and \$1500 in the first through fourth years, consecutively. Three tractors earn \$3000 the first year, \$2650 the next, etc.

Table 7–7 Hypothetical Data Illustrating a Production Function for Tractors

	Added Number of Tractors					
Year	*1*	*2*	*3*	*4*	*5*	*Discount Factor*
1	\$1500	\$2500	\$3000	\$ 3250	\$ 3375	$(1.10)^1 = 1.1$
2	1300	2200	2650	2875	2990	$(1.10)^2 = 1.21$
3	1100	1900	2300	2500	2600	$(1.10)^3 = 1.331$
4	900	1500	1800	1950	2025	$(1.10)^4 = 1.464$
Four-Year Total	\$4800	\$8100	\$9750	\$10,575	\$10,990	
VMP of tractors	\$4800	\$3300	\$1650	\$ 825	\$ 415	

Table 7–8 Discounting at a Rate of 10 Percent

	Number of Tractors				
Year	*1*	*2*	*3*	*4*	*5*
1	$1363.63	$2272.72	$2727.27	$2954.54	$3068.18
2	1074.38	1818.18	2190.08	2376.03	2471.07
3	826.45	1427.50	1728.02	1878.29	1953.42
4	614.75	1024.59	1229.51	1331.97	1383.20
Discounted Total	$3879.21	$6542.99	$7874.88	$8540.83	$8875.87
Discounted *VMP*	$3879.21	$2663.78	$1331.89	$ 665.95	$ 335.04

Source: Data from Table 7–7.

Over the four-year period, the value of the marginal product of the first tractor is $4800, the second $3300, the third $1650, the fourth $825, and the fifth $415. But these marginal earnings do not accrue at one point in time but represent a stream over the four years. Hence they must be discounted. Assuming a discount rate of 10 percent, the discounted earnings of tractors are presented in Table 7–8. A graph of the discounted *VMP*s is presented in Figure 7–4.

The discounted *VMP*s represent what the tractors will earn on the farm and are therefore called the *use* value. But all durables have three prices or values: the purchase price, the use value, and the salvage

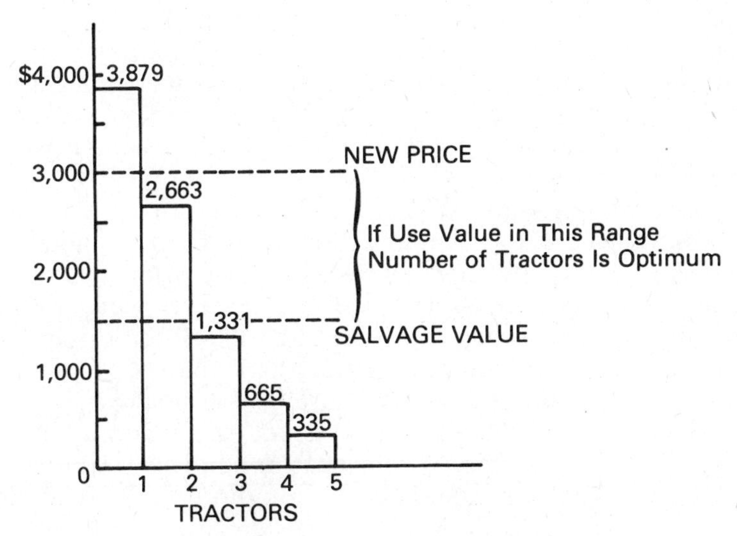

Figure 7–4. Discounted value of the marginal product of tractors.

value.[8] Any decision regarding the optimum amount of durables must consider all three. The following rules are important:

1. If the use value exceeds the purchase price, more of the durable will be purchased. If the farmer has one tractor and new tractors cost $2500, he will buy a second because the use value of the second, $2663.78 (see Table 7–8), exceeds the purchase price.
2. If the use value of the durable is less than the salvage value, some of the durables will be sold. In Figure 7–4, if the farmer owns three tractors and the salvage value for tractors is $1500, then one tractor will be sold because its sale value exceeds the present value of its future earnings.
3. If the use value of the durable exceeds the salvage value and is smaller than the purchase price, the optimum amount of durable resource is being used and the resource should not be purchased or sold. As illustrated in Figure 7–4, if tractors cost $3000 new (rather than the $2500 suggested above), have a salvage value of $1500 and the farmer owns two, he will neither buy nor sell. He has the optimum amount. Tractors are fixed in quantity because profits cannot be increased by varying them.

The importance of the distinction between purchase costs and salvage values for durable inputs can be illustrated using short-run cost curves. In this case suppose the only fixed input is a machine with salvage value lower than the purchase price. Assuming that variable costs do not change with the value of the machine, average total cost, *ATC*, will depend only on fixed costs, which in turn depend on the value placed on the machine. Fixed costs, and thus *ATC*, will be the highest when the machine is valued at the purchase price. As the machine depreciates, fixed costs and *ATC* will fall to a lower level determined by the salvage value. Therefore, in Figure 7–5, the appropriate *ATC* curve for a farmer with a newly purchased machine is marked *ATC*—PURCHASE while a farmer with a machine depreciated out to salvage value would have the total costs denoted by *ATC*—SALVAGE. *ATC* curves for other values of the machine would fall between these extremes.

As illustrated in Figure 7–5, if the product price is above *OZ*, the farmer without a machine could afford to purchase a new machine because it will earn above normal profits. If the price falls below *OW*, the farmer who owns a machine will lose money even when his machine is valued only at the salvage price, and he will sell the machine if that price persists. When the price is between *OW* and *OZ*, the farmer will neither buy nor sell the machine because he is covering costs when the machine is valued at salvage value but is not earning enough to justify purchase of another machine. From a beforehand view, when the price is less than *OZ*, he also is not earning enough to purchase a machine.

[8]For nondurables, the purchase price and salvage value are identical.

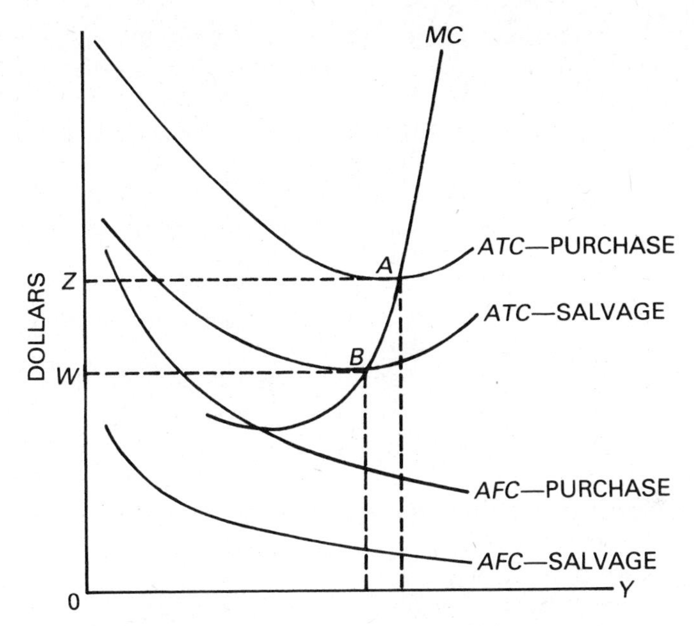

Figure 7–5. Effects of valuation of a durable on enterprise profit.

Depreciation and Obsolescence

Production is the transformation of goods and/or services into goods and/or services of a different form. In agricultural production, tractor fuel, labor, seed, and other inputs are essentially transformed into bushels of corn, pounds of pork, beef, or whatever the end product may be. If all inputs were nondurable—that is, completely transformed to product in one production period—the earnings of the production process could easily be assessed. Profit would be the cash remaining at the end of the period after actual and imputed costs are paid.

Services embodied in durable inputs are forthcoming over several production periods. Thus, at the end of any particular production period, the manager possesses some cash and the unused services of his durable inputs. It is in this context that depreciation becomes important. Assuming no changes in market prices or technology (obsolescence), depreciation is the value of the services of a durable input that are transformed into product during a production period. Each production period a portion of the value of the durable is transformed into product, and this reduction in value is equal to the depreciation. This type of depreciation is often called *use* depreciation.

The cash receipts from a production period must be of sufficient magnitude to cover costs of nondurable inputs, actual labor costs, imputed family labor and management costs, depreciation of durables, and interest on capital invested in production, including the capital invested in durables at the beginning of the period. Profit is any surplus left after paying these costs and, of course, may be positive, negative, or zero.

If the amount attributed to depreciation is set aside at the end of each production period into what is often termed a sinking fund, then when the services of the durable input are depleted, the sinking fund should contain enough capital to purchase another durable input. The unprofitable business will not be able to maintain a sinking fund large enough to replace durable inputs and thus will not survive in the long run. Businesses which survive only until their durable inputs are worn out are said to be living off of depreciation.

Depreciation is a cost because it is a payment for services rendered during a particular production period. The investment in a durable input (value of the unused services) is not a cost as such, but the interest imputed to that investment is a cost. The present production process should yield cash receipts sufficient to pay the interest on these stored-up future services, but the cost of the service is incurred only when the service is used.

In practice, the actual depreciation of a durable input is not known. Between the times a durable input is bought and sold, its value is largely unknown. If the input is of the type that is commonly bought and sold, a market price might be available. But many times such market prices are not available or do not indicate the true value to the firm. The manager must, of course, determine a value for depreciation which is as accurate as possible. If the input is "depreciated out" too fast, profits computed for the period are artificially depressed at the expense of profits for a later period. If depreciation is too slow, profits for early periods appear too large. A sinking fund must be established to replace the input when its services are depleted.[9]

Suppose that a machine costs \$1200 and sells for \$200 at the end of its useful life, five years. The \$200 salvage value is affected by the use depreciation of the machine, its obsolescence, and changing market values. (Whether the latter two are costs will depend on the situation. The machine may or may not be obsolete, and market values may increase or decrease.) The \$1000 depreciation must be allocated in some manner among the five years of use.

One method of estimating depreciation is called the *straight line*

[9]The durable input may be purchased either with borrowed funds or owned funds. When borrowed funds are used, the depreciation payment is used to make an annual payment on the principal of the loan. Interest on capital invested becomes actual on that part of the loan outstanding and imputed on owned capital. As the loan is repaid, more and more of the "total" interest becomes imputed. When the input is depreciated out and the loan is repaid, the firm is unencumbered and may incur another loan. Setting aside a sinking fund from profits is known as "internal" financing as compared to credit or "external" financing. One of the reasons for the tremendous expansion in farm production in recent decades has been due to a shift from internal to external financing on farms. From an economic point of view, the same costs are incurred either way. In fact, internal financing may be less risky.

or *linear* method. To use this method, the amount of depreciation, \$1000, is divided by years of use, five, to give an annual depreciation of \$200. The machine is therefore assumed to be worth \$1200 at the beginning of the first year, \$1000 beginning the second, \$800 the third, \$600 the fourth, \$400 the fifth, and \$200 at the end.

A second method of estimating depreciation is called the *declining balance* method. This method makes use of the assumption that depreciation is a proportion, K, of the actual value of the durable input. If P is the purchase price, the value of input at the end of the first production period is KP, the value at the end of the second period is $K(KP)$, or K^2P, etc. For the example given above, $K = 0.7$.[10] Using this method, the machine is worth:

\$1200	beginning of the first period
\$1200(0.7) = \$840	beginning of the second period
\$840(0.7) = \$588	beginning of the third period
\$588(0.7) = \$411.6	beginning of the fourth period
\$411.6(0.7) = \$288.12	beginning of the fifth period
\$288.12(0.7) = \$201.68	end of the fifth period

The straight line and declining balance methods of depreciation are two common methods of depreciation estimation. Other methods are also possible. Besides the establishment of a sinking fund, depreciation estimation is necessary in the estimation of profits. Many a farmer has thought he was making a sound profit until the services of his durable inputs were depleted. Equally as important, accurate estimates of profit will indicate when further investments in durables are justified.

The concept of depreciation in and of itself is reasonably simple. But computation of depreciation is confounded by obsolescence and other market effects.

Obsolescence is in essence a rating of the efficiency of an input in operation. A machine is technically obsolete when another machine can do the job more efficiently. Usually this efficiency means doing the same job at lower cost, but other management considerations, such as reduction of time or physical effort required, may also make one machine more desirable than another. For example, a grain auger is more expensive than a scoop shovel, but farmers buy the auger to save labor during busy times and to ease their workday. Power steering on tractors saves no time but reduces the physical effort required to drive tractors.

Increases or decreases in market value also confound the apparent

[10]Let S = salvage, P = purchase price, K = the proportion, and t = time; then $S = K^tP$. This can be solved for K using the method of footnote 5.

effects of true depreciation. When prices in the nation are generally on the rise, farmers often find they can sell durable inputs for more than the original cost or at least for more than they expected to gain from sale at the time of purchase. This phenomenon has been particularly true for farmland, but it is also true to a lesser extent of other durables. In times of rising prices, trade-in allowances for old tractors can equal or exceed the original cost, particularly if traded on a larger unit.

Obsolescence and effects of market values of depreciation are sometimes called *time* depreciation. Time and use depreciation are often both included in the meaning of the general term "depreciation." When buying durable inputs, the manager must weigh the possible effects of obsolescence and market prices, neither of which he can know beforehand, as well as the machine's productive value on the farm. Thus he becomes, whether or not he likes it, something of a market speculator as well as the manager of a business.

Buying Versus Renting Durable Inputs

As noted earlier, the present value of discounted revenues is not as high as the sum of the undiscounted returns. The higher the discount rate, the greater the difference between discounted and undiscounted returns. Similarily, compounded cost (F) grows over time. In some instances, the cost of owning a machine may exceed the cost of renting the service that machine provides.

As a crude rule of thumb, the estimated fixed cost of owning a machine (depreciation, interest, taxes, insurance, housing) averages about 17 percent of its new value each year. For example, a $6000 feedwagon will cost approximately $1020 per year to own ($6000 · 0.17). If it handles 500 tons of silage per year, the estimated average fixed cost would be approximately $2 per ton ($1020 ÷ 500T) plus variable costs. If the silage handled is doubled to 1000 tons, the average fixed costs would be reduced by half to approximately $1 per ton ($1020 ÷ 1,000T).

The fixed costs of a $20,000 combine will be approximately $3400 per year using the same 17 percent estimate. If this combine harvests 600 acres each year, the estimated average fixed costs would be approximately $6 per acre. Estimates of variable costs must be added to this to determine total machinery costs. By increasing the acreage harvested to 900 acres, the estimated average fixed costs would be reduced to approximately $4 per acre. As the use of combines is spread over more acres, the average fixed cost per acre declines.

As machinery costs increase, operators should consider alternative methods of obtaining machinery services. Leasing is an alternative that has often appeared feasible because it would enable farmers to share the services embodied in lumpy inputs. But to the present time, leasing has not been widely used.

A lease is an agreement granting use of equipment by an owner (lessor) to a user (lessee) for a specified period of time for an agreed price. The length of the lease may be for a day, a year, or several years. Leases of less than one year are generally referred to as *short term* and are used for machinery, such as tractors, combines, and hay swathers. Rates vary with the type of machinery involved and the machinery dealer.

Another method used is to hire the services of expensive machinery on a custom basis. This seems feasible in situations where the size of the farming operation does not justify the purchase of a complete set of equipment. One of the characteristics of farmers who are financially successful is that their farms are well equipped but not over-equipped.

Figure 7–6 shows, as an example, the custom combining rates for wheat, sorghum grain, corn, and soybeans in Kansas. In 1982, the average custom combining rate for wheat was $13.70 per acre. Thus, if the fixed cost of owning a combine plus the variable cost of its operation exceeded the custom rate, the operator would be financially better off to hire the work on a custom basis rather than purchase the equipment and do the work himself.

Decisions on whether to lease, custom hire, or own equipment include such considerations as: (1) the rate of return, if any, to be earned on the capital that would be freed when machines are not owned, (2) the number of hours the machine would be used, (3) the effects of leasing or custom hiring on the timeliness of the operation, and (4) the rate of depreciation normally used on owned machinery.

Long-run Production Possibility Curves

The addition of time causes shifts in input-output relationships. In farming, as in other sectors of the economy, rapid changes have occurred in production technology, in the structure of production costs and prices, and in quality standards for farm products. These changes have profound effects on the size and organization of farms, farm enterprises, and management practices (Chapters 6 and 10).

Farming has changed from an industry based largely on land and labor to one that also requires large amounts of capital. These changes have led to larger and more specialized enterprises. Specialization is the division of productive activities among individual farms and regions. Ordinarily, specialization means increasing productivity from the resosources and lowering the unit cost of production. But specialization also leads to interdependence, so that no farm or region is self-sufficient. On some farms, 75 percent or more of the inputs are purchased, and some of these inputs are uniquely adapted to production of single products.

These specialized inputs contribute to the heterogeneity of today's farms. The types and amounts of available resources differ widely

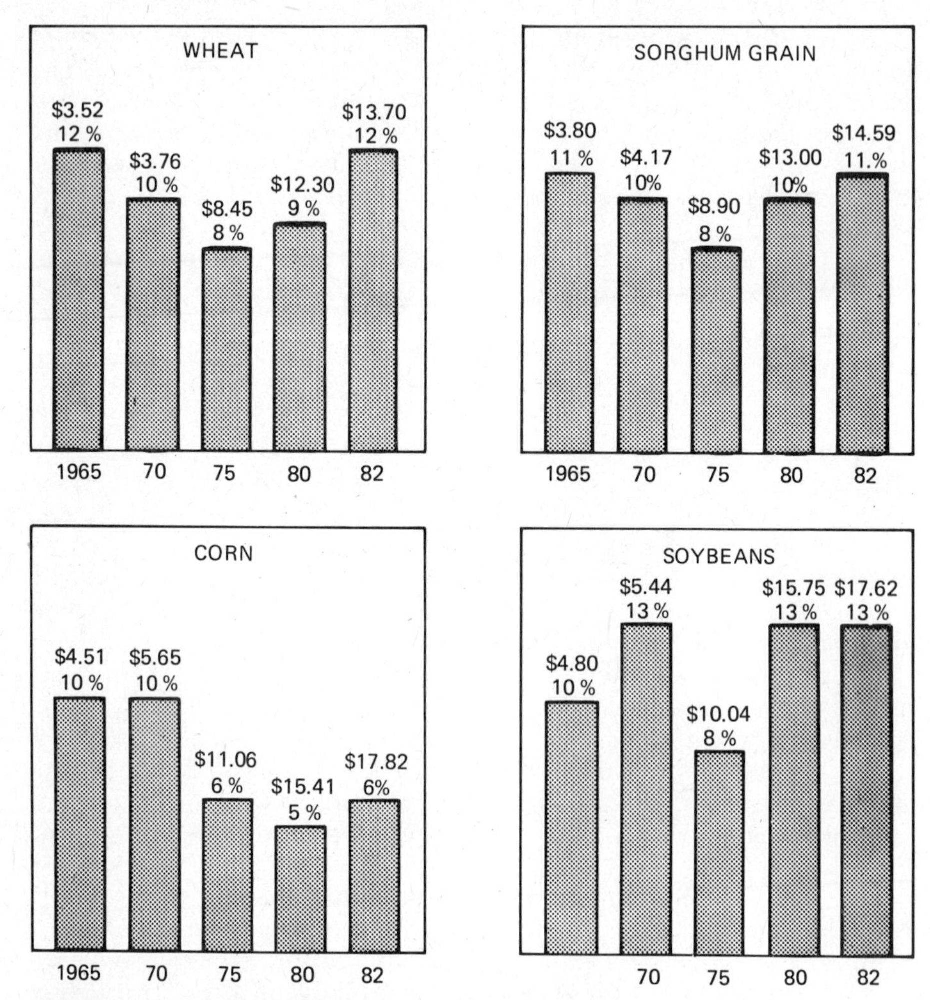

Figure 7–6. Average custom combining in dollars and as a percent of crop value per acre for selected years and crops in Kansas. (*Source:* Holland, T. E. *et al., Kansas Custom Rates, 1982,* Kansas Crop and Livestock Reporting Service, USDA and Kansas Department of Agriculture, January 1983.)

among farms. For this reason the degree of adjustments in response to changing farm prices also differ in the short run. This is because specialized resources are much more efficient in the production of one particular product. The rates at which resources or inputs are transformed into different products, as measured by their marginal products in different uses, will vary. For example, durable inputs (buildings and equipment) constructed for a specialized dairy enterprise can be utilized for the production of hogs, but the efficiency of these inputs in pork production would be lower than the efficiency of buildings and equipment designed especially for the production of hogs.

Table 7–9 and Figure 7–7 illustrate the differences in resource adaptability among two products, Y_1 and Y_2. The opportunity or production possibility curve *LR* in Figure 7–7 depicts the production alternatives in the long run. It can be viewed as a planning curve showing the combination of the two products that can be produced with a given outlay of funds or investment. Once a particular product combination is chosen and investment funds are committed to that combination, only one particular short-run production possibility curve, such as S_1R or S_2L, will apply. In this example, if the investment funds are committed to a specialized set of durables for enterprise Y_1, 35 units of Y_1 can be produced; similarly a specialized set of durables for enterprise Y_2 would result in 29 units. Other combinations of Y_1 and Y_2 on the long-run opportunity curve could be chosen (Table 7–9, columns 2 and 3).

However, potential output of product Y_2 is reduced where specialized resources are purchased to produce Y_1 or vice versa, even though the short-run opportunity curves are based on the same amount of investment as the long-run opportunity curve. Resources purchased especially to produce enterprise Y_1 would produce only 19 units of Y_2; 10 less than if the investment had been used to purchase resources adapted to Y_2. Other examples can be derived from Table 7–9 and Figure 7–7.

The use of inputs specialized in the production of certain products makes enterprise output more stable. When the manager has a specialized set of durables, small changes in relative product prices will not cause a shift to a different enterprise. Investment in durable inputs represents fixed costs. These durables embody services that can be transformed into product or products only over long periods of time.

Table 7–9 Production Functions for Y_1 and Y_2 with Adapted and Unadapted Durable Inputs

Units of Variable Input	*Production functions with durable inputs adapted to the specific product.*		*Production functions with durable inputs adapted to the opposite product.*	
	Y_1	Y_2	Y_1	Y_2
0	0	0	0	0
1	10	8	6	1
2	18	15	11	9
3	25	20	15	13
4	31	24	18	16
5	34	27	20	18
6	35	29	21	19

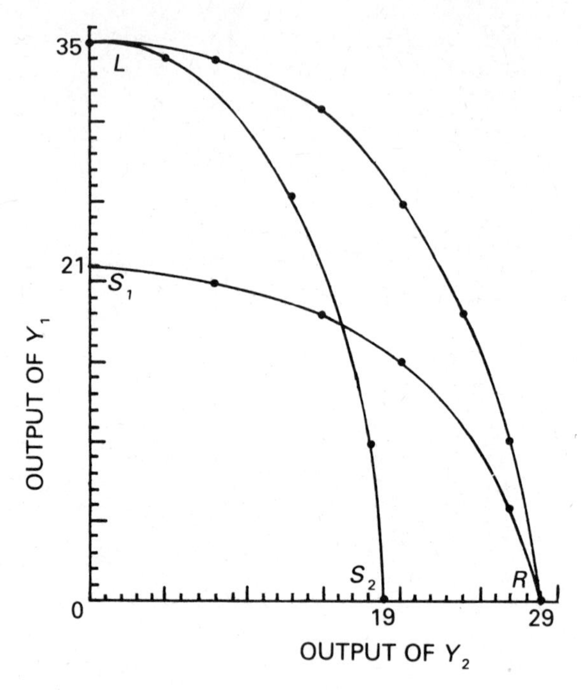

Figure 7–7 Long-run and short-run production possibility curves.

Enterprises using such durables are less responsive to relative input and output price change than are enterprises using inputs that are transformed into products within short periods of time.

VALUING AGRICULTURAL LAND

The Land-Price Paradox

One of the most important, and certainly the most unique, inputs in agricultural production is farmland. Not only an important input, farmland also is the most important asset in terms of value—land value represented approximately 80 percent of the estimated trillion dollars of farm assets in 1982, an average of nearly $350,000 of real estate assets per farm. Because of the significance of its role, the value of farmland has been a concern of agricultural economists.

In this section, we will review some of the literature pertaining to the valuation of farmland. The subject provides a useful case study for applying the techniques discussed earlier in this chapter. Further, we hope to provide some insights into an issue of extreme importance in United States agriculture today.

As explained in Chapter 6, the classical theory of production suggests that in the long run an input in a perfectly competitive market should earn a return equal to the value of its marginal product. Because long-run factor shares are difficult to determine in practice, researchers traditionally used alternative short-run measures such as net farm income per acre or earnings imputed to land as a fixed asset. These measures of earnings were compared to the current market price of farmland to determine if that price could be justified by the productivity of the land.

Historically, land prices and farm income appeared to move together. But in the Fifties and into the Sixties, land prices began to diverge from income trends. In 1960, Scofield observed this divergence and called it the "land-price paradox." This price-earnings gap has subsequently become an important research topic and policy issue.

The index of the real value of farmland per acre, defined to be the current or "nominal" dollar value of farmland divided by the Consumer Price Index, is shown in Figure 7–8. Because the CPI provides a measure of the rate of inflation in the United States, the index in Figure 7–8 represents value increases over and above the inflation rate. Increases in real land values have been steady since 1945, and particularly dramatic since 1965. As an investment, farmland has clearly provided a good hedge against inflation. Reference to Figure 10–10 (p. 392), however, suggests that, except for a peak in 1973, net farm income in constant dollars has been slowly declining since 1965. The sharp drop in net farm income in 1981 and 1982 was accompanied by a decline in the real value of farmland in those same years, the first such decline since 1970. With the exception of these two years, the divergence of farmland value and farm income illustrate the general nature of the land-price paradox, although most research studies used more refined data.

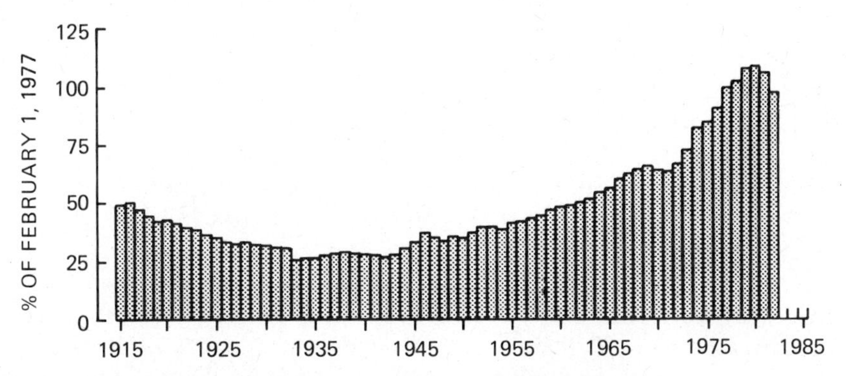

Figure 7–8 Index of real value per acre of U.S. farmland. Reported as of March 1, 1913–75, February 1, 1976–81, and April 1, 1982. Excludes Alaska and Hawaii. The indexes of real farmland value computed by dividing the nominal land value indexes by the Consumer Price Index.

The traditional valuation formula presented above, $V = R/i$, was in general use to determine the present value of farmland, where V is now used to represent that value. Researchers and other land market observers were therefore inclined to use this formulation to estimate farmland value. In the search for the causes of the land-price paradox, the applicability of this formula was questioned and a host of other explanations was offered. The issues considered can be classified into four general questions: (1) Is the formula $V = R/i$ appropriate for valuing land? (2) What forces are increasing annual earnings, resulting in higher land values? (3) What is the proper measure for R? (4) Are market forces shifting the supply or demand curves for farmland, therefore setting market prices that are not totally dependent on farmland earnings?

In the sections to follow, we will concentrate on the valuation formulas, which are of primary interest given the purpose of this chapter. But for completeness, a short discussion of the other effects will be presented now.

It was hypothesized that several factors were causing the annual earnings, R, to increase. Technological change decreased costs while government farm programs maintained favorable prices and reduced price uncertainty. Technological change also favored large-scale farming operations, thereby increasing earnings on present holdings and enhancing the value of additional acreage to the farmer with adequate modern equipment. As a result, expansion demand, that is, purchases to increase farm size compared to the purchase of a whole farm by a "new" owner, increased. In addition to the increased value for the marginal increment of land, expansion demand was supported by the favorable financing terms available to owners with large equities in existing farms.

The use of net farm income per acre as an appropriate measure of R was questioned. Melichar analyzed the arguments involved and suggested a measure of "returns-to-assets" could better be used to determine the value of assets.

The major force seen shifting the supply curve for land was nonfarmland use. Highways, airports, urban expansion, and recreational uses all reduced land available for farming. Two new effects were suggested that could shift the demand curve for land to the right. Hathaway, in 1957, noted that capital appreciation made farmland an attractive investment. The continued existence of real capital gains would attract investors to the land market. A second component, mentioned by Martin and Jefferies, was that some owners regarded farmland as a consumption good and were willing to accept reduced returns for the privilege of ownership. Thus, farmland simultaneously produces utility as consumption good and value as a production asset. A more complete discussion of these factors may be found in Doll and Widdows.

Extensions of the Valuation Equations

The formula for valuing an asset that produces a return into perpetuity, $V = R/i$, was presented by Fisher in his classic book *Theory of Interest.* When applied to farmland, the assumption of a constant, no-growth stream of income forever, while providing a very simple mathematical equation, is unrealistic. Reinsel and Reinsel argued that the formula cannot explain land values and proposed instead the erratic earnings formula

$$V = \frac{R_1}{(1 + i_t)} + \frac{R_2}{(1 + i_t)^2} + \frac{R_3}{(1 + i_t)^3} + \ldots = \sum_{t=1}^{\infty} \frac{R_t}{(1 + i_t)^t}$$

In this formula, both earnings, R_t, and discount rates, i_t, may change each year.

It may be that the investor does not have an infinite planning horizon and may prefer to consider the land investment for some finite period. Then, Crowley proposed the modification

$$V = \sum_{t=1}^{n} \frac{R_t}{(1 + i_t)^t} + \frac{P_n}{(1 + i_p)^n}$$

where the land is to be held for n years and sold for price P_n. Although i_t and R_t vary by years, they could be assumed constant for simplicity. Also, the sale price, P_n, is discounted at an interest rate, i_p, which could equal i_n.

Both Crowley and the Reinsels argue that the simple, constant-stream formula, $V = R/i$, leads to the erroneous conclusion that agricultural land has been overvalued in recent decades. They believe the erratic growth model is better able to capture the effects of increasing returns on land values.

Melichar proposed two variations on the land valuation problem. First, as mentioned above, he replaced net income measures with a measure of returns to assets. Second, he argued that returns to farmland are increasing (although net farm income is not) and suggested the use of the following model to permit growth rates to be incorporated into the valuation process. Assuming farmland earnings grow at a constant rate of g per year, then the value of the discounted earnings over n period is

$$V_n = \frac{(1 + g)R_0}{(1 + i)} + \frac{(1 + g)^2R_0}{(1 + i)^2} + \frac{(1 + g)^3R_0}{(1 + i)^3} + \ldots + \frac{(1 + g)^nR_0}{(1 + i)^n}$$

where R_0 is the current earnings of the asset, in this case farmland, and $R_1 = (1 + g)R_0$, $R_2 = (1 + g)^2R_0$, etc. But at the end of n periods, the land will be sold and the discounted sale price should be added to the value, V_n, as shown above for the erratic earnings formula. But because returns grow at rate g, the sale value at the end of period n will be based on the same growth formula as V_n, beginning at time $(n + 1)$ and extended into the future. Since this is true for each potential sale and the farmland produces into perpetuity, then the actual values of the asset is given by

$$V = \sum_{t=1}^{\infty} \frac{(1 + g)^t}{(1 + i)^t} R_0$$

which is an infinite series of discounted earnings. When $0 < g < i$, then it can be shown (see Van Horne) that the equilibrium value of the asset is given by

$$V_e = \frac{(1 + g)}{(i - g)} R_0$$

Comparing this to the traditional formula, $V = R/i$, $(1 + g)R_0$ is in a sense equivalent to R, but $(i - g)$ replaces i in the denominator. Thus, if the constant growth model is appropriate for valuing land, but the traditional model is used, then what appears to be the annual return or discount rate, i, is actually the discount rate, i, minus the anticipated growth rate, g. Put another way, if land earnings are divided by land values to estimate an annual rate of return, the result is not the discount rate, i, but rather the difference, $i - g$. To find the true interest or discount rate in this case, an estimate of the growth rate, $\hat{g}$, must be added to earnings-value ratio. An anticipated high growth rate increases the value of an asset and has the apparent (but not actual) effect of decreasing the return to the asset.

Calculus can be used to investigate the effects of changes in i, g, and R_0 on V_e. Thus,

$$\frac{\partial V_e}{\partial R_0} = \frac{(1 + g)}{(i - g)} > 0$$

$$\frac{\partial V_e}{\partial g} = \left(\frac{1}{i - g}\right) R_0 > 0$$

and

$$\frac{\partial V_e}{\partial i} = -\frac{1 + g}{(i - g)^2} R_0 < 0$$

Increases in expected earnings or growth rates will increase farmland value while increases in the discount rate will decrease value. Thus, the land buyer can be regarded as revising his estimates of i, g and R_0 each year. In so doing, he will consider not only the absolute level of current earnings, represented by R_0, but also forces that promise to enhance future earnings. For example, new technologies will increase R_0 but may also cause the investor to adjust g upward (i.e., expect even higher earnings in the future). Present values are therefore increased in two ways.

Basing his interpretations on this model, Melichar concluded that asset earnings did generally support asset values in agriculture for the 1950–78 time period. Doll, Widdows, and Velde tested his hypothesis further, with mixed results. Both studies were conducted for time periods immediately preceding the high interest rates and low farm product prices of the early Eighties.

Inflation and the Value of Land

The valuation formulas can be modified to reflect the effects of inflation. We will discuss the constant growth earnings model; the others can be similarly adjusted. Dobbins *et al.* present a more complete discussion along with an empirical analysis for Indiana farmland.

If the real rate of growth in earnings is represented by δ and the annual inflation rate by π, then the nominal growth rate is the real rate compounded forward by the inflation rate. Thus,

$$1 + g = (1 + \delta)(1 + \pi)$$

Similarly, if the discount rate in real terms is r, then the inflated discount factor is

$$1 + i = (1 + r)(1 + \pi)$$

The present-value constant growth formula can then be rewritten

$$V = \frac{(1 + \delta)(1 + \pi)}{(1 + r)(1 + \pi)} R_0 + \frac{(1 + \delta)(1 + \pi)^2}{(1 + r)(1 + \pi)} R_0 + \dots .$$

where π is assumed constant for each year. But the rates of growth due to inflation are offsetting in each fraction so that the equilibrium value ultimately becomes

$$V_e = \frac{(1 + \delta)}{(r - \delta)} R_0$$

assuming, as before, that $0 < \delta < r$. In this formulation, the effects of inflation cancel and, as Dobbins *et al.* explain, the equilibrium value of land is unchanged from the nominal formulation. Furthermore, starting with the same expression, if investors believe the only source of income growth is inflationary, δ is zero and the equilibrium value reduces to $V_e = R_0/r$, the traditional formula in real terms.

In fact, the effects of inflation are probably not neutral. The markets for farm products are distinct in many ways from the markets for farm inputs. In practice, each farm product and each input price has a unique inflation rate, none of which are adequately represented by the Consumer Price Index, a nationwide index based on a bundle of consumer goods. When inputs and outputs have different rates of inflation, it is likely that the highest rates will exist for purchased inputs. If so, then inflation would not have a neutral effect on land values; real returns would fall over time because costs would rise faster than revenues and V_e would be shifted downward.

Inflation has some other important effects. Historically, farmland has been financed with mortgages of fixed length and fixed interest rates. Inflation in such cases redistributes real wealth from the lenders to the borrowers. Simply put, inflation enables borrowers to repay debts with dollars that have less real value each year. This wealth transfer increases with the rate of inflation, the size of the mortgage, and the length of the mortgage. Because mortgages on farmland have been readily available, farmland has become a convenient method of buying debt and thereby obtaining a real wealth transfer.

Inflationary trends that cause costs and revenues to increase at differing rates could create cash flow problems for farmers. First, revenue flows could be reduced relative to cost flows. This squeeze could affect resource use and productivity, further reducing revenues—although the effects of inflation on productivity are not well defined. Second, the reduced cash flows create difficulties for farmers committed to meeting sizable mortgage payments on land. This is especially true if land buyers hold the newer, variable-interest-rate mortgages. In this case, the interest payments increase with inflation while farm revenues may not.

Determination of the effects of inflation therefore requires a more sophisticated modeling process. In addition to effects on cost and revenue streams, inflationary trends may effect property valuation through differential effects on income, capital gains, and inheritance taxes, as well as a number of other factors. Recently, researchers have suggested methods of analyzing these as well as other determinants of land values. The next sections contains a short review of these models.

Newer Valuation Models

Models developed in the Seventies are complex compared to the traditional valuation formulas and therefore are not presented in detail here. Instead, we will sketch their general nature.

Most methods proposed are based on some variation of cash flow and/or capital budgeting models. As such, they represent a micro rather than a macro approach to the valuation problem. The models are tested on case studies or are applied to scenarios most appropriate for investment decisions by individual farmers. Designed specifically for computer simulation or solution, the analyses determine a maximum bid price or maximum value for farmland, given a specific set of assumptions.

Valuation models of this type are specified to include the variables known to affect decisions of land investors. That is, they include more than the potential flow of land earnings, and consider variables that can affect both income flows and wealth over a manager's lifetime. These variables include inflation rates, marginal income tax rates, capital gains taxes at time of sale, finite planning horizons, down payments and financing terms, owner's equity, farm size and efficiency, income variance, and the manager's risk aversion. Analyses of this type have been presented by Lee and Rask, Plaxico and Kletke, and Harris and Nehring. The effects of income taxes on land valuation, as incorporated in these and other valuation models, have been discussed by Adams.

The models presented enable researchers to capture the heterogeneity of the land market—not only is farmland not homogeneous, but a given tract of land does not have the same value to all potential buyers. Thus, while they provide estimates of the price that could be paid for farmland in individual instances, they do not answer the question of whether land in the aggregate is overpriced relative to other assets in the economy.

These models can be visualized as separating land value into a set of components that are interrelated and not as easily separated as we will regard them here. Some components represent values, others costs, and still others represent institutional arrangements. One value component is traditional: the discounted present value of the future stream of earnings from the land. But this future income is subject to income taxes, which can vary from zero to 50 percent.

A second value component that can exist over a finite time horizon is real capital gains. The investor can realize these either by selling the farmland or borrowing against its increased value. When sold, the increase in value will be subject to a capital gains tax, which also will vary with the individual's income tax rate.

A third value component that has been suggested represents a wealth transfer resulting from holding debt during inflationary periods. Described above, this permits the land buyer to accrue wealth simply by being in debt.

A cost component relates to the methods of financing the farmland debt. Important considerations here are the amount of the down payment or owner's equity, the mortgage interest payments, which are deductible expenses, and the length of the mortgage period. A second cost component can be included to represent the farmer's efficiency in production. Management efficiency increases the income flows and hence the value of land to the purchaser.

Other components affecting land values include off-farm income, tenancy, total wealth, farm programs, etc. All may help determine the price a buyer may place on a parcel of land.

In sum, the classic method of valuing farmland, using the valuation formula $V = R/i$ or one of its variants, is thus seen to be quite different from the price a buyer might be willing to pay for the same farmland. The classic model considers only the first value component. But value is ultimately determined in the marketplace; price is set by what buyers are willing to pay and sellers are willing to accept. In a given area, sellers will rationally seek the highest bid price, perhaps that offered by the most efficient and well-financed farm operators. To compare aggregated imputed earnings to aggregated imputed value of farmland seems to neglect the varied nature of demand for farmland. When land is purchased for expansion, the value to the firm can be determined by estimating the increment in profits resulting from that increment in land. If land is purchased in part for consumption, as some researchers imply, then the purchaser must decide how much of his discretionary income he wishes to apply to farmland, compared to other consumer goods. If land is purchased solely as an asset in an investment portfolio, then this investment must be compared to other potential assets, such as gold, bonds, etc. This aspect of land ownership, largely neglected in the literature, has been studied by Feldstein. Finally, in ending this section, we note that the above models are intended to determine the prices land buyers can offer. Little, if any, research has been directed toward the supply side of the market.

Suggested Readings

Adams, Roy D. "The Effect of Income Tax Progressivity on Valuation of Income Streams by Individuals." *American Journal of Agricultural Economics,* Volume 54, August 1977, pp. 538–542.

Allen, R. G. D. *Mathematical Analysis for Economists.* New York: Macmillan and Company, 1956, Chapter 11.

Barish, Norman. *Economic Analysis for Engineering and Management Decision Making.* New York: McGraw-Hill Book Company, 1962, Chapters 5, 6, 7, and 9–17.

Boulding, K. E. *Economic Analysis,* Revised Edition. New York: Harper & Bros., 1948, Chapters 35 and 36.

Carlson, Sune. *A Study on the Pure Theory of Production.* New York: Kelley and Millman, Inc., 1965, Chapter 4.

Crowley, William D. "Actual Versus Apparent Rates of Return on Farmland Investment." *Agricultural Finance Review*, Volume 35, October 1974, pp. 52–58.

Dillon, John L. *The Analysis of Response in Crop and Livestock Production*, Second Edition. Oxford: Pergamon Press, 1977, Chapter 3.

Dobbins, C. L., Baker, R. B., Dunlop, L., Pheasant, J. W., and McCarl, B. A. *The Return to Land Ownership and Land Values: Is There an Economic Relationship*? Purdue Agricultural Experiment Station Bulletin No. 311, February 1981.

Doll, John P. and Widdows, Richard. *A Critique of the Literature on U.S. Farmland Values.* Paul Velde, ed. ERS Staff Report No. AGES830124, Washington, D.C., January 1983.

Doll, John P., Widdows, Richard, and Velde, Paul D. "The Value of Agricultural Land in the United States: A Research Report." *Agricultural Economics Review*, Volume 35, Washington, D.C., April 1983, pp. 39–44.

Faris, J. E. "Analytical Techniques Used in Determining the Optimum Replacement Pattern." *Journal of Farm Economics*, Volume 42, November 1960, pp. 755–66.

Feldstein, Martin. *The Effects of Inflation on the Prices of Land and Gold.* Working Paper No. 296, Cambridge, Mass.: National Bureau of Economic Research, November 1978.

Fisher, Irving. *The Theory of Interest.* New York: Kelly and Millman, Inc., 1954.

Frisch, Ragnar. *Theory of Production.* Chicago: Rand McNally and Company, 1965, Chapter 4.

Gisser, Micha. *Intermediate Price Theory.* New York McGraw-Hill Book Company, 1981, Chapter 16.

Gordon, Myron J. "Optimal Timing of Capital Expenditures." In Levy H., and Sarnat, M., eds. *Financial Decision-Making under Uncertainty*, New York: Academic Press, 1977.

Gould, J. P., and Ferguson, C. E. *Microeconomic Theory*, Fifth Edition. Homewood, Ill.: Richard A. Irwin, Inc., 1980, Chapter 17.

Harris, Duane G., and Nehring, R. F. "Impact of Farm Size on the Bidding Potential for Agricultural Land." *American Journal of Agricultural Economics*, Volume 51, May 1976, pp. 161–169.

Hathaway, Dale E. "Agriculture and The Business Cycle." Policy for Commercial Agriculture, Joint Economic Committee, 85th Congress, U.S. Government Printing Office, Washington, D.C., 1967.

Heady, Earl O. *Economics of Agricultural Production and Resource Use.* New York: Prentice-Hall, Inc., 1952, Chapter 13.

Johnson, G. L. "The State of Agricultural Supply Analysis." *Journal of Farm Economics*, Volume 42, May 1960, pp. 435–52.

Lee, Warren F., and Rask, Norman. "Inflation and Crop Profitability: How Much Can Farmers Afford to Pay for Land?" *American Journal of Agricultural Economics*, Volume 58, December 1976, pp. 986–989.

Martin, W. E., and Jefferies, G. L. "Relating Ranch Prices and Grazing Permit Values to Ranch Productivity." *Journal of Farm Economics*, Volume 48, May 1960, pp. 233–242.

Melichar, Emanual. "Capital Gains Versus Current Income in the Farming Sector." *American Journal of Agricultural Economics,* Volume 61, December 1979, pp. 1085–1092.

Nelson, Kenneth E. and Purcell, Wayne D. "A Quantitative Approach to the Feedlot Replacement Problem." *Southern Journal of Agricultural Economics,* Volume 4, July 1972, pp. 143–149.

Nelson, Kenneth E. and Purcell, Wayne D. "A Comparison of Liveweight, Carcass and Lean Meat Criteria for the Feedlot Replacement Decision." *Southern Journal of Agricultural Economics,* Volume 5, December 1973, pp. 99–107.

Plaxico, James S. and Kletke, Daniel D. "The Value of Unrealized Farm Land Capital Gains." *American Journal of Agricultural Economics,* Volume 61, May 1979, pp. 327–330.

Quirk, James P. *Intermediate Economics.* Chicago: Science Research Associates, Inc., 1976, Chapter 11.

Reinsel, R. D. and Reinsel, E. I. "The Economics of Asset Values and Current Income in Farming." *American Journal of Agricultural Economics,* Volume 61, December 1979, pp. 1043–1097.

Scofield, William H. "Returns to Productive Capital in Agriculture." *Farm Real Estate Market Developments,* Washington, D.C., CD-54, February 1960.

Van Horne, James C. *Financial Management and Policy,* Fourth Edition. Englewood Cliffs, N.J.: Prentice-Hall, Inc., 1977.

Widdows, Richard and Doll, John P. *Four Econometric Models of the U.S. Farmland Market: An Updating with Comparisons.* Paul Velde, ed., ERS Staff Report No. AGES820702, Washington, D.C., July 1982.

INTRODUCTION TO DECISION THEORY

CHAPTER 8

Chapters 2 through 6 present an economic analysis of the production process under the assumptions of perfect certainty and timelessness. While the concepts presented in these chapters provide the foundation needed for a basic understanding of the theory of the firm, they do not embody all or even most of the important characteristics of a dynamic, ever-changing firm. But just as algebra must be mastered before calculus, the fundamentals of economic logic must be mastered before more complex concepts can be attacked.

Chapter 7 includes a discussion of the problems encountered when time is introduced into the elementary timeless analysis. In this chapter, the assumption of perfect certainty will be dropped and an introduction to the theory of decision making under conditions of risk and uncertainty will be presented. Many of the concepts developed in earlier chapters will be utilized but, as was seen in Chapter 7, relaxing a key assumption ultimately requires development of some new concepts.

A GENERAL PICTURE

Perfect certainty, the ability to predict exactly the future outcome resulting from present actions, does not usually exist for the manager, at least in situations of consequence. When a farmer buys a red tractor, he knows it will stay red for some time to come. But that is at best trivial knowledge. He is far more interested in its power, fuel consumption, and dependability. Some of these unknowns may be determined to a more or less certain degree, depending on the conditions of purchase. If the tractor is new, the farmer will expect a high degree of dependability, and he will be able to determine estimates of fuel consumption and drawbar power from demonstrations, the experience of neighbors, or published information. If the tractor is second-hand, the farmer may have some estimate of these factors, based on the reputation of the previous owner or seller, but will face a larger degree of uncertainty than if he had purchased a new machine.

Of the many uncertainties facing the farmer, the uncertainties of machinery quality are relatively minor. Indeed, the efficiency of modern farm machinery has been the primary reason for the rapid mechanization of farms in the last five decades. Price and yield fluctuations of farm products comprise a far more important source of uncertainty.

When the dryland farmer of the Great Plains plants winter wheat each fall, he can only speculate on the yield of next summer's harvest. As the season passes, the degree of uncertainty surrounding the expected yield diminishes until immediately before harvest, when the farmer will be able to determine expected yields with reasonable accuracy. He cannot be sure of harvesting the crop even then, because a hail storm may blow up quite suddenly. Thus, yields are known for certain only after the crop is safe in storage.

Price variations create uncertainty for hog farmers of the Corn Belt. Uncertainties arising from disease, feed quality, and heredity can be controlled by the competent husbandryman; in contrast, the price of hogs is determined by factors beyond his control. Prices may be high at the beginning of the production period and fall steadily until marketing time, or may do the reverse. Again, as the time of sale approaches, uncertainty surrounding the price to be received diminishes.

Many other examples of uncertainty may be found in agriculture. The vineyard owner in Washington faces the hazard of early frost. An asparagus crop in the same area may go unharvested or grow too large because of labor scarcity. The problem common to all these situations in that inputs must be allocated at one point in time while returns accrue at a later date. Because the future cannot be foretold, the inputs must be allocated without perfect assurance of a specific return.

Knowledge about the future exists in different degrees. That the color of the tractor would be red was certain but unimportant. That there will be a price for hogs or a yield of wheat is certain (if prices and yields of zero are permitted), but the exact magnitude of that price or yield is uncertain. Thus, some classification of outcomes is possible even for situations involving less than perfect certainty.

AN EMPIRICAL EXAMPLE

Estimates of the magnitude of price and yield variability facing farmers in southwestern Oklahoma have been developed by Mapp *et al.* and are contained in Table 8–1. Gross margins are presented on a per acre basis for three crops, wheat, cotton and grain sorghum, and on a per head basis for two livestock systems, winter and summer steers. The estimates in Table 8–1 are based on yields from experimental plots and other budget data. Yields and prices were adjusted to remove trends in technology and inflation. Prices and costs were estimated in 1977 dollars. Gross margins are defined as total revenue minus total variable cost. Fixed costs should be subtracted to obtain net farm income, but these costs are assumed constant for the 13-year period and therefore do not affect variability.

Because of the method of estimation, the gross margins in Table 8–1 represent both price and yield variability. The annual margins show considerable variation about their mean values. Cotton earned a

Table 8–1 Estimated Annual Gross Margins for Five Farm Enterprises in Southwestern Oklahoma, 1965–1977

	Dollars per acre			*Dollars per head*	
Year	*Wheat Sale at Harvest*	*Cotton*	*Grain Sorghum*	*Winter Steers*	*Summer Steers*
1965	23.03	64.51	36.21	31.50	53.77
1966	76.62	−55.84	17.74	69.28	8.08
1967	35.55	72.55	38.03	20.14	27.03
1968	−29.57	−6.35	38.71	35.25	24.14
1969	43.95	87.56	43.04	56.86	11.61
1970	17.10	39.86	34.21	73.06	26.30
1971	−29.89	85.54	19.65	34.95	60.49
1972	−32.30	−16.66	38.41	64.66	90.63
1973	35.86	−61.90	130.91	134.35	35.20
1974	135.24	22.53	106.71	−21.14	−45.85
1975	57.77	24.77	98.72	6.86	69.52
1976	86.37	89.12	35.70	61.75	−22.38
1977	48.56	113.37	29.08	44.38	15.96
Mean	36.02	35.31	52.86	47.07	27.27
Standard deviation	48.79	56.99	34.92	37.57	36.78
Coefficient of variation	135.44	161.38	66.05	79.82	134.87

Source: Mapp, Harry P., Jr., Hardin, M. L., Walker, O. L., and Persaud, T., "Analysis of Risk Management Strategies for Agricultural Producers," *American Journal of Agricultural Economics,*Volume 61, December 1979, pp. 1071–1077.

negative return in four of the 13 years and only grain sorghum earned a positive return each year. The relative variability of the enterprises can be determined by examination of the standard deviations of the gross margins. The coefficient of variation (*CV*), found by dividing the standard deviation by the mean, provides a measure of variation relative to earnings.[1] Grain sorghum and winter steers display the highest average margins and lowest coefficients of variation. Cotton has the most variability per dollar of earnings (*CV* equal to 161 percent) while summer steers have the lowest average margin and a high variability, with a *CV* approximately equal to wheat and exceeded only by cotton.

Interpretation of these data would suggest that grain sorghum would be an important crop in southwestern Oklahoma while summer steer enterprises would appear risky. The role of the other enterprises, both intermediate in returns and variation, is not clear. Further, prices and yields for the enterprises (and, hence, variability) cannot be regarded as independent of each other. That is, yields may move up and

[1]For this chapter, the reader must be familiar with the basic concepts for statistics: means, variances, probability, random variables, expected values, etc.

down together while prices, influenced by quite different forces than weather, may exhibit different patterns. Finally, livestock prices and production may display cycles and trends different from the crops. This suggests the covariance of earnings among the enterprises is also important. Clearly, a more rigorous analysis is required.

A HISTORICAL VIEW

Uncertainty is often trivially defined as the lack of certainty. Examination of that definition leads to the further refinement that uncertainty is the lack of knowledge about the state of the world at some future time, where "world" is meant to represent both economic and natural events. But lack of knowledge could be regarded as the cause of uncertainty and not the phenomenon of uncertainty itself. The final answer cannot be provided here, but the issue suggests the intriguing questions that can be asked about uncertainty. For example, how is uncertainty measured? Can it be said there are types and degrees of uncertainty? Is uncertainty a good that provides utility and is preferred by some managers?

From the viewpoint of the theory of the firm, the most important consideration is the impact of uncertainty on the manager's decision making. How does uncertainty affect the traditional equilibrium position of the firm, the optimal amounts of inputs and outputs, and thereby the firm's demand for factors and supply of product? Is it possible that uncertainties surrounding such future events as prices, production, technology, and government programs not only create changes in equilibrium but cause disequilibrium? These are, in fact, some of the questions towards which analyses of decision making under uncertainty are directed.

The first attempts to categorize production processes subjected to conditions of less than perfect certainty were patterned after the pioneering work of Frank H. Knight. Writing in 1921, Knight created two subclasses to describe lack-of-knowledge situations: risk and uncertainty. The manager was defined as facing *risk* in a production process when he was aware of all possible outcomes that could result from the process and could attach a probability to each outcome. Thus, a risky situation existed when the outcome of an enterprise, say, corn production in Ohio, was a random variable with a known probability distribution. Continuing with Knight's definitions, the manager was said to face *uncertainty* when he was unable to associate probabilities with the outcomes of the production process. This classification could be made even more ethereal by extension to situations for which the manager is unable even to enumerate the possible outcomes of the process. The first homesteaders in the Great Plains might have faced this type of uncertainty; they not only were unaware of the yield distributions of crops but, as time has shown, often selected inappropriate crops.

Modern decision theory has evolved beyond the basic definitions of Knight. The focal point of the departure from Knight's analysis revolves around the nature of the probabilities to be used in decision-making processes. As originally conceived, probabilities were regarded as the relative frequency of events that occur in nature. To determine the probability of an event the observer recorded the outcome of repeated experiments, such as flipping a coin or recording weather variables. *Objective* measures of probability were obtained by increasing the sample size until the desired accuracy was obtained; thus, the mathematical or axiomatic development of probability theory regarded probabilities as the limit values of relative frequences as the number of experimental observations increased (became "infinite"). Estimating such probabilities was viewed primarily as an empirical or data-gathering problem. But choices must often be made in economic or natural settings that are not repeatable. Empirical data required to estimate the long-run relative frequencies are therefore not available. Yet, observation of farm managers suggests that they must and do make satisfactory decisions in such situations. For example, decisions to buy or sell land are uniquely determined by the economic conditions existing at a point in time, and most of us would probably agree that these conditions will never be repeated in history, given the dynamics of technical change, world demand for food, etc. Nonetheless, participants in the land market must formulate appropriate actions regarding land transactions. In such situations, modern decision theorists have argued that decision makers will formulate *subjective* estimates of the probable occurrence of each outcome.[2] These subjective estimates are personal in nature but must be consistent with the axioms of probability. When faced with a risky situation where a decision must be made, the modern decision theorists argue that the manager will use all available information, including historical records, expert opinion, and personal experience, to formulate the subjective probabilities appropriate for the decision-making process. The culmination of this view is perhaps most succinctly stated by Anderson *et al.* (page 18):

> Subjective probability is the only valid concept for decision making, just as (purposes of communication aside) decision making is the only valid use for probabilities. The probabilities used in decision making ideally should be those of the person who bears responsibility for the decision. . . . subjective probabilities cannot be 'right' or 'wrong' although a rational person would presumably wish to refine his probability judgements, eliminating as far as possible any

[2]See Chapter 2 of Anderson et al., the readings by Savage, and the book edited by Kyberg and Smokler listed at the end of this chapter.

biases arising from misconceptions or misinterpretations of the information available to him.

Subjective probabilities are degrees of belief held about future outcomes. They obey the laws of probability, but they do not represent the results of experimentation and can vary among individuals. Thus, two wheat farmers may disagree about the effects of a Russian wheat sale on prices, but if both act in a manner consistent with their beliefs (as expressed by their subjective probability distributions), both can be regarded as rational even though their actions differ.

To summarize, uncertainty can be considered as the lack of information. As the amount of information increases, uncertainty will decrease. Perfect certainty exists when the future outcome of a production process is known. When confronted with a lack of information, managers rely on their subjective evaluations to determine the selection of an action appropriate to the uncertain situations. Because decisions must be made, the subjective probabilities will be determined by the manager. Viewed in this manner, Knight's dichotomy becomes unimportant and writers in modern decision theory often use either of the terms *risk* and *uncertainty* to refer to situations where complete information is lacking. All decision analysis is placed in the context of risk with subjective probabilities.

If lack of information can be measured in degrees and placed on a spectrum with perfect certainty on one end and risk based on subjective probabilities in the middle, is there a boundary opposite perfect certainty which could be regarded as Knight's uncertainty? Undoubtedly there are such phenomena. Catastrophic events do occur, but the farmer may be emotionally unable to evaluate outcomes or place subjective probabilities on them. Such situations are probably not amenable to the logical analyses to be developed in this chapter.[3]

ANALYZING RISKY PRODUCTION PROCESSES

Decision making under certainty, as developed above in Chapters 2 through 7, is a logically straightforward process that leads to the selection of various optima to minimize costs or maximize profits. Although the details may seem technically complex, the principles developed lead to a set of rules that enable the manager to determine

[3]It is difficult to conjure up textbook examples. Death or injury resulting from farm accidents might be an example, but the fact that farmers and their families operate equipment suggests they have assessed the situation and decided to proceed. A somewhat similar instance arises in utility theory designed to lead to the derivation of consumer demand functions. The utility resulting from threshold amounts of necessities, such as food, water and shelter, are not considered within the limits of conventional theory. What price will be paid for survival? Shackle has suggested a theoretical construct that might be deemed to fall between Knight's uncertainty and the modern subjective analysis, but his rather esoteric analysis is too abstract to be pursued here.

the actions that will achieve his goals. In a similar vein, decision making in risky situations is also conceptually straightforward. Because risk is created by the lack of information, decision making for risky situations requires more abstraction of the general decision-making process which will, as a matter of fact, subsume perfect-certainty analysis as a special case.

The explicit recognition of risk as a lack of information creates a new variable input in the decision process: information. The manager must decide how to structure the decision-making process given the lack of complete information and must also decide how to acquire and use additional information. To answer the first question, he must have available a logical process that leads to the optimal decision. To answer the second, he must be able to determine the value of additional information. Because information is usually not costless, the framework to analyze decision making under risk must provide an estimate of the value of information. Ultimately, the marginal cost of information must be compared to its value.

To begin the analysis of decision making under risk, decision problems are divided into the following components: (1) actions available to the manager, (2) the states of nature which could occur, (3) the probabilities the manager attaches to the occurrence of each state, (4) the outcomes or consequences that result from each action under each state of nature, and (5) the objective of the manager, such as maximizing profits. Using these five components, the manager may formulate two additional ones: (6) a decision to seek new information about the probable occurrence of the states of nature (forecasts) and (7) the formulation of an optimal strategy for using these forecasts. The general properties required for each component are as follows:

Actions The manager must be able to select from a set of alternative actions. If he has no alternatives, then he has no decision problem. The determination of actions that could potentially be undertaken in a complex production process is obviously not an easy task and will require considerable skill on behalf of the manager. The set of actions must be defined so that they are mutually exclusive and exhaustive, in the sense that they include all feasible acts. Actions are regarded as discrete and are denoted a_1, a_2, a_3 . . . , etc.[4]

States of Nature The states of nature represent the decision maker's estimate of the events that could occur in the natural or economic world. The states of nature to be considered are those that can affect the production process. Definition of states of nature usually requires judgment about such variables as rainfall, prices, governmental actions regarding farm programs, grain exports to Russia, etc. States of nature are regarded as independent of actions; the manager's choice of an action is assumed to have no effect on the occurrence of the states of nature. States of nature are regarded as discrete and denoted θ_1, θ_2, θ_3, . . . etc.

[4]The analysis presented will be the usual one, which assumes discrete actions, states, etc. Extensions to continuous analysis is possible and can be found in more advanced literature.

Probabilities The subjective probabilities attached to each state of nature are denoted $P(\theta_j)$ where j is the jth state of nature. As noted, these probabilities represent the manager's degree of belief in the occurrence of a particular outcome and can be formulated in a variety of ways. They do have the axiomatic properties ascribed to probability and must sum to one when added over all states. The $P(\theta_j)$ are called *prior* probabilities because they represent the manager's initial degree of belief about the outcomes of the production process. These prior probabilities might eventually be modified by new information. In the theory of decision making under risk, it is interesting to note that while actions, states of nature, and prior probabilities all depend on the extent and accuracy of initial information, only the prior probabilities are regarded as being subject to change after new information is gathered. Actions and states do not change because they are assumed to be complete in the beginning. In fact, the manager would reformulate the total problem in any manner suggested by new knowledge.

Consequences Each action will result in a consequence that depends on the state of nature. In order for a meaningful decision problem to exist, the consequences must vary depending on the action taken and the state of nature that occurs. The outcome or consequence must be measured in some manner, such as profit or, more generally, something called "utility." For most agricultural production problems, profits can be determined by budgeting techniques, assuming the necessary data are available. But dollars may not be the appropriate measure of whatever the farmer is seeking to maximize (translation of money into this other measure, utility, will be discussed later in this chapter). The consequence resulting from taking action a_i when state of nature θ_j occurs will be denoted $U(a_i/\theta_j)$, regardless of whether the outcome is measured in utility or dollars.

Choice Criterion As mentioned in Chapter 3, the farmer must have a basis for selecting among alternatives: a choice indicator or objective function. This choice criterion relates to the manager's objective, that is, what he wishes to maximize. Modern utility theory makes the assumption he wishes to maximize utility, which is related uniquely to dollars and wealth. This assumption will be discussed further below in the section entitled "Utility in Risky Situations."

Experiments and Strategies The above components provide sufficient information to determine expected utility and select an appropriate action. But after evaluating the problem based on the information initially available, the manager may decide to seek additional information. This new data may take a variety of forms, from reading the newspaper and visiting with neighbors to reading research bulletins and utilizing sophisticated computer models, but all are classified under the general heading of "experiments." As a result of this new data, the decision maker may decide to revise his initial estimates of the probabilities of the states of nature. Where the initial probability estimates were called *prior* probabilities, the revised estimates are called *posterior* probabilities. Our view of the decision process for a risky situation is similar to a road map in the sense that all routes must be determined beforehand. Thus, to complete the picture, the manager must determine how to react to each possible bit of new information. If the information comes from a forecast, then he must determine appropriate actions to be taken for each possible forecast. Such a set of actions is called a strategy.

The decision-making process, as it is structured for risky situations, need not result in the same optimal action for all managers. Because the selection of an action depends on the manager's subjective probabilities, the decision process only guarantees that the action selected is consistent with the manager's beliefs. If two managers hold different beliefs about the probable outcomes, they will choose different courses of action and both will be correct a priori. Thus, the good risky decision does not guarantee an ultimately favorable outcome—only that the action selected was rational given the manager's initial knowledge, plus the information added by "experiment." The decision-making process outlined is highly personal; the managers must structure their problem, formulate their beliefs, choose an action, and accept the consequences.

The decision process has been likened to a chain by Anderson *et al.* (page 4), in the sense that the components represent links that interlock and none can be said to be the most important. In our view, managers are continually involved in collecting information and reformulating decision problems. The information available at the end of the last growing season is the information available at the beginning of this growing season. Thus, in the long run, the chain is a circular one.

A Hypothetical Example

Marketing problems are common in agriculture. Consider the following hypothetical example. In January, a farmer must decide whether to sell wheat immediately or hold it until April. He knows prices usually increase throughout the winter months, and in this particular year he believes that the price rise will enable him to make a profit (above storage costs) in April. On the negative side, prices could fall and cause him to suffer a loss if he chooses to wait until April to sell.

To quantify this example, which is admittedly oversimplified but contains the components described above, suppose that the January price is \$3.50 per bushel and the two possible prices in April are \$2.50 or \$5.00. The April prices represent the states of nature. The farmer's subjective belief is that the chances are 3 out of 10 that the price will be \$2.50 and 7 out of 10 that the price will be \$5.00. These are the prior probabilities of the states of nature and are presented in Table 8–2. Using θ_1 to denote an April price of \$2.50, the probability of θ_1 occurring, $P(\theta_1)$, is 0.30. Similarly, defining θ_2 to be an April price of \$5.00, the probability of θ_2 occurring, $P(\theta_2)$, is 0.70. The two probabilities sum to one.

Table 8–2 Prior Probabilities, $P(\theta_j)$, for Wheat Prices

	State of Nature: Price of Wheat in April	
	θ_1 = *\$2.50*	θ_2 = *\$5.00*
Prior Probability	$P(\theta_1) = 0.30$	$P(\theta_2) = 0.70$

The farmer is holding 1000 bushels of wheat. He must choose among two actions: (a_1) sell 1000 bushels in January or (a_2) sell 1000 bushels in April. Depending on the outcome—which will vary with the state of nature that occurs—the farmer will incur a profit or a loss. The possibilities are tabulated in Table 8–3; the result is called a utility function.[5] Referring to Table 8–3, if the farmer sells in January at \$3.50, his revenue is \$3500, and he neither gains nor loses from price variations in April. If he holds and sells in April, he will lose \$1000 if the price drops to \$2.50 and will gain \$1500 if the price increases to \$5.00. Utility resulting when action a_i is taken and state of nature θ_j occurs is denoted $U(a_i,\theta_j)$, where i may take the values 1 or 2 and j the values 1 or 2.

The choice criterion in this case is to maximize expected utility. To compute expected utility from the two actions:

$$\begin{aligned} E(\text{Utility } a_1) &= P(\theta_1)\, U(a_1,\theta_1) + P(\theta_2)\, U(a_1,\theta_2) \\ &= 0.3\ (\$0) + 0.7\ (\$0) \\ &= 0 \\ E(\text{Utility } a_2) &= P(\theta_1)\, U(a_2,\theta_1) + P(\theta_2)\, U(a_2,\theta_2) \\ &= 0.3\ (-\$1000) + 0.7\ (\$1500) \\ &= \$750 \end{aligned}$$

Selling 1000 bushels in April, a_2, maximizes expected utility and is the optimal action, given the farmer's subjective prior probabilities.

The decision problem can be analyzed to determine its sensitivity to the prior probabilities. By setting the $E(\text{Utility } a_1) = E(\text{Utility } a_2)$, the prior probabilities that will make the manager indifferent between a_1 and a_2 can be determined. Solving the expression

$$\$0 \cdot P(\theta_1) + \$0 \cdot P(\theta_2) = -\$1000\ P(\theta_1) + \$1500\ P(\theta_2)$$

will determine the odds or probabilities that will return equal expected utility from each action; they are $P(\theta_1) = 3/5$ and $P(\theta_2) = 2/5$. Because

Table 8–3 Utility Function, $U(a_i,\theta_j)$, For Wheat Problem

	State of Nature: Price of Wheat in April	
Possible Actions	$\theta_1 = \$2.50$	$\theta_2 = \$5.00$
a_1 Sell in January	0	0
a_2 Sell in April	−\$1000	\$1500

[5]For this example, the dollar amount of gain or loss will be used to represent utility, assuming the decision maker is risk-neutral. This may not be true in general, as will be explained in the next section of this chapter.

these "break-even" probabilities represent odds of 3 to 2 while the farmer evaluated the odds to be 3 to 7, the farmer would have to be willing to revise substantially upward his probability of a low April price before he would change his action. Therefore, new information about the price of wheat in April would have to (1) lead to probabilities different from the farmer's priors and, in addition, (2) be quite accurate to cause the farmer to change his prior action, a_2.

As an initial step, the farmer might like to know the value of a perfect forecast. A perfect forecast would enable him to anticipate the state of nature and always take the correct actions by (a_1) selling in January when the April price is \$2.50 and (2) selling in April when the price is \$5.00. Because he expects these events to happen with the frequencies 3/10 and 7/10, the most he can expect to earn over time from this speculative activity is

$$0.3(\$0) + 0.7(\$1500) = \$1050$$

The optimal action based on the prior probabilities returned \$750. Therefore, the expected value of the perfect forecast is \$1050 − \$750 = \$300. The farmer could pay up to \$300 for additional information.

Our technique permits the manager to include new information in his decision process. The farmer selects a_2 because he believes the probability of a price of \$2.50 in April is low, 0.3 or 30 percent. If he were to revise this estimate above 60 percent, he would choose a_1. At this juncture in the process he may decide to seek new information and revise his subjective prior probabilities. As noted above, this new evidence is usually in the form of a forecast or other timely experimental data. When analyzing weather problems, the forecast might be issued by the local weather service.

For our price forecasting problem, assume that the farmer's daughter is an agricultural economist working at a land grant university. The daughter is home for New Year's Day and discusses the price outlook with her father. She has a model to forecast prices for April, but the problem is that the model doesn't forecast perfectly. Experience has shown that when the price will be \$5.00, the model's forecast is correct 80 percent of the time and wrong 20 percent of the time, and when the price will be \$2.50, the forecast is correct 90 percent of the time and wrong 10 percent. These probabilities, conditioned on the state of nature and called the likelihood function, are summarized in Table 8–4.

The probabilities for the likelihood function are expressed using the notation for conditional probabilities. A forecast of \$2.50 for the April wheat price is denoted F_1, and a forecast of \$5.00 is denoted F_2. If the April price is eventually determined to be \$2.50, the probability that the model forecast \$2.50 back in January is given by the conditional

Table 8–4 The Likelihood Function, $P(F_j|\theta_i)$, for the Wheat Price Problem

April Price Forecast	*Conditional Probabilities*	
	State of Nature: Price of Wheat in April	
	θ_1 = *\$2.50*	θ_2 = *\$5.00*
F_1 = \$2.50	$P(F_1\|\theta_1) = 0.90$	$P(F_1\|\theta_2) = 0.20$
F_2 = \$5.00	$P(F_2\|\theta_1) = 0.10$	$P(F_2\|\theta_2) = 0.80$
Sum	1.00	1.00

probability $P(F_1|\theta_1) = 0.90$, and the probability that the model forecast \$5.00 in January is given by the conditional probability $P(F_2|\theta_1) = 0.10$. The two probabilities sum to one because a forecast of either \$2.50 or \$5.00 will always be issued. Similar interpretations can be made for the probabilities contained in Table 8–4 for θ_2 = \$5.00. Notice that the model's forecast of the higher price is less accurate; forecasts are not required to be equally accurate for each state of nature.

The farmer would like to use the forecast of the agricultural economist because its accuracy is good in both cases, but he is loath to abandon his subjective prior information based on years of experience. $P(\theta_1)$ and $P(\theta_2)$ represent information to the farmer; rather than discard it, he would prefer to temper it with the new information provided by the forecast. The daughter suggests he combine the two sets of probabilities using Bayes' theorem.

The result is obtained intuitively as follows: The question the farmer must ask is, "Of all the times that an April price of \$2.50 is forecast, what will be the actual frequency of its occurrence?" The decision maker needs to have the outcomes conditioned on forecasts (rather than forecasts conditioned on outcomes). $P(\theta_1)$, $P(\theta_2)$, $P(F_1|\theta_1)$, and $P(F_2|\theta_2)$ are known, but the farmer needs to know $P(\theta_1|F_1)$, $P(\theta_2|F_1)$, $P(\theta_1|F_2)$, and $P(\theta_1|F_2)$.

Consider a forecast of F_1, a price of \$2.50. The probabilities that this forecast will occur jointly with the states of nature, θ_1 and θ_2, can be determined by using probability formulas:

$$P(F_1,\theta_1) = P(\theta_1)P(F_1|\theta_1) = 0.3 \cdot 0.9 = 0.27$$
$$P(F_1,\theta_2) = P(\theta_2)P(F_1|\theta_2) = 0.7 \cdot 0.2 = 0.14$$

These are the usual probability formulas given in any elementary statistics text; intuitively, they weight the farmer's subjective priors by the forecast accuracy. By addition, the marginal probability of F_1, $P(F_1)$ is

$$P(F_1) = P(F_1,\theta_1) + P(F_1,\theta_2) = 0.41$$

Now, after an April price of \$2.50 is forecast, one of the two states of nature, either (F_1,θ_1) or (F_1,θ_2) must actually occur. The April price will either be \$2.50 or \$5.00. To help the farmer make his decision, we must find $P(\theta_1|F_1)$ and $P(\theta_2|F_2)$. But these conditional probabilities can be computed from $P(F_1,\theta_1)$, $P(F_1,\theta_2)$, and $P(F_1)$ as follows:

$$P(\theta_1|F_1) = \frac{P(F_1,\theta_1)}{P(F_1)} = \frac{P(\theta_1)P(F_1|\theta_1)}{P(F_1)} = \frac{0.27}{0.41} = 0.66$$

$$P(\theta_2|F_1) = \frac{P(F_1,\theta_2)}{P(F_1)} = \frac{P(\theta_2)P(F_1|\theta_2)}{P(F_1)} = \frac{0.14}{0.41} = 0.34$$

These conditional probabilities, called posterior probabilities, are summarized in Table 8–5. Note in Table 8–5 the probabilities sum across (for forecasts) rather than down (for prices or states of nature) as in Table 8–4.

The Bayesian technique for computing posterior probabilities can be illustrated using a diagram. The states of nature (April prices) and new information (price forecasts) can be regarded as having a bivariate probability distribution. This is shown in Figure 8–1 where the area of the rectangle is assumed equal to one and divided into four areas representing each of the four outcomes. The farmer's prior probabilities divide the probability space vertically into two parts, as shown by the heavy lines around the rectangles representing $P(\theta_1) = 0.3$ and $P(\theta_2) = 0.7$. The forecasts subdivide each of these probability spaces into two smaller areas. If the forecast is \$2.50, then the probability space is restricted to the shaded area. By examining those areas, it can be seen that the appropriate posteriors are $P(\theta_1|F_1) = 0.27/0.41 = 0.66$ and $P(\theta_2|F_1) = 0.14/0.41 = 0.34$. If the forecast is \$5.00, then the unshaded area applies and the posteriors are $P(\theta_1|F_2) = .03/.59 = 0.05$ and $P(\theta_2|F_2) = 0.56/0.59 = 0.95$.

The farmer now has two sets of posterior probabilities. One for a forecast of \$2.50 and the other for a forecast of \$5.00. Depending on which forecast is made, one of these sets would be used to compute

Table 8–5 Posterior Probabilities, $P(\theta_j|F_i)$, for the Wheat Problem

April Price Forecast	*Conditional Probabilities* — *State of Nature: Price of Wheat in April* θ_1 = *\$2.50*	θ_2 = *\$5.00*	*Sum*
F_1 = \$2.50	$P(\theta_1\|F_1) = 0.66$	$P(\theta_2\|F_1) = 0.34$	1.00
F_2 = \$5.00	$P(\theta_1\|F_2) = 0.05$	$P(\theta_2\|F_2) = 0.95$	1.00

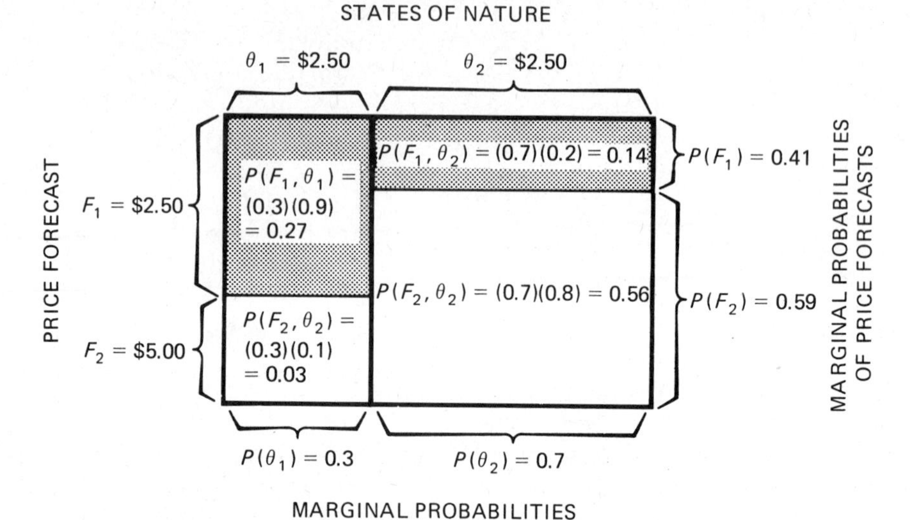

Figure 8–1. A graphic presentation of the determination of posterior probabilities.

expected profits and thereby lead to the selection of the appropriate action.

We now return to comparing expected profits. The results, using the subjective priors and the two sets of posteriors, are presented in Table 8–6. If the farmer were to use his subjective priors, he would choose action a_2 and expect to earn \$750. If he chooses to utilize the forecasts and the model forecasts an April price of \$2.50, then expected utility from each action would be

$$E(\text{Utility } a_1) = 0.66(0) + 0.34(0) = 0$$
$$E(\text{Utility } a_2) = 0.66(-\$1000) + 0.34(\$1500) = -\$150$$

He would choose a_1, sell in January, and earn no profit from speculation. When the model forecasts an April price of \$5.00, then expected

Table 8–6 Expected Profits From Two Actions For Wheat Problems

	Expected Profits		
Possible Actions	*Using Subjective Priors*	*Using Posteriors for Forecast of \$2.50*	*Using Posteriors for Forecasts of \$5.00*
a_1 Sell in January	\$ 0	\$ 0	\$ 0
a_2 Sell in April	\$750	−\$150	\$1375

utility would be

$$E(\text{Utility } a_1) = 0.05(\$0) + 0.95(\$0) = 0$$
$$E(\text{Utility } a_2) = 0.05(-\$1000) = 0.95(\$1500) = \$1375$$

and he would choose a_2 with an expected earning of $1375. The farmer's *strategy* is to choose a_1 when F_1 is forecast and a_2 when F_2 is forecast.

Why does the forecast cause the farmer to select forecast strategies that differ from the prior optimal action, a_2? As mentioned above, the key is the magnitude of the farmer's prior probabilities relative to the forecast accuracy. The farmer is optimistic about the chances of a $5.00 price in April, and thus will sell in January only when his optimism is offset by an accurate forecast. As examples, two cases will be presented.

First, suppose the farmer is not optimistic—reduce his priors to 0.50 for each outcome. (This is the Laplace rule of ignorance: When empirical information is completely lacking, assign equal prior probabilities to outcomes.) Then the above analysis will determine that the farmer should sell in January when the price forecast for April is $2.50 and hold to sell in April when the price forecast is $5.00 any time $P(F_1|\theta_1) \geqq 0.50$ and $P(F_2|\theta_2) \geqq 0.50$. In this case, the accuracy of the forecast outweighs the priors, which now reflect indifference.

Second, suppose the farmer's prior probabilities remain at 0.3 for an April price of $2.50 and 0.7 for an April price of $5.00. Then, the farmer will choose a_2 and always hold for sale in April whenever $P(F_1|\theta_1) < 0.70$. Because the farmer is optimistic, the forecast must be sufficiently accurate to affect his optimism. When the forecasts are perfect, the farmer should ignore his priors and base his actions only on the forecasts.

The strategies selected as a result of reevaluating the prior probabilities using Bayes' theorem are called *Bayesian strategies*. We might note that the manager could use other methods of revising prior information, but the Bayesian method does provide a workable scheme that combines both new and old information without discarding either.

UTILITY IN RISKY SITUATIONS

When describing the overall structure of the abstract decision process, the consequences of the manager's actions were measured in terms of the utility resulting from those actions. We noted that utility might be represented by the monetary outcome and, in fact, used money as the measure of outcome in the forecasting example just presented. In general, the monetary outcome of an action will not be the same as the utility associated with the outcome of the action; the two are the same only in one special instance. In this section, a brief introduction to the concept of utility as developed for risky decision processes will be presented. In the terminology developed in Chapter 2, we are seeking a more general "choice indicator" to replace money, and this choice indicator is called, for lack of a better name, "utility."

The problem inherent with using dollars of expected income as an indicator of utility is often demonstrated by presenting a table of consequences of the following type:

States	*Probabilities*	*Actions*	
θ_j	$P(\theta_j)$	a_1	a_2
θ_1	1/2	$1000	−$20,000
θ_2	1/2	$1000	+$22,000
Expected Earnings		$1000	$ 1000

Both actions in this example return the same expected earnings, $1000. If the choice criterion were to maximize expected earnings, the manager would be indifferent between the two actions. But if the manager chooses a_2, then he will be incurring a loss of $20,000 half of the time. To be indifferent to a_1 and a_2, the manager must be indifferent between large fluctuations in earnings and constant earnings. He must not be concerned about the possible frequent loss of $20,000. Thus, the probability of loss and amount of loss are masked by the use of expected earnings. The use of utility rather than dollar earnings permits this additional factor to be included in the decision process. To continue with this simple example, if the manager prefers a_1 to a_2 (or a_2 to a_1) in the above example, then this preference suggests the existence of a "return" other than expected money outcome. This return is called "utility" in the literature on decision making. The decision maker is regarded as having a "utility function" which assigns utility to each consequence of his actions.

If the utility function of the manager is to be useful, it should assign utilities in such a way that the rules of decision making ensure that the manager is acting according to his wishes. If the decision maker maximizes expected utility, then he should be able to review the process after the outcome and be satisfied that he made the correct choice, even if the outcome was not to his liking. To express it another way, maximizing expected utility should lead to the choice of actions consistent with preferences. How do we know such a function exists and, if it does, what does it look like?

The logical foundation required to verify the existence of a utility function is beyond the scope of this text. It is developed in many places in the literature—including Anderson *et al.*, Hey, and Halter and Dean, as well as in the more advanced sources cited by them. Starting from a basic set of axioms, a utility function can be constructed to reflect the preferences of the individual in such a way that if the utility of one action exceeds the utility of a second, the manager prefers the first action to the second. Such a utility function is called a *von Neumann–Morgenstern utility function*, after the writers who popularized the concept in a 1947 book entitled *The Theory of Games and Economic Behavior*. The von Neumann–Morgenstern utility function is derived logically from their axioms. If, upon study, those axioms are deemed reasonable, then it follows that the utility function exists; if the axioms cannot be accepted, then the existence of the utility functions described below can be questioned.

Given the existence of the utility function, what does it look like? In practice, utility is often expressed as a function of wealth; more general cases are possible but cannot be developed here. The utility of a given action is then considered in context of the resulting increase or decrease in wealth. Three possible utility functions are presented in Figure 8–2. In each case, utility, U, is regarded as a function of wealth, W, measured around some initial level of wealth, W_1. The functions drawn are regarded as segments of a much larger utility

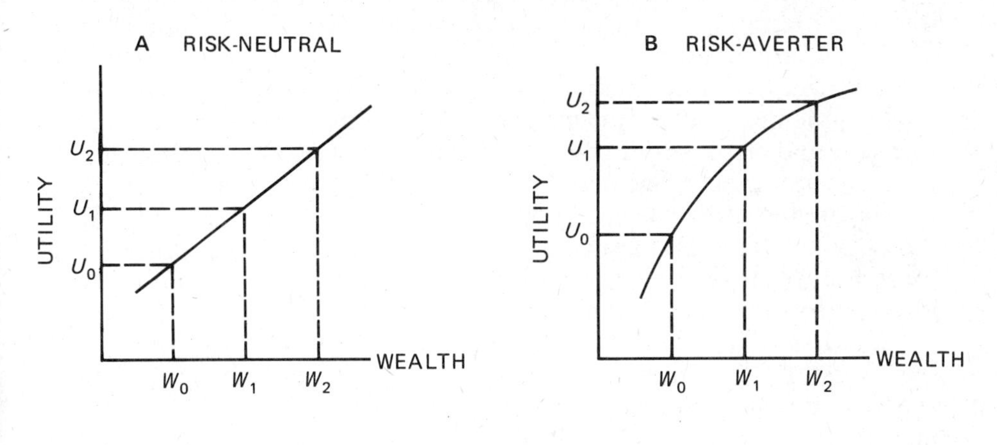

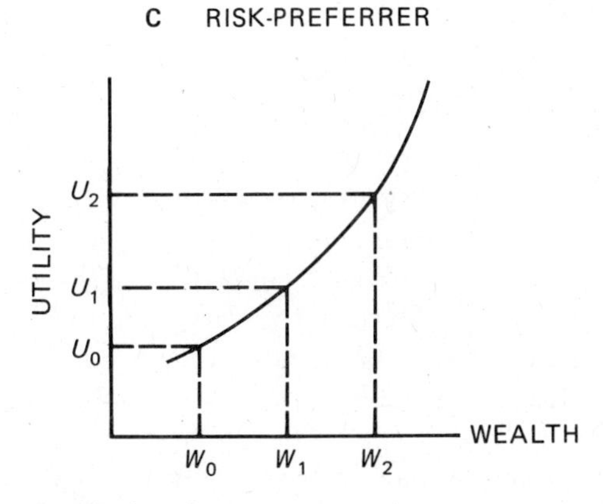

Figure 8–2. Examples of utility functions.

function which might appear quite different if it could be seen in its entirety.[6] The three functions presented represent three cases typical of a manager's response to a given situation.

In Figure 8–2A, the manager's utility function is a linear function of wealth. Each added dollar of wealth has the same marginal utility to him. Put another way, the first derivative of his utility function with respect to wealth is constant and the second derivative is zero. In this case the manager is said to be risk-neutral.

In Figure 8–2B, the manager has a concave utility function. Each added dollar of wealth adds less utility so that marginal utility is positive but decreasing. In this case the first derivative is positive while the second derivative is negative. Added wealth, the increment from W_1 to W_2, increases utility less than an equal decrease in wealth, from W_1 to W_0, would reduce utility. Therefore, losses reduce utility relatively more than gains and the manager is said to be risk-averse.

Finally, in Figure 8–2C, the manager's utility function is convex. As wealth increases, the marginal utility of money also increases. Both the first and second derivatives of the utility function are positive. Because additions to wealth increase utility more than reductions in wealth decrease utility, the manager is said to be a risk-preferrer.

As a reminder, note that the examples presented represent only small segments of utility functions and only include one variable, wealth. Decision makers often must take actions on decisions affecting different areas of business and personal life. At any particular time, a person may be buying life insurance, betting on the Super Bowl, playing high-stakes poker, taking vitamin pills, and driving on freeways in a small foreign import. One elementary analysis cannot reconcile all the differences found in production and consumption habits.

It is useful to consider further why the linear, concave, and convex segments of the utility function correspond to the risk-neutral, risk-averse and risk-preferrer categories. Consider the example presented in Figure 8–3. Given $1000 of wealth, all individuals have a resulting utility level of 10; at $2000, all individuals have a utility level of 20. At a wealth level exactly halfway between these two extremes, $1500, the risk-averter has a resulting utility of 17.5, the risk-neutral person has a utility level of 15, and the risk-preferrer has a utility level of 12.5. Suppose there are two actions and two equally likely states of nature. Action one, a_1, returns $1500 for each state of nature while action two, a_2, returns $1000 for the first state of nature and $2000 for the second.

Because a_1 always returns $1500, that action's utility for each person can be read directly from the graph and will be 17.5, 15.0, and 12.5 for the risk-averter, risk-neutral, and risk-preferrer, respectively. Because this is the certain event, we need only to compare these utilities

[6]For an extension, see the suggested readings by Friedman and Savage.

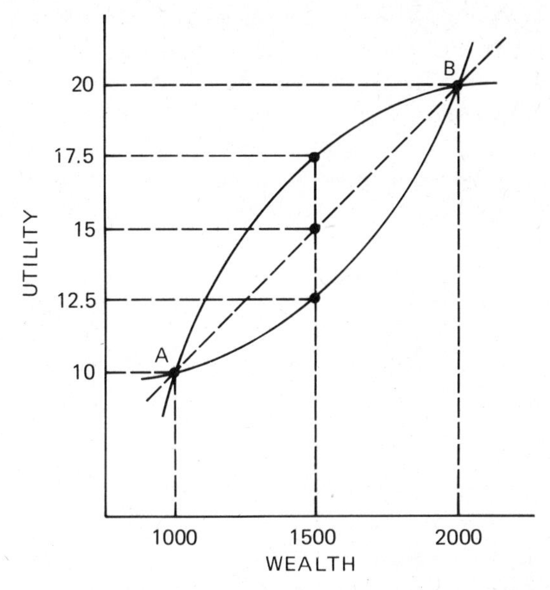

Figure 8–3. A comparison of attitudes toward risk.

to those resulting from the risky event, which, in each case, will be given by

$$
\begin{aligned}
E(U) &= P(\theta_1)U(a_2,\theta_1) + P(\theta_2)U(a_2,\theta_2) \\
&= \frac{1}{2}\,[U(\$1000)] + \frac{1}{2}\,[U(\$2000)] \\
&= \frac{1}{2}\,(10) + \frac{1}{2}\,(20) = 15
\end{aligned}
$$

which is 15 for each person. But the expected wealth in this example is \$1500; furthermore, the expected wealth and expected utility from a_2 falls on a line segment connecting the endpoints A and B. This is true in general. When the probability of state one is one ($P(\theta_1) = 1$), the expected utility and wealth are \$1000 and 10, respectively, located at point A; as the probability of state one decreases to zero, expected utility moves along the linear segment to point B.[7]

Thus, the expected utilities computed for the risky problem fall exactly on the utility function of the risk-neutral person; he is indifferent between the gamble and the certain event. The expected utilities fall below the utility function of the risk-averter; he prefers to have the certain return of \$1500 than the same wealth (\$1500) when he

[7]Let α be the probability of the first state of nature. Then, $E(U) = 10\alpha + (1 - \alpha)\,20 = 20 - 10\alpha$, which is linear and will be 10 when $\alpha = 1$ and 20 when $\alpha = 0$. The same is true of expected wealth.

must take risks to get it. Finally, the utility function of the risk-preferrer drops below the straight line representing expected utility.

The example in Figure 8–3 is also shown in Table 8–7. Here, expected utilities are computed in the usual fashion. The risk-neutral person would have the same expected utility from either action; the risk-averter maximizes expected utility from choosing the riskless action a_1; while the risk-preferrer maximizes expected utility by choosing a_2. For him, risk is a good to be consumed.

The utilities of the three cannot be compared. It cannot be said that the risk-averter gets more utility from a_1 than either of the other managers. Her utility, 17.5, serves only as a choice indicator for her and does not suggest that she is in any way happier than the others, who have expected utilities of 15 and 12.5. The utility indexes serve only as an ordering scheme for a single decision maker; the magnitude of the values assigned to utility is not important as long as the ordering of preferences is not changed. If U represents a manager's utility function, then $U^* = c + dU$ will also, where $d > 0$. Obviously, an infinite number of such functions exist.

The acceptance of the existence of risk aversion poses some interesting new problems. How is the degree of risk aversion to be compared among managers and, if differences exist, what measures can be provided for them? What does it mean to say one production process is riskier than another? Again, detailed answers are beyond the scope of this discussion; some general directions can be suggested.

The concavity of the utility function provides a reasonable starting point. From above, it was seen that if the second derivative of the utility is positive, the manager is a risk-preferrer, when the second derivative is zero, the manager is risk-neutral, and when the second derivative is negative, the manager is a risk-averter. But second derivatives do not provide a useful means for comparing risk aversion among managers because the utility functions are not unique. By using the linear transformation shown in the last paragraph, the magnitude of the second derivative of the utility function can be changed while leaving the

Table 8–7 Effect of Risk Attitudes on Utility

Probabilities		*Wealth for Each Action and State of Nature*		*Utility From Each Action*					
				Risk-Averse		*Risk-Neutral*		*Risk-Preferrer*	
$P(\theta_j)$	θ_j	a_1	a_2	$U(a_1)$	$U(a_2)$	$U(a_1)$	$U(a_2)$	$U(a_1)$	$U(a_2)$
1/2	θ_1	$1500	$1000	17.5	10	15	10	12.5	10
1/2	θ_2	1500	2000	17.5	20	15	20	12.5	20
Expected Utility				17.5	15	15	15	12.5	15

ordering of preferences unchanged. This problem has been resolved by using a ratio of derivatives of the utility function, $(-U''/U')$. This measure, called the Pratt–Arrow measure of absolute risk aversion, is unaffected by linear transformations of the utility function.

A second measure of risk aversion is called the risk premium. The risk premium is defined as the amount of expected wealth required to make the manager indifferent between a risky outcome and a certain outcome. Determination of the risk premium for a risk-averse manager is shown in Figure 8–4. For the situation illustrated, the manager is considering a risky action that returns $1000 or $2000, each with probability 1/2. The expected wealth from this action is $1500 and the expected utility is 30 units [30 = 0.5(20) + 0.5(40)]. But, as explained above, the expected wealth and expected utilities for the action always lie on a line between *A* and *B*, falling on *A* when the probability of $1000 is one and moving towards *B* as the probability of $2000 increases. Thus, the expected utilities from the risky situation always lie below the utility function. In this case, expected utility is 30; the amount of certain wealth that would result in 30 units of utility for the manager is called the "certainty equivalent" of the risky process. This can be determined using the utility function. A horizontal line 30 units above the wealth axis intersects the utility function at *C*, where wealth is $1250. The manager receives the same utility from $1250 certain dollars or $1500 expected dollars; $1,500 − $1,250 = $250 is the manager's risk premium for this decision.

When the risk premium is positive, zero, or negative, the manager is risk-averse, risk-neutral, or a risk-preferrer, in that order. As the utility function becomes more concave, the manager becomes more risk-averse. Note also that whenever wealth is used directly as a choice

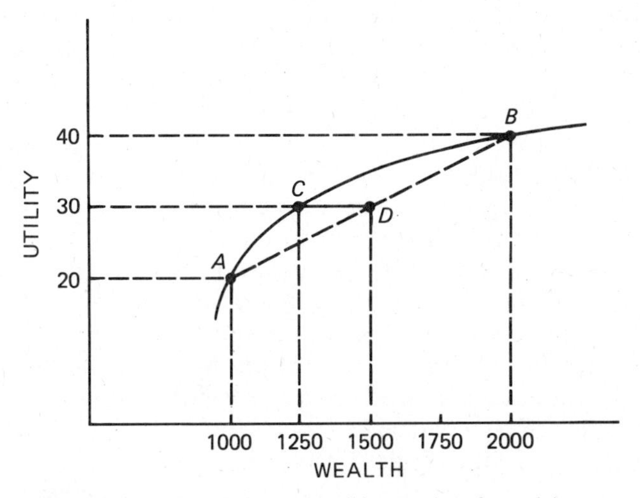

Figure 8–4. Determining the risk premium for a risk-averse manager.

criterion, the underlying assumption necessarily is that the manager is neutral to risk. This criterion was used throughout the early chapters of this text; we now see it as a special case of a more general theory.

Both the measures of risk discussed are defined only for one point or the immediate neighborhood of one point on the utility function. Such measures are called "local" rather than "global." Thus, while risk aversion among managers might be compared for a particular decision, no general statements can be made on the basis of those measures. Further, the risk premium depends on the concavity of the utility function and the probabilities attached to each outcome. Because the probabilities are subjective, if the manager places a probability of one on either outcome, then his risk premium is zero. It will be a maximum for those probabilities that maximize the distance between C and D in Figure 8–4.

When does one probability distribution imply more risk than a second? The question is not easily answered; interested readers should pursue the subject in our suggested readings. In practical applications, researchers have tended to use the variance of a distribution as a direct measure of risk, assuming that increasing variance implies increasing risk. While the use of the variance for this purpose has been subjected to increasing criticism and is, in fact, largely discredited by recent theoretical work, it remains a common measure of risk in applied research.

The Quadratic Utility Function

The existence of utility functions that imply the maximization of something other than income would probably not be disputed by many researchers in agricultural economics. A significant area of research has been the determination of utility functions of managers. But, in practice, applied researchers often use a formulation called the "quadratic" utility function. The quadratic equation is tractable computationally and lends itself nicely to empirical studies. The quadratic formulation is appropriate, of course, when the manager does in fact have a quadratic utility function (which can be an assumption) or when the risky event being considered has associated with it a normal distribution, which permits the distribution to be described by its mean and variance. The assumption of a quadratic utility function may not be inappropriate for practical work when the decision involves reasonably moderate changes in wealth and, consequently, small movements along the utility function. For many natural phenomena, the normal distribution is adequate; it might often apply in agriculture because it was originally developed to describe the chance outcomes of biological events.

The quadratic utility function has the form:

$$U(x) = x + bx^2$$

where x can be interpreted to be the variable of concern, such as wealth. [Any linear function of $U(x)$ would serve as well, such as $U^*(x) = c + dU = c + d(X + bx^2)$ where $d > 0$.] The sign of b is determined by the risk taker's attitude toward risk. For the three cases, risk aversion implies b is negative; risk-neutral implies b is zero; and risk-preferrer implies b is positive. When used in practice, $U(x)$ is assumed to have a positive slope.

When x is a risky outcome with a normal probability distribution, $U(x)$ can be expressed as a function of the expected values of x and x^2. Thus,

$$U(x) = E(x) + bE(x^2)$$

but $E(x^2) = \sigma^2 + (E(x))^2$ so that the utility of the expected value of the risky outcome is

$$U(x) = E(x) + b\{\sigma^2 + [E(x)]^2\}$$
$$U(x) = E(x) + b[E(x)]^2 + b\sigma^2$$

The convenience of such a utility formulation is clear; utility becomes a function of the first and second moments about the mean of the probability distribution. To apply the quadratic utility function in practice, researchers need to determine estimates of the mean and variance for the risky outcome being considered, with b adjusted to reflect the manager's preference for risk.

When the notation used above is simplified somewhat by letting $U(x) = U$, $E(x) = E$, and $\sigma^2 = V$, then the quadratic utility function can be written

$$U = E + bE^2 + bV$$

which can be recognized as the three-dimensional surface expressing $U = f(E, V)$, exactly as those developed in Chapter 4. By setting U equal to a fixed value, U_0, an isoutility curve can be derived which would have the same interpretation as an isoquant. The typical isoutility curve is derived by starting with

$$U_0 = E + bE^2 + bV$$

and taking the easy solution

$$V = \frac{1}{b}U_0 - \frac{1}{b}E - E^2$$

By regarding U_0 as a parameter to be varied, this equation defines a family of isoutility contours, any one of which traces the combinations of E and V which result in the same level of utility, U_0, to the manager. This is called (E, V) analysis.

Although the algebraic expression for isoutility curves are most easily derived expressing V as a function of E, graphical analysis of the problem has traditionally placed V or $\sqrt{V}$ on the horizontal axis. To find the derivative of E with respect to V, note that

$$\frac{dE}{dV} = -\frac{\partial U/\partial V}{\partial U/\partial E} = -\frac{b}{1 + 2bE}$$

where the denominator $(1 + 2bE)$ must be positive in order for the quadratic utility function to be appropriate. The slope of the (E, V) isoutility curve will therefore depend on the sign of b. For a risk-averter, b will be negative, dE/dV will be positive, and the isoutility curves will slope upward. For a risk-preferrer, b will be positive and the (E, V) isoutility curves will slope downward. Examination of the second derivatives suggests they are concave up and concave down, respectively. For the risk-neutral manager, b is zero and the isoutility curves are horizontal lines ($b = 0$ implies $U = E$). The risk-neutral person is indifferent to the variance of returns.

The three situations described are sketched in Figure 8–5. When V is zero, the isoutility curves intersect the E axis; these intercepts represent the expected certain return that is equivalent to all the (E, V) combinations on that isoutility curve. Because more E is preferred to less, high isoutility curves (measured vertically) are preferred to lower curves. The slope of the isoutility curves suggests the degree of aversion or attraction for risk. The steeper the slope, the stronger is the feeling.

COMPARISON OF TRADITIONAL AND MODERN ANALYSES

Modern decision theory has evolved beyond the definitions of Knight. Nevertheless, the early literature of agricultural economics has been greatly enriched by applications of the traditional analysis of Knight. In this section, the classic approach resulting from Knight's definitions will be compared to the modern decision theory.

Risk

The analysis of risk in the classic case is similar to that used by modern theorists. Now, subjective probabilities are substituted for "objective" probabilities. In both cases, expected returns are computed and the manager selects the action that maximizes expected returns. Modern theory substitutes the von Neumann–Morgenstern definition of utility

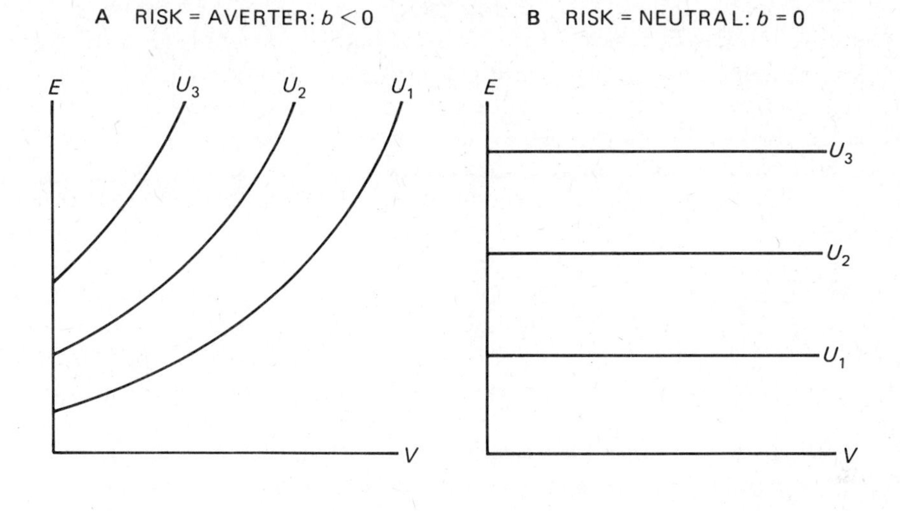

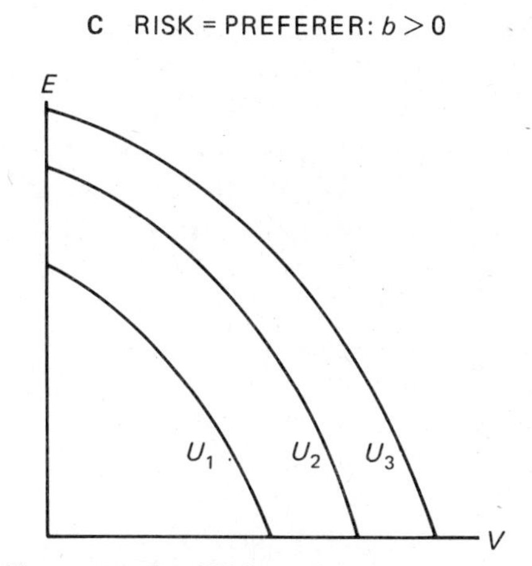

Figure 8–5. Isoutility curves for (E,V) analysis.

for the traditional measure of success, money, but profit remains an important choice indicator in much applied research in agricultural economics.

In addition, modern theory has clarified the role of information as a factor in reducing risk (Chapter 7 in Hey is useful reading on this topic). As a result, the value of new information can be determined and compared to the cost of that information. Using Bayes' theorem, the new information is introduced into the decision process in a method that is analytically rigorous. But this extension can be applied as well to Knight's traditional risk analysis. "Objective" information need only be substituted for "subjective" information.

Uncertainty

The differences between modern and traditional analyses are more clearly defined when approaching the problem Knight defined as uncertainty: when the manager does not know the probabilities associated with the various outcomes of the production process.

To combat uncertainty as defined by Knight, a number of well-known courses of action were recommended to farmers. These actions, which might better be called strategies, include crop and livestock diversification; insurance against hail, fire, and crop failure; contracts for specified price levels, such as forward contracts for the sale of a crop or cash rent for the landlord; and flexibility when planning farm enterprises and farm investments (see Heady, Chapters 9 and 15–18). These were often applied to both "risk" and "uncertainty" situations and often recommended without careful study of the manager's goals.

The strategies just listed might be regarded as informal. In addition, a number of more formal strategies were suggested for decision making in uncertain situations. These strategies have been identified in the literature by rather eclectic titles, such as The Maximin Rule, The Minimax Rule, The Minimax Regret Rule, and The Principle of Insufficient Reason. All purport to instruct the manager how to act in the absence of knowledge of the probability distributions of the random events. Each contains implicit assumptions concerning the manager's utility function and, hence, risk aversion. Hey, in his Chapter 5, presents a numerical example for which each of the above rules leads to a different course of action.

Modern decision theorists argue that the above rules are in fact equivalent to similar problems using subjective probability distributions. Halter and Dean (pp. 92–93) show this equivalence for each of the above rules and state the case for subjective probability:

> Each criteria has been shown to be equivalent to expressing a subjective probability distribution over the states of nature. Thus, if we admit subjective probability then we need only consider the criteria of maximizing expected value. Furthermore, we believe that no decision maker is so ignorant of his domain of interest that he cannot propose a subjective probability distribution.

EMPIRICAL RESEARCH ON RISK SITUATIONS

A rich and diverse literature has been created in the area of applied decision theory. Although other classifications are possible, existing research could be divided into four areas: (1) verifying that farmers do maximize utility rather than money income and investigating the properties of utility functions; (2) applying the theory to problems of resource allocation within individual enterprises: (3) using decision theory to evaluate the usefulness of forecasts; and (4) determining farming systems that are compatible with the manager's utility preferences. To end this chapter, we will briefly sketch some applications in the four areas and then present two research examples. The first example will be a rather traditional analysis of an enterprise, fertilizing a crop in a risky situation, and the second will be a more modern example of whole farm planning.

A Short Summary of the Literature

Derivation of Utility Functions. Halter and Dean, in their appendix to Chapter 3, discuss methods of determining utility functions for decision makers using questionnaires and interview techniques. Managers are asked to respond to hypothetical choices, and the responses are analyzed to generate utility functions. Anderson *et al.*, Chapter 4, present some practical methods of determining preferences in direct interview. In 1968, Officer and Halter compared three methods of utility estimation and concluded (p. 275) that ". . . utility analysis can be more successfully used than conventional economic analysis in making risky decisions." In 1974, Lin, Dean, and Moore studied a sample of California farms to determine whether farmers were profit maximizers or utility maximizers. They concluded that utility-maximization models explained farmers' behavior more accurately than profit-maximizing models. But none of the models they used predicted actual behavior well. A comprehensive review of research efforts to determine utility functions and risk aversion is presented by Young.

Resource Allocation Within an Enterprise. Effects of risk on resource allocation is outlined in Chapter 6 of Anderson *et al.* and Chapters 17 to 23 of Hey. Because even simple examples quickly become complex, we will only outline the general approach. Using notation from Chapter 3, the profit from an enterprise can be written

$$Profit = P_Y Y - P_X X - TFC$$

where X is the variable input and Y is the resulting yield, P_Y and P_X are their unit prices, and TFC represents total fixed costs. This traditional profit equation is converted to the risky situation by assuming that one or more of the variables (P_Y, P_X, Y, X) are random. (If X is random, then Y, which is a function of X, will also be random.) Utility is introduced by defining it as a function of profit; thus, Utility = U(Profit). The manager is assumed to maximize expected utility.

In the most elementary cases, P_X is assumed fixed and X can be controlled by the manager. Output, Y, is a random variable with a price, P_Y, that is also random. Even this simple example is complex—Anderson *et al.* show the necessary mathematics. However, the conclusion is straightforward and we will use the interpretation of Young (p. 1066). Maximizing utility for the above situation results in the expression

$$E(VMP_X) = P_X + RI$$

where $E(VMP_X)$ represents the expected value of the marginal product. Because output price and output response are random, this expected value can be interpreted intuitively as a long-run return. This expression equates the expected value of the marginal return to input price, P_X, plus or minus a risk factor. R represents the manager's risk aversion for this enterprise and I represents the marginal contribution to risk of an additional unit of input. I is usually positive, so the effect on input use will depend on R, which is termed the "risk coefficient." When the manager is risk-neutral, $R = 0$ and the manager will use the (long-run) profit-maximizing amount of input. When she is risk-averse, $R > 0$ and the amount of input used will be less than the optimum. This result is not surprising.

Analyses similar to the above example can be conducted for any farm enterprise. Just and Pope have studied a related problem for production functions and present a useful reference list. Young also contains a review of recent work in the area.

Estimating the Value of Forecasts. The future is intriguing to the manager and of concern to the researcher. Because of the unique manner forecasts are integrated into the decision process, decision theory provides a useful framework for analyzing the usefulness of forecasts. The accuracy of the forecast can be varied by changing the probabilities in the likelihood function (Table 8–4), and the value of the forecast can be determined for any given set of prior probabilities held by the manager.

A significant amount of the literature in agricultural economics has dealt with applications of decision theory to forecasting. Bullock and Logan applied it to cattle marketing. Byerlee and Anderson to long-run weather forecasts and fertilizer strategies, Carlson to crop disease control, and Eidman, Dean, and Carter to price forecasting for turkey production. Baquet, Halter, and Conklin estimated the value of frost forecasts. An important feature of the decision model, when used for the evaluation of forecast problems, is that the forecast itself need not be technically feasible. The value of the forecast to the manager can still be determined. Doll provides estimates of the value of a forecast of a "growing season" to Missouri corn farmers.

Planning the Farm. The introduction of utility analysis into agricultural economics research has resulted in a new emphasis on whole-farm planning under risk. As noted above, traditional analysis suggested such general strategies as diversification and insurance, without regard to the farmer's overall objectives. Modern theory, based on maximizing utility and "risk" programming, created an entirely new (at the time) area of research.

Farm planning under risk is based on the pioneering work of Markowitz, who was concerned with selecting a stock portfolio to minimize the variance of stock earnings. The solutions are based on (E, V) analysis. When this technique is applied to farm planning, the researcher must know the expected returns for all enterprises and, hence, the farm. The variances and covariances of expected returns among all enterprises must also be known. The objective is to find the farm plan that maximizes the farmer's utility; for (E, V) analysis, this goal is achieved by a farm plan that minimizes the variance of the expected income desired by the farmer. The expected returns, variances, and covariances can be substituted into the quadratic utility function and the resulting expression maximized. Unfortunately, this results in an objective function that is quadratic and must be solved using quadratic programming techniques (see Appendix II).

The farmer's utility function is not usually known; or, if the plan is to be offered to a group of farmers, it is reasonable to suppose that all do not have identical utility functions. As an alternative, quadratic programming can be used to minimize the variance of returns for each level of expected returns. This technique generates an efficient set of solutions from which individual farmers can select their preferred plans.

A large body of literature is available on quadratic risk programming. Anderson *et al.* present a systematic development in their Chapter 7. Hazell developed a procedure known as MOTAD, used by Mapp *et al.* in the example presented below, which permits the use of linear programming to approximate the risky solutions of quadratic programming. The articles by Johnson, Paris, Holt and Anderson, Brink and McCarl, Scott and Baker will lead the reader to more advanced concepts and references.

The Future. Several articles on risk preference and management may be found in the December 1979 Proceedings Issue of the *American Journal of Agricultural Economics*. In addition to summarizing the literature, authors of these articles address some interesting policy questions concerning risk-related research. Arguments are presented concerning research methods and applications of results. Students interested in pursuing the subject of risk research will find these articles useful.

Fertilization in a Risky Situation

Commercial fertilizers have become increasingly important in American agriculture. Chapter 3 contains an example of the marginal analysis used to determine optimum rates of fertilization when the production function is known with perfect certainty. In practice, of course, production response to fertilizer is not known with certainty. An example will now be presented to demonstrate the use of production functions and probability distributions to determine the expected earnings from the use of commercial fertilizer.[8]

Two experiments were conducted in Tennessee to determine the response of pearl millet to nitrogen under irrigated conditions. For one experiment, five rates of nitrogen applications ranged from zero to 240 pounds per acre. Four "levels" of irrigation were used; that is, irrigation water was applied in the appropriate amounts and at the proper times to maintain soil-moisture levels in each set of plots at four predetermined saturation levels. The second experiment was similar but used four rates of nitrogen and three levels of irrigation.

The unit used to measure the effects of weather on plant growth during the growing season was called a "drouth day." Computation of a drouth day is explained in detail in the reference cited below; it is sufficient here to define a drouth day as a day during the growing season when the plant is unable to obtain moisture from the soil. The use of the drouth day as a measure of moisture enables the researcher to consider the water-holding capacity of the soil, the contribution of rainfall and irrigation to soil moisture, the drying effects of sunlight, and other climatic factors affecting plant growth. The number of drouth days increases with the severity of drouth during the growing season.

The results of the experiments were used to determine the response of pearl millet yields to nitrogen and drouth days. The researchers were able to obtain an estimate of a production function with two variable inputs, similar to those described in Chapter 4.[9] Using this production function, they derived yield isoquants for millet (Figure 8–6). Notice that as the number of drouth days increases, the isoquants become more steeply sloped, particularly at high nitrogen levels. Thus, for any given level of nitrogen (on any horizontal line), the marginal product of nitrogen decreases as the number of drouth days increases.

One objective of the research was to determine the most profitable amount of nitrogen to use on pearl millet, given that the production function could not be predicted with perfect certainty. This problem could be approached by conducting the experiment with nitrogen applications at the same location for many years. The data from these experiments could then be analyzed to determine the nitrogen rate most profitable in the long run. However, by using irrigation, the

[8]The research results presented here are taken from Knetsch, J. L. and Parks, W. L., *Interpreting Results of Irrigation Experiments—A Progress Report,* TVA Report No. T-59-1 AE, Knoxville, Tennessee, August 1958.

[9]The production function presented by Knetsch and Parks is given by $Y = 3.07 + 0.1506N + 0.0010D - 0.0023N^2 - 0.0007D^2 - 0.0005ND$ where Y is the estimated yield of forage in tons, N is ten-pound units of nitrogen, and D is the number of drouth days in the June–September growing period. The production function is not presented in graphs or tables here due to space limitations. It is similar to those presented in Chapter 4.

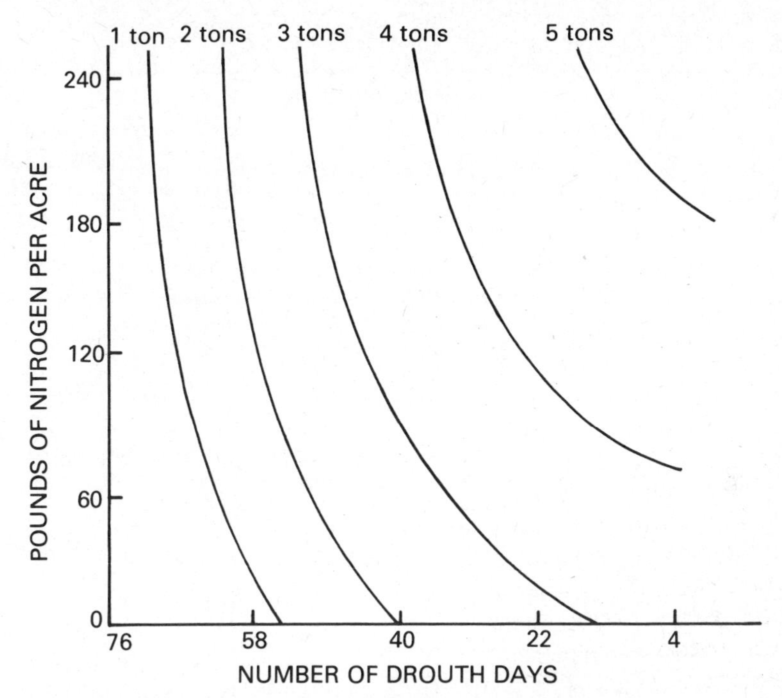

Figure 8–6. Yield isoquants showing combinations of nitrogen and drouth days needed to produce specified yields of pearl millet.

researchers were able to simulate yield response under many different moisture conditions from one year's data. The experimental data were used to determine yield response to a wide range of nitrogen rates and drouth days while rainfall and other meteorological data collected by the National Weather Service were used to estimate the number of drouth days per season for as many years back in time as the researchers deemed relevant. Thus, all the information needed to determine the expected profit from nitrogen fertilizer was available: (1) a probability distribution of drouth days, based on historical climatological data, and (2) a production function relating millet yield to nitrogen, the controlled input, and drouth days, the random weather input. The results of the analysis are presented in Table 8–8.

For purposes of the analysis, the drouth days occurring at the experimental location (Ashwood, Tennessee) for the period 1927–1956 were classified under six headings: 19, 29, 38, 47, 56, and 73 drouth days. This classification simplified the analysis, yet provided enough different growing seasons to be meaningful. The drouth classifications were further selected so that they occurred with equal probability, 1/6.

From Table 8–8, it can be seen that none of the nitrogen appli-

Table 8–8 Estimated Returns Above Fertilizer Costs for Nitrogen Applied to Millet at Ashwood, Tennessee[a]

	Number of Drouth Days[b]						
Nitrogen Pounds/Acre	*73*	*56*	*47*	*38*	*29*	*19*	*Expected Profit*
0	$ 0	$23.34	$39.34	$52.50	$62.83	$70.98	$41.50
30	−5.96	26.06	42.40	55.90	66.56	75.09	43.34
60	−9.48	29.74	46.42	60.26	71.26	80.16	46.39
90	−7.46	32.39	49.40	63.58	74.92	84.19	49.50
120	−6.44	34.00	51.35	65.86	77.54	87.19	51.58
150	−6.56	34.58	52.27	67.12	79.13	89.16	52.62
180	−7.65	34.12	52.15	67.34	79.69	90.09	52.62
210	−9.78	32.63	50.99	66.52	79.21	89.98	51.59
240	−12.95	30.10	48.80	64.66	77.69	88.84	49.52

Source: Knetsch, J. L. and Parks, W. L., *Interpreting Results of Irrigation Experiments—A Progress Report*, TVA Report No. T-59-1 AE, Knoxville, Tennessee, August 1958.

[a]Assumed prices are forage $25 per ton, nitrogen $0.132 per pound, and an application rate of $2.00 per acre was charged.
[b]The drouth days were classified so that each occurs with equal probability, 1/6.

cations returned a profit at the high drouth level, 73 days, but profits from nitrogen applications varied from $70 to $90 per acre at the low drouth level, 19 days. The probability of a profit was the same, 5/6, for all nitrogen amounts. The probability of a loss is 1/6 for all nitrogen amounts except zero.

The expected profits from each amount of nitrogen are presented in the right-hand column of Table 8–8. The nitrogen rates 150 and 180 pounds per acre yielded the same expected profit, about $11 per acre more than would result if no nitrogen were used. The farmer wishing to maximize $E(P)$ would use 150 pounds of nitrogen per acre and would expect to suffer a loss one year in six. The more conservative farmer, or one limited in capital, might note that the long-run expected gain from increasing nitrogen use from 90 to 150 pounds is only $3 per acre, $3 gained from the use of $8 of additional fertilizer.

This analysis is based on the assumptions that the probability distribution of drouth days based on historical data does represent the farmer's subjective prior distribution and also that the farmer has a linear utility function. A change in the prior probability distribution can easily be accommodated by changing the probabilities assigned to the six categories. As shown above, converting expected profits to utility requires knowledge of the farmer's utility function. If a quadratic utility function is assumed, then the variance of the expected returns must also be estimated. Thus, application of decision theory into a problem such as this requires a substantial increase in information; although that information may be quite useful, it may not always be available.

Farm Planning in a Risky Situation

An example of whole-farm planning using modern decision theory will now be presented. Table 8–1, at the beginning of this chapter, contains estimated gross margins, along with their means and standard deviations, for five farm enterprises in southwestern Oklahoma. This information was utilized by researchers at Oklahoma State University to determine the impact of the expected variability of gross margins on enterprise selection on the farm.

A "typical" farm was defined for study. The farm was assumed to be 1500 acres in size with 1200 acres of cropland and 300 acres of native pasture. The enterprise data in Table 8–1 were assumed to be representative for growing conditions in the area. Assets were specified to be land and buildings valued at $855,000 and machinery valued at $70,000. Equity was assumed to be $654,372 and real estate debt equal to $270,628.

Farm enterprise plans were developed for this model farm using a version of linear programming known as MOTAD. Developed by Hazell in 1971, MOTAD permits linear programming models to be used to develop risk-efficient farm plans. The farmer is assumed to be a risk-averter with a quadratic utility function. This suggests that the farmer's utility function can be expressed in terms of expected income and the variance of income. Estimates of this information are contained in Table 8–1 for the farm in southwestern Oklahoma. The problem thus becomes one of using the MOTAD linear programming technique to derive the farmer's efficient E-V boundry, that is, to select the farm plan that minimizes the *variance* of expected returns for each *level* of expected returns. Using MOTAD, the mean absolute deviations in gross margins are used as a substitute for the variance in gross margins. The concept is straightforward, but the programming details are not. More complete explanations can be found in Hazell, Mapp *et al.* and Anderson *et al.*

Three enterprise plans for the Oklahoma farm are presented in Table 8–9. The profit-maximizing plan is a standard linear programming solution that assumes the manager selects a set of enterprises to maximize gross margins, assuming sale at harvest time. Risk is not considered. In the second plan, MOTAD-I, risk is introduced. It is assumed the manager wishes to minimize total negative deviations from the expected gross margins, subject to receiving a specified level of income. The third plan, MOTAD-II, is similar to MOTAD-I but permits storage, delayed sale of storable commodities, and forward contracting of wheat.

The three plans illustrate the effects of considering risk in farm plans. The first plan represents the maximum point on the E-V frontier; $66,340 is returned without regard to possible income variations, which would average $41,360. MOTAD-I illustrates the effect of considering risk. Steers are increased, grain sorghums are reduced, and

Table 8–9 Enterprise Combinations for an Oklahoma Farm When Risk is Considered

Activities in the Optimum Solution	*Units*	*Profit-Maximizing Solution—Sale at Harvest*	*MOTAD-I Solution—Sale at Harvest*	*MOTAD-II Solution—Sequential Marketing and Forward Contracting*
Steers	Head	171	233	222
Grain sorghum	Acres	1172	918	866
Alfalfa	Acres	28	28	28
Cotton	Acres		65	150
Wheat—sell in June	Acres		189	
Wheat—sell in December	Acres			60
Wheat—forward contract in March for June sale	Acres			96
Gross margins	Dollars	66,340	65,000	65,000
Standard deviation	Dollars	41,360	36,100	33,605

Source: Mapp, Harry P., Jr., Hardin, M. L., Walker, O. L., and Persuad, T., "Analysis of Risk Management Strategies for Agricultural Producers," *American Journal of Agricultural Economics*, Volume 61, December 1979, pp. 1071–1077.

additional enterprises, cotton and wheat, are added. Diversification is increased and the gross margin, $65,000, represents a sacrifice of only $1340 of expected income. But the standard deviation of this income, $36,100, represents a reduction of $5260. A risk-averter might find this trade-off attractive. The third plan, MOTAD-II, reduces grain sorghum and steers, increases cotton, and adds decisions to forward contract wheat and store wheat for December sale. Diversity in marketing as well as production causes the standard deviation of gross margins to drop even farther, $33,605, while maintaining the income level of $65,000.

The Oklahoma researchers note that additional risk management strategies could be used, including crop insurance, alternative land rental arrangements, futures markets, and federal disaster programs. In an extension of their analysis, which will not be presented here, they use a computer to simulate each solution over a 20-year period for a variety of scenarios involving the product prices, input prices, land values, land equity, and other costs. From the simulations, the chance of business failure was found to increase substantially when the farmer's beginning equity fell as low as 45 percent.

THE MANAGER AS A PRODUCTION PROCESS

The manager, as he faces uncertain situations, is similar in many respects to any production process. Inputs, in the form of experience, gathered information, inherent knowledge, and intuition, flow in and management decisions flow out. Costs as well as returns are associated with the decision process. The manager eventually reaches the point where the (expected) added cost of delaying a decision exceeds the (expected) returns added by the delay. Several reasons exist for this.

First, any immediate decision is suboptimal in time. Chapter 7 discussed some of the reasons for this. Another reason is the uncertainty of the future. We do not go to extremes to select an optimum now because unforeseen future events might negate our best efforts. The manager has many decisions to make. He cannot ignore all others while lingering over one.

Second, the number of possible actions are usually so numerous that the individual cannot consider all with care. As pointed out above, the costs would eventually exceed the returns. Many alternatives are ignored. Most farmers, for example, do not consider the possibility of moving to Australia when faced with a farm program they abhor. Simon has suggested that decision makers do not attempt to maximize but rather select from a set of alternatives they consider satisfactory.

Finally, in most important decision situations, many of the determining factors cannot be controlled by the manager. This has been discussed in detail. Nonetheless, when faced with such occurrences as war, peace, accidents, disease, depression, and overkill, as well as many other unknowns, the decision maker attempting to carefully weigh all factors would probably find himself with an insoluble dilemma.

It is probably true that people do not have an irrational passion for dispassionate rationality.[10] On the other hand, decisions must be made and, as evidenced by the world around us, can be made successfully. Managers may not be ultimate maximizers but they do attempt to act rationally in the attainment of their goals.

Suggested Readings

Anderson, Jock R., Dillon, John L., and Hardaker, J. Brian. *Agricultural Decision Analysis.* Ames: Iowa State University Press, 1977.

Baquet, A. E., Halter, A. N., and Conklin, Frank S. "The Value of Frost Forecasting: A Bayesian Appraisal. "*American Journal of Agricultural Economics,* Volume 58, August 1976, pp. 511–520.

Black, Roy. "Weather Variations as a Cost-Price Uncertainty Factor as It Affects Corn and Soybean Production." *American Journal of Agricultural Economics,* Volume 57, December 1975, pp. 940–944.

[10]To paraphrase Miller and Starr, who paraphrase John Monroe Clark. This discussion is based on ideas presented by Miller and Starr, Chapter 3.

Borch, Karl. *The Economics of Uncertainty.* Princeton, N.J.: Princeton University Press, 1968, Chapter 5.

Brink, Lars and McCarl, Bruce. "The Tradeoff Between Expected Return and Risk Among Cornbelt Farmers." *American Journal of Agricultural Economics,* Volume 60, May 1978, pp. 259–263.

Bross, I. D. J. *Design for Decision.* New York: MacMillan Company, 1959.

Bullock, J. Bruce and Logan, Samuel H. "An Application of Statistical Decision Theory to Cattle Feedlot Marketing." *American Journal of Agricultural Economics,* Volume 52, May 1970, pp. 234–241.

Byerlee, D. R. *A Decision Theoretic Approach to the Economic Analysis of Information.* Department of Farm Management, Farm Management Bulletin 3, University of New England, Armidale, 1968.

Byerlee, D. R. and Anderson, J. R. "Value of Predictors of Uncontrolled Factors in Response Functions." *Australian Journal of Agricultural Economics,* Volume 13, December 1969, pp. 113–127.

Carlson, Gerald A. "A Decision Theoretic Approach to Crop Disease Prediction and Control." *American Journal of Agricultural Economics,* Volume 52, May 1970, pp. 216–223.

Chernoff, H. and Moses. L. E. *Elementary Decision Theory.* New York: John Wiley and Sons, Inc., 1959.

Colyer, D. "Fertilizer Strategy Under Uncertainty." *Canadian Journal of Agricultural Economics,* Volume 17, November 1969, pp. 144–149.

Doll, John P. "Obtaining Preliminary Bayesian Estimates of the Value of a Weather Forecast." *American Journal of Agricultural Economics,* Volume 53, November 1971, pp. 651–655.

Doll, John P. "A Comparison of Annual versus Average Optima for Fertilizer Experiments." *American Journal of Agricultural Economics,* Volume 54, May 1972, pp. 226–233.

Eidman, Vernon, Dean, Gerald W., and Carter, Harold O. "An Application of Statistical Decision Theory to Commercial Turkey Production." *Journal of Farm Economics,* Volumne 49, November 1967, pp. 852–868.

Friedman, M. and Savage, L. J. "The Expected Utility Hypothesis and the Measurability of Utility." *Journal of Political Economics,* Volume 60, December 1952, pp. 463–474.

Halter, A. N. and Dean, G. W. *Decisions under Uncertainty with Research Applications.* South-Western Publishing Company, 1971.

Hazell, P. B. R. "A Linear Alternative to Quadratic and Semivariance Programming for Farm Planning Under Uncertainty." *American Journal of Agricultural Economics,* Volume 53, February 1971, pp. 53–62.

Heady, E. O. *Economics of Agricultural Production and Resource Use.* Englewood Cliffs, N.J.: Prentice-Hall, Inc., 1952.

Hey, John D. *Uncertainty in Microeconomics.* New York: New York University Press, 1979.

Holt, John and Anderson, Kim B. "Teaching Decision Making Under Risk and Uncertainty to Farmers." *American Journal of Agricultural Economics,* Volume 60, May 1978, pp. 249–253.

Johnson, S. R. "A Re-examination of the Farm Diversification Problem." *Journal of Farm Economics,* Volume 49, August 1967, pp. 610–621.

Just, Richard E. and Pope, Rulan D. "Production Function Estimation and Related Risk Considerations." *American Journal of Agricultural Economics,* Volume 61, May 1979, pp. 276–284.

Knetsch, J. L. and Parks, W. L. *Interpreting Results of Irrigation Experiments—A Progress Report.* TVA Report No. T 59-1 AE, Knoxville, Tennessee, August 1958.

Knight, Frank H. *Risk, Uncertainty, and Profit.* Boston: Houghton-Mifflin, 1921.

Kyberg, H. E., Jr., and Smokler, Howard E., eds. *Studies in Subjective Probability.* New York: John Wiley and Sons, Inc., 1964.

Langham, Max R. "Game Theory Applied to a Policy Problem of Rice Farmers." *Journal of Farm Economics,* Volume 45, February 1963, pp. 151–162.

Lin, W., Dean, G. W. and Moore, C. W. "An Empirical Test of Utility vs. Profit Maximization in Agricultural Production." *American Journal of Agricultural Economics,* Volume 56, August 1974, pp. 497–508.

Luce, R. Duncan and Raiffa, H. *Games and Decisions,* New York: John Wiley and Sons, Inc., 1957.

Mapp, Harry P., Jr., Hardin, J. L., Walter, O. L., and Persand, T. "Analysis of Risk Management Strategies for Agricultural Producers." *American Journal of Agricultural Economics,* Volume 61, December 1979, pp. 1071–1077.

Markowitz, Harry M. "Portfolio Selection." *Journal of Finance,* Volume 12, March 1952, pp. 77–91.

McKinsey, J. C. C. *Introduction to the Theory of Games.* McGraw-Hill Book Company, Inc., 1952.

McQuigg, J. D. and Doll, John P. *Weather Variability and Economic Analysis.* Missouri Agricultural Experiment Station Bulletin 771, June 1961.

Miller, D. W. and Starr, M. K. *Executive Decisions and Operations Research.* Englewood Cliffs, N.J.: Prentice-Hall, Inc., 1960.

Officer, R. R. and Halter, A. N. "Utility Analysis in a Practical Setting." *American Journal of Agricultural Economics,* Volume 50, May 1968, pp. 257–277.

Paris, Quirino. "Revenue and Cost Uncertainty, Generalized Mean-Variance, and the Linear Complementarity Problem." *American Journal of Agricultural Economics,* Volume 61, May 1979, pp. 268–275.

Savage, L. J. "Bayesian Statistics." In Robert E. Machal and Paul Gray, eds. *Recent Developments in Decision and Information Processes.* New York: Macmillan Company, Inc., 1966, pp. 161–194.

Scott, John R., Jr., and Baker, Chester B. "A Practical Way to Select an Optimum Farm Plan under Risk." *American Journal of Agricultural Economics,* Volume 54, November 1972, pp. 657–660.

Shackle, G. L. S. *Decision, Order and Time*, Second Edition. Cambridge: Cambridge University Press, 1969.

Simon, Herbert A. "Behavioral Model of Rational Choice," *Quarterly Journal of Economics,* February 1955, pp. 99–118.

von Neumann, John and Morgenstern, Oskar. *Theory of Games and Economic Behavior.* Princeton, N.J.: Princeton University Press, 1944.

Young, Douglas. "Risk Preferences of Agricultural Producers: Their Use in Extension and Research." *American Journal of Agricultural Economics.* Volume 61, December 1979, pp. 1003–1069.

LINEAR PROGRAMMING CHAPTER 9

Since the early 1950s, linear programming has become one of the important research techniques in agricultural economics. Applications have been made to farm management problems, feed-mix analyses, spatial location, and transportation models, to name a few. These applications would be sufficient to justify the need to study linear programming. However, linear programming also provides a synthesis for many of the basic concepts of economics presented in previous chapters. Its use in research on farm planning has provided additional insights into management problems and, as such, has enhanced decision making on farms.

The study of resource allocation on an individual farm when resources are limited—that is, when available resources will not permit the manager to use the profit-maximizing amounts—is basic to production economics. In Chapter 4, decision criteria were developed to determine the combination of inputs that would minimize the cost of a given quantity of output. In Chapter 5, criteria were developed to determine the profit-maximizing combinations of output that could be produced given a limited amount of input. The last section of Chapter 5 presented the general equimarginal principles and discussed the optimal allocation of several inputs among several enterprises.

Those two chapters contain the economic logic underlying linear programming. Linear programming, of course, does involve some unique features. One is quite basic. The discussion of equimarginal principles in Chapter 5 was based on the assumption that input use would be in Stage II of the production function. Other than that, no restrictions were placed on the form of those functions. Linear programming requires that all relevant production and revenue functions be linear. This implies that the tools of calculus used in previous chapters cannot be applied in linear programming. Thus, while the economic principles are the same, the mathematical techniques needed for a solution are different.

This chapter will start with a general discussion of the linear programming model and will emphasize the economic logic of the model. Next, the simplex solution will be presented in elementary form. Although linear programming problems are now solved on electronic computers, a review of the simplex method is traditional in classroom presentations of linear programming and does underline some of the unique features of linear programming models. The remainder of the chapter will contain a discussion of the uses of linear programming analyses. (Appendix II contains an introduction to nonlinear programming; the concepts in Appendix II are more general but require a good working knowledge of calculus.)

THE LINEAR PROGRAMMING MODEL

The Production Function

The production functions assumed for linear programming are, as mentioned, linear, but of a very special type. They do not permit input substitution, and they must exhibit constant returns to scale. Such production functions are often called *fixed-proportion* production functions.

The fixed-proportion production function used in linear programming should not be confused with the continuous linear production function that is also characterized by constant returns to scale. Any linear function that passes through the origin will possess the property of constant returns to scale. To compare the two, consider the linear production function

$$Y = a_1X_1 + a_2X_2$$

This function shows constant returns to scale; it is linear and homogeneous, so that doubling input amounts will double output, but the inputs do substitute for each other. The isoquant equation for any constant output level, Y_0, is given by

$$X_1 = \frac{Y_0 - a_2X_2}{a_1} = \frac{Y_0}{a_1} - \frac{a_2}{a_1}X_2$$

and the marginal rate of substitution is given by

$$-\frac{MPP_{X_2}}{MPP_{X_1}} = -\frac{a_2}{a_1}$$

In this case, the expansion path lies on one input axis or the other, or in the event that $-(a_2/a_1) = -(P_{X_2}/P_{X_1})$ would consist of the entire plane (see Chapter 5).

The fixed-proportion production function is written

$$Y = \text{minimum}\ (a_1X_1, a_2X_2)$$

which means that output is determined by the limiting input. For example, if $a_1 = 2$, $a_2 = 3$, $X_1 = 2$, and $X_2 = 5$, output would be

Y = minimum $(2X_1, 3X_2)$ = minimum $(2 \cdot 2, 3 \cdot 5) = 4$. X_2 could be increased to 100 or decreased to 4/3 and output would remain at 4. However, if the amount of X_2 dropped below 4/3, X_1 remaining at 2, then X_2 would be limiting and output would drop and equal a_2X_2.

Unfortunately, the algebraic expression for the fixed-proportion production function does not clearly express the fact that both inputs are required to produce the minimum amount of Y. For example, when Y = minimum $(2X_1, 3X_2)$ and two units of X_1 are used, then 4/3 units of X_2 must also be used to produce four units of output. When $X_2 = 5$ and X_1 is the limiting input, X_2 will be in surplus by the amount equal to $(5 - 4/3 = 11/3)$ or [the amount of X_2 available $-$ (output/a_2)]. Thus, the amounts of inputs actually *used* to produce four units of output will be $X_1 = 2$ and $X_2 = 4/3$.

Graphs of these functions are presented in Figure 9–1. The production surface for the fixed-proportion function would look like one corner of a pyramid. The rate of incline of the corner would depend on the magnitudes of a_1 and a_2. The isoquants would appear as right angles with their vertex located on the corner. On the isoquant map for the fixed-proportion production surface, the expansion path is a straight line passing through the origin because the firm will always expand up the corner of the pyramid. Regardless of prices, the same combinations of inputs will always be used to produce a given level of output—economic efficiency requires that the minimum amount of input be used to produce a given level of output. Thus, on the isoquant map of Figure 9–1, point A dominates both points B and C because B requires more X_1 and C requires more X_2. Input use will always be in a fixed ratio that also determines the slope of the expansion path. The arrow tips on the axes and expansion paths are intended to suggest constant returns to scale.

Along the expansion path, the inputs can be termed technical complements (Chapter 4). Therefore, for movements off the expansion path, such as from A to B or A to C in Figure 9–1, the marginal product of the input in excess is zero.

For purposes of contrast, the linear production function that permits substitution is also presented in Figure 9–1 along with its isoquant map. This production function would appear as a flat surface, such as a sheet of plywood, with its corner at the origin and with its inclination determined by the marginal products of the inputs, which are constant and nonzero. Functions of this type were discussed in Chapter 5.[1]

[1]The traditional methods of calculus cannot be used to locate the extreme points of linear functions. Consider the linear and homogeneous function $Y = 2X$. If this function is defined to exist for all values of X (everywhere on the X axis), then it has no extreme points. If it is defined for a particular interval, say $1 \leqq X \leq 3$, then the first derivative, $dY/dX = 2$, is of no help because it is never zero on the interval. In this case, graphing the function on the interval will demonstrate that its minimum on the interval, 2, occurs when $X = 1$ and its maximum on the interval, 6, occurs when $X = 3$. A more general discussion of these problems is included in Appendix II.

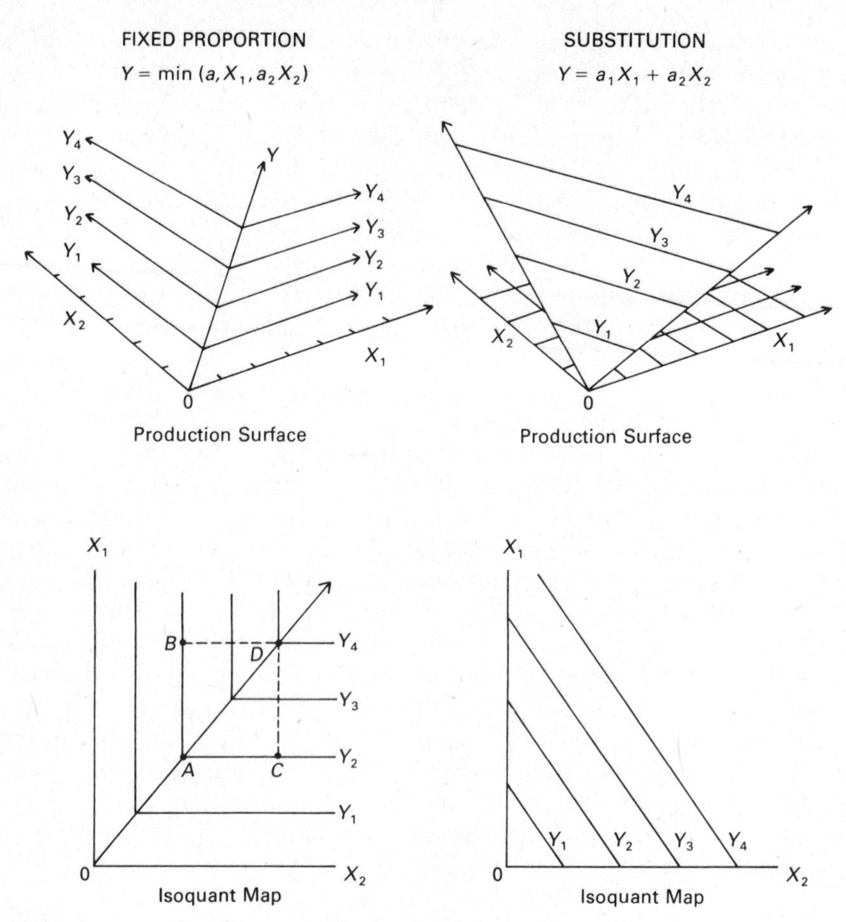

Figure 9–1. Linear homogeneous production functions.

The Maximization Model

The typical linear programming problem must start with a set of fixed-proportion production functions for the enterprises to be considered. In farm management research, the production functions are usually derived from a budgeting procedure. The farmer, who wants to formulate a farm plan consisting of several different enterprises, first must develop a budget for each enterprise. These budgets are then used to determine input requirements and profits per unit of output for the enterprise. The coefficients of the fixed proportion production functions are derived from the budgets.

For example, consider the data in Table 9–1. Three inputs are used to produce two outputs. From budgeting, it is determined that a unit of Y_1 requires 3 units of X_1, 1 unit of X_2, and 3 units of X_3 while a unit of Y_2 requires 2, 2, and 1 unit of the same inputs, respectively. Using the notation presented above, Y_1 = minimum $(3X_1, 1X_2, 3X_3)$

Table 9–1 Linear Programming: The Budget for a Hypothetical Maximization Problem

	A Input Requirements and Availability		
Input	*Units of Input/Units of Output*		*Total Amount of Available Input*
	Output Y_1	*Output Y_2*	
X_1	3	2	36
X_2	1	2	30
X_3	3	1	30
	B Profit per Unit of Output		
	\$1	\$1	

and Y_2 = minimum ($2X_1$, $2X_2$, $1X_3$); that is, the fixed proportions needed are (3:1:3) for Y_1 and (2:2:1) for Y_2. Y_1 and Y_2 both return a profit of \$1 per unit, and the farmer would like to maximize profits. He has limited quantities of inputs: 36 units of X_1, 30 units of X_2, and 30 units of X_3.

The total quantity of the inputs will limit the amount of output that can be produced. The 36 units of X_1 can be used to produce either Y_1 or Y_2 or a combination of Y_1 and Y_2. Thus, $3Y_1 + 2Y_2 \leqq 36$. That is, for any values of Y_1 and Y_2 chosen to be the possible solutions to the problem, the total amount of X_1 used cannot exceed 36. It could be less than 36—hence, the inequality. There will be three such equations, one for each input, and all three can be written

$$3Y_1 + 2Y_2 \leqq 36$$
$$1Y_1 + 2Y_2 \leqq 30$$
$$3Y_1 + 1Y_2 \leqq 30$$

In linear programming terminology, these equations are called *constraints*. Notice that the columns of numbers to the left of the inequalities are the coefficients of the production functions; the column on the right represents input amounts. Two additional requirements must now be added: $Y_1 \geqq 0$, $Y_2 \geqq 0$. These *nonnegativity restrictions* are dictated by mathematical considerations but have a practical interpretation. In practice, farmers cannot produce minus 10 cows or minus 150 acres of corn. Linear programming is a mathematical technique that could, without these last two restrictions, have solutions that include negative outputs.

The final equation needed to set up the linear programming problem is called the *objective function*. The farmer operates the farm for profit, and the linear programming solution is based on the assumption that he seeks to maximize profits. Profit is given by

$$\text{Profit} = P_{Y_1}Y_1 + P_{Y_2}Y_2 = \$1\ Y_1 + \$1\ Y_2$$

Thus, the objective of the linear programming model is to maximize the objective function (profit) subject to the three input constraints and the two nonnegativity restrictions.

Granted some changes in terminology, the linear programming problem can be seen to be a special case of the problems analyzed in Chapter 5. In the discussion of the general equimarginal principle, the profit equation was not presented as an equation and called the objective function; and although limited capital was specified, the capital constraint was not written in equation form. Yet the objective of the analysis was to maximize profits by equating the ratio $(VMP/P_X) \geqq 1$ for all inputs in all enterprises, given that capital available to purchase inputs was limited.

The linear programming problem lends itself to a graphic solution. The principles are identical to those developed in Chapter 5. (See Figure 5–2 and the discussion pertinent to it.) First, the production possibility curve must be derived. This can be done by graphing the three constraint equations, ignoring the inequalities (for graphing purposes only). This is shown in Figure 9–2A. The three constraints are given by three linear equations graphed in the first quadrant only, because $Y_1 \geqq 0$ and $Y_2 \geqq 0$.

Because of the properties of the fixed-coefficient production function, the constraints can be considered in this rather unique way: the available amount of input X_1 will permit the farmer to produce (1) any amount of Y_1 and Y_2 on the line $3Y_1 + 2Y_2 = 36$, or (2) any amounts of Y_1 and Y_2 that fall to the southwest of $3Y_1 + 2Y_2 = 36$, including those on the axes. The farmer could produce zero quantities of Y_1 and Y_2 if he wished. Similarly, the amounts of X_2 and X_3 will permit the production of output combinations on or below the lines $Y_1 + 2Y_2 = 30$ and $3Y_1 + Y_2 = 30$, respectively. Taken together, the three constraints define a production possibility "curve" for Y_1 and Y_2. Between points $C = (0, 15)$ and $D = (3, 13\frac{1}{2})$, X_2 is the limiting input; between D and $E = (8, 6)$, X_1 is the limiting input; between E and $F = (10, 0)$, X_3 is the limiting input. Thus, the production possibility "curve" is traced out by the connected line segment CDEF. On or below this line all three constraints are satisfied simultaneously. Above the line, one or more are violated. The importance of the inequalities in the constraints is now apparent. Without them a solution would be possible only if all the three constraint lines intersected at the same point.

The production possibility curve along with the axes define the set of solutions for the problem. That is, the profit-maximizing combi-

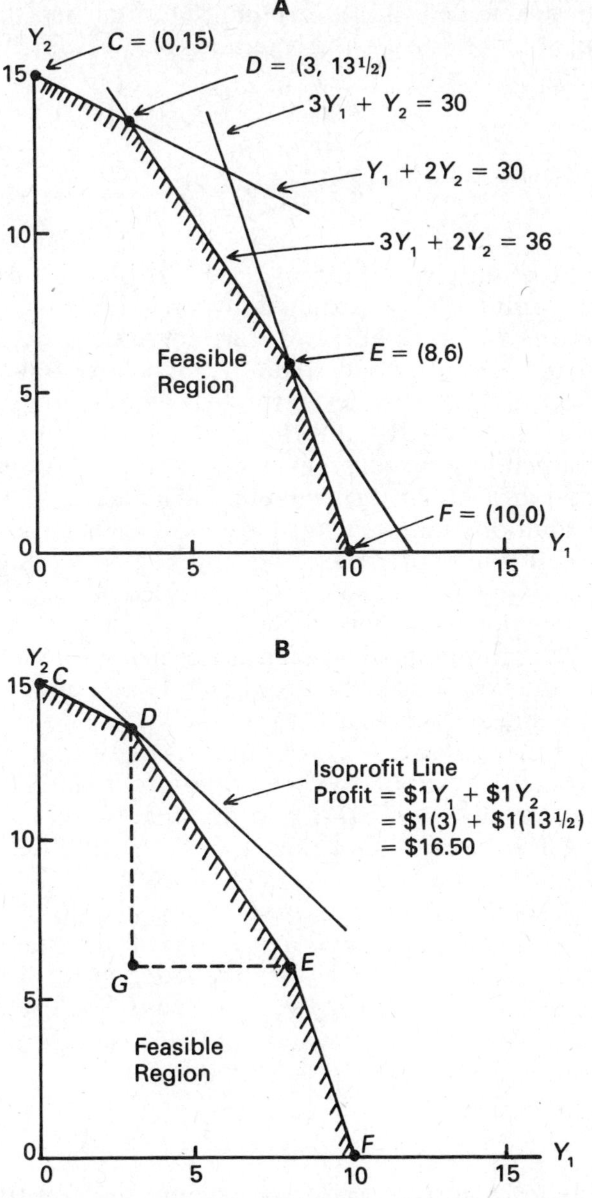

Figure 9–2. Graphic solution to the maximization problem.

nation of Y_1 and Y_2 must be one (or more) of the points on the axes, on *CDEF*, or within the region bounded by these lines. This set of points is called the *feasible* region.

The profit-maximizing combination of outputs will be one of the points in the feasible region. The profit function ($P_{Y_1}Y_1 + P_{Y_2}Y_2 = \text{Profit}$)

is identical in concept to the isorevenue line discussed in Chapter 5. For a given profit level, say P_0, the equation for the isoprofit line would be

$$Y_2 = \frac{P_0}{P_{Y_2}} - \frac{P_{Y_1}}{P_{Y_2}} Y_1$$

A family of these lines with the slope $(-P_{Y_1}/P_{Y_2}) = -1$ can be drawn on the same graph with the production possibility curve. As the isoprofit line moves to the right, profits are increased. Because the objective is to maximize profit, the solution will be given by the point on the highest isoprofit line that lies in the feasible region. Thus, point D is the solution (Figure 9–2B) and the maximum profit is \$16.50. Because the production possibility curve is not smooth (does not have a continuous first derivative) at D, tangency is not attained, but the maximum profit solution is nonetheless located. This is an important difference between linear programming and classic methods of maximization. It is somewhat similar to solutions determined in Chapter 5 for linear production possibility curves.

In linear programming, the profit-maximizing solution will always fall on the production possibility "frontier." This is because the objective function is linear and always increases when Y_1 and Y_2 increase. Consider point G in Figure 9–2B. By increasing output of Y_2 until all X_1 and X_2 is used, the outputs at D could be produced. Because D has more Y_2 and the same Y_1 as G, D clearly dominates G. The same is true of G and E; E has more Y_1 and the same Y_2 as G. Therefore, E also dominates G.

This suggests that the constraint equations could be solved algebraically (ignoring the inequalities) in a search for the optimal solution. At point C, the constraint for X_2 is effective. Setting $Y_1 = 0$ in that constraint gives $Y_2 = 15$. Solving the constraints for X_1 and X_2

$$3Y_1 + 2Y_2 = 36$$
$$1Y_1 + 2Y_2 = 30$$

simultaneously leads to $D = (3, 13\frac{1}{2})$. Solving the constraints for X_1 and X_3

$$3Y_1 + 2Y_2 = 36$$
$$3Y_1 + Y_2 = 30$$

simultaneously gives $E = (8, 6)$, and setting $Y_2 = 0$ in the constraint for X_3 yields $F = (10, 0)$. These four points can be substituted into the profit equation, Profit $= P_{Y_1} Y_1 + P_{Y_2} Y_2$, as follows:

$$\begin{aligned} \$1\ (0) + \$1\ (15) &= \$15 \\ \$1\ (3) + \$1\ (13\tfrac{1}{2}) &= \$16.50 \\ \$1\ (8) + \$1\ (6) &= \$14 \\ \$1\ (10) + \$1\ (0) &= \$10 \end{aligned}$$

This again shows $D = (3, 13\frac{1}{2})$ to be the optimal solution.

The Minimization Model

The economic problem often involves minimizing the cost of a given level of production; linear programming techniques are equally applicable to problems of minimization. A common example used in linear programming texts is called the "diet" or "feed-mix" problem. The problem of the farmer is to feed an animal at minimum cost and yet supply the daily minimum requirements of nutrients needed to maintain and fatten the animal.

Hypothetical data for a feed-mix problem are given in Table 9–2. Two feed ingredients have varying quantities of calcium, protein, and calories. The minimum daily requirement from these two sources is assumed to be 10, 15, and 15 units of calcium, protein, and calories, respectively. A unit of ingredient X_1 costs \$1 and a unit of ingredient X_2 costs \$2. Sufficient information is available to set up the following linear programming model:

$$\text{Minimize Cost} = \$1\ X_1 + \$2\ X_2$$

subject to

$$\begin{aligned} 1X_1 + 1X_2 &\geqq 10 \\ 3X_1 + 1X_2 &\geqq 15 \\ 1X_1 + 6X_2 &\geqq 15 \end{aligned}$$

and

$$X_1 \geqq 0,\ X_2 \geqq 0$$

The constraints are again expressed as inequalities but now ensure that the minimum requirements are met. Thus, $X_1 + X_2 \geqq 10$ states that at least 10 units of calcium are required; more is acceptable but less is not. The same interpretation holds for $3X_1 + 1X_2 \geqq 15$ (protein) and $X_1 + 6X_2 \geqq 15$ (calories). The graphic solution is presented in Figure

Table 9–2 Linear Programming: The Budget for a Hypothetical Minimization Problem

	A Nutrient Availability and Requirements		
Nutrient	*Unit of Nutrient/Unit of Feed*		*Minimum Daily Requirement*
	Ingredient X_1	*Ingredient X_2*	
Calcium	1	1	10
Protein	3	1	15
Calories	1	6	15
	B Cost per Unit of Feed		
	\$1	\$2	

9–3. The feasible set now consists of all points above or beyond the lines determined by the minimum requirements. These are all the points on or above the point (0, 15) on the X_2 axis, the protein constraint between (0, 15) and (2.5, 7.5), the calcium constraint between (2.5, 7.5) and (9, 1), the calorie constraint between (9, 1) and (15, 0), and the X_1 axis beyond the point (15, 0).

Because the isocost function is linear and decreases with shifts to the left, the minimum cost point will again be on the boundary of the feasible region but now will be on the lower boundary. (There is no upper boundary specified; in fact, the upper boundary would be represented in some manner by the animal's stomach capacity.) By computing the cost of each corner of the lower boundary of the feasible region, it can be determined that $X_1 = 9$ and $X_2 = 1$ leads to the minimum cost solution, \$11. The isocost line will have a slope of

$$-\frac{P_{X_1}}{P_{X_2}} = -\frac{1}{2}$$

A line with this slope would touch the boundary of the feasible region at (9, 1). A lower cost line would not touch the feasible region.

The cost-minimizing solution is again not located by a tangency, but rather by an intersection of the feasible set and lowest attainable isocost line. At the solution point (9, 1), the slope of the isocost line ($-1/2$) is between the *MRS* of X_1 for X_2 along the constraint line for calcium and the *MRS* of X_1 for X_2 along the constraint for calories. That is, $-1/6 > -1/2 > -1$. This is characteristic of any linear programming corner solution.

Additional Considerations

The different types of constraints can be combined in the same model. Thus, constraints representing upper bounds, lower bounds, and equality

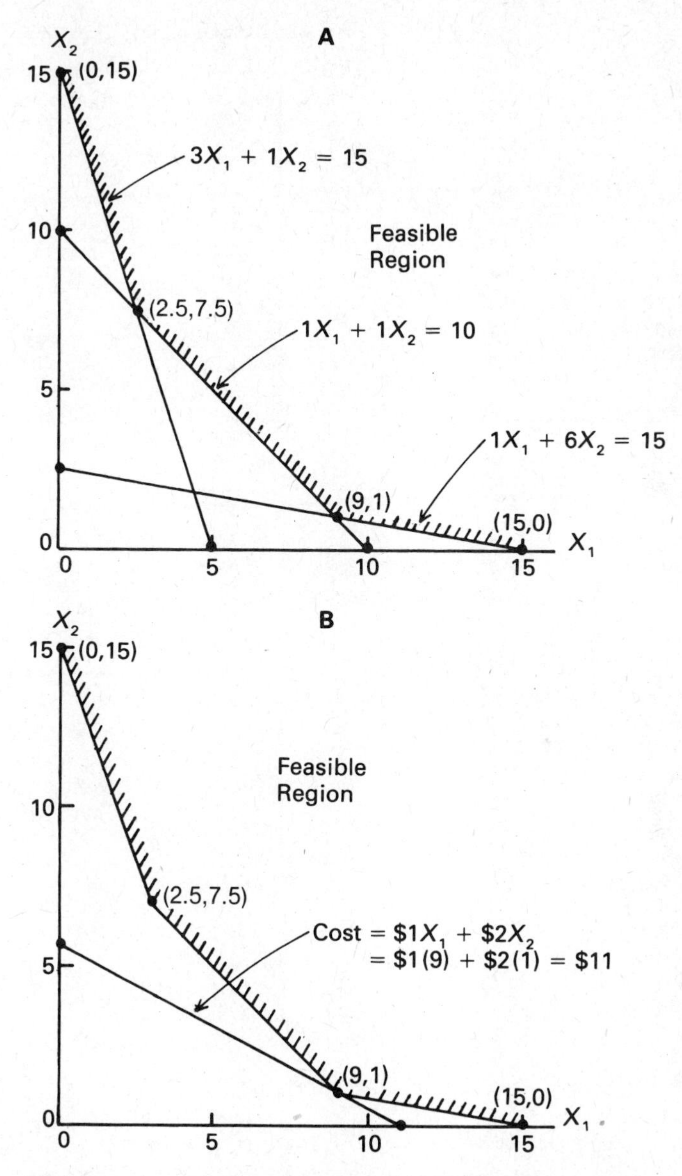

Figure 9–3. Graphic solution to the minimization problem.

requirements can be used in the same problem. When different types of constraints are utilized, care must be taken to ensure that they are consistent. Inconsistency is always a danger to be aware of, but the hazard is increased when different types of constraints are combined.

Figure 9–4 illustrates some common constraint problems. Figure 9–4A depicts a case in which the constraints all have positive slopes and define a feasible set that has no maximum. For example, the two constraints might be $Y_2 - 2Y_1 \leqq 5$ and $-Y_2 + Y_1 \leqq 6$. The negative

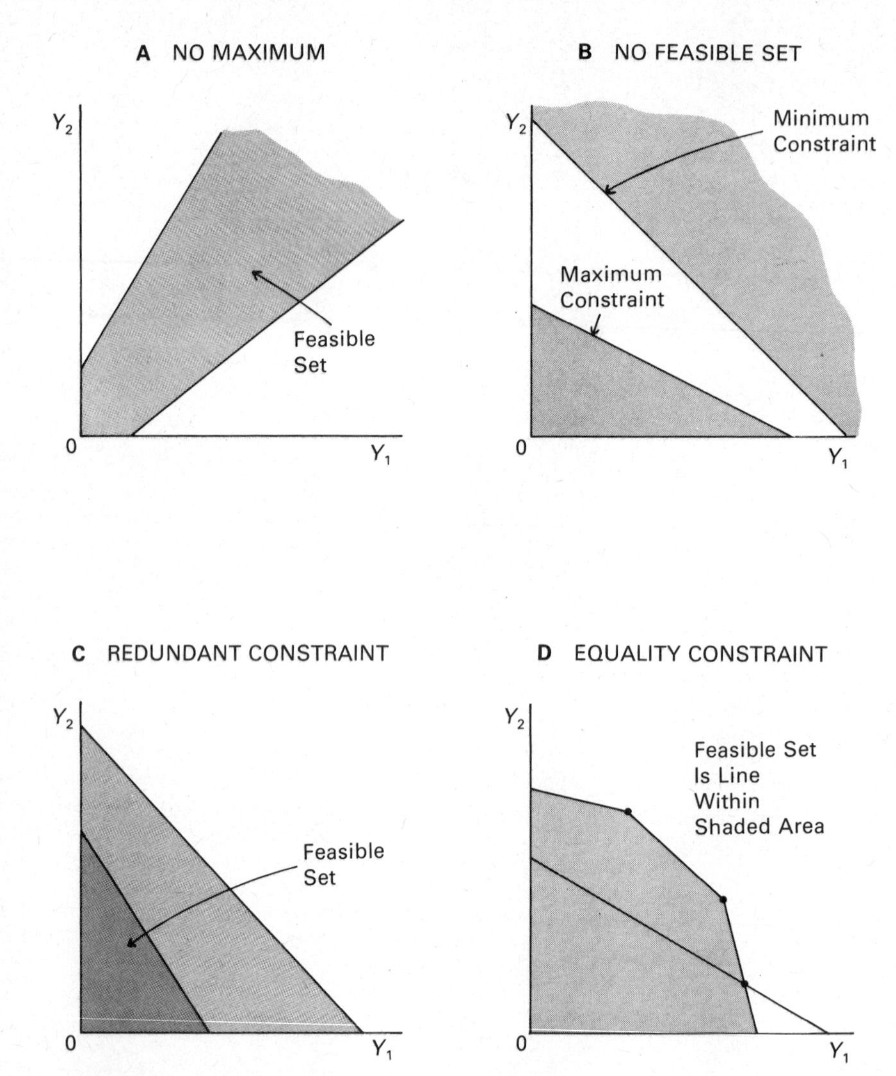

Figure 9–4. Examples of constraint problems.

coefficients imply that a particular production activity uses one input (with the positive coefficient) and produces the other (with the negative coefficient). The activities supply inputs to each other, allowing production to be expanded without limit, and define an economic perpetual motion machine. Figure 9–4B illustrates maximum and minimum constraints that are inconsistent. The minimum constraint lies above the maximum constraint. Hence, there are no points in the first quadrant that will satisfy both. Figure 9–4C contains a redundant constraint in a maximizing problem. The lower constraint dominates the upper constraint. The redundant constraint causes no problem because the optimal solution is unchanged in the presence of the constraint. It is simply not needed.

Finally, an equality constraint is shown in Figure 9–4D, along with three inequality constraints. The equality constraint would appear, for example, as $3Y_1 + 2Y_2 = 10$. It causes no problems for determining the profit-maximizing solution, but forces the solution to lie on the line within an area defined by the other constraints.

The Dual

Problems in economics can often be divided into a dichotomy. For example, the most profitable output of a production process could be determined by finding the optimum input or the optimum output. In Chapter 3, it was shown that the optimum input, when substituted into the production function, would produce the optimum output. A similar dichotomy, called the *dual*, exists in linear programming. Every linear programming problem has a dual.

Consider the typical maximization problem. The farmer's goal is to maximize profits from his fixed bundle of resources. The value of the objective function of the linear programming problem, which is usually some measure of profit, also represents the value of his stock of resources—the ones that give rise to the constraints of the model—when the resource services are converted to output services and sold at prevailing prices in the output markets. Thus, in a very real sense, it can be said that the maximization process imputes a value to the farm's resources. These imputed values represent the worth of the inputs when used within the firm, not their value when sold directly in the input markets outside the firm. But this imputation is at best indirect.

The dual of the maximization problem is designed to determine the imputed value of the resources in a direct fashion. Each resource is assumed to have an imputed value, called a *shadow* price. The shadow price is defined as the number of dollars that would be added to profit by an additional unit of the resource. Suppose that a simple model includes two products, Y_1 and Y_2, and two resources, X_1 and X_2, and that each resource has a shadow price, λ_1 for X_1 and λ_2 for X_2. Because the production functions in question are linear with constant returns to scale, the total imputed value of the resources can be written

$$\text{Total imputed value} = b_1\lambda_1 + b_2\lambda_2$$

where b_1 and b_2 represent the available amounts of the resources X_1 and X_2, respectively. This is the objective function of the dual of the maximization problem. b_1 and b_2 are given or fixed input quantities; λ_1 and λ_2 are measured in units of dollars.

The goal of the dual solution is to minimize the objective function!

If the total imputed value of the inputs in fact equals total profit to the farm, why does the dual seek to minimize it? The answer lies in the nature of the constraints.

Let us write out the constraints for the dual and then explain them. For the two input-two output example, the constraints would be

$$a_{11}\lambda_1 + a_{21}\lambda_2 \geqq P_{Y_1}$$
$$a_{12}\lambda_1 + a_{22}\lambda_2 \geqq P_{Y_2}$$

The a_{ij} are the production coefficients explained above in more general notation. a_{11} is the amount of X_1 used to produce a unit of Y_1. a_{21} is the amount of X_2 used to produce a unit of Y_1. The product $(a_{11}\lambda_1)$ is the imputed cost of using X_1 to produce a unit of Y_1 and $(a_{21}\lambda_2)$ is the imputed cost of using X_2 to produce a unit of Y_1. P_{Y_1} is the profit from the unit of Y_1. Thus, the constraint says that the total imputed cost (or value) of the resource quantities used to produce a unit of Y_1 must equal or exceed the profit received from that unit of Y_1. To restate this proposition: The imputed value to the farm of all inputs used to produce a unit of Y_1 must at least equal the profit earned from a unit of Y_1.

How are the values of λ_1 and λ_2 determined? Ultimately from the solution of the programming problem. They will represent the return the inputs can earn when put to their most profitable use in the farm business. Thus, an inequality can never hold for an enterprise in the final solution. If it did, the farmer would be losing money on that particular enterprise and could reallocate the resources to another enterprise and increase revenue. This implies that if Y_1 is to be produced, only the equality for Y_1 can hold in the final solution. When the equality holds, resources used to produce Y_1 will earn exactly what they could earn elsewhere on the farm, and the imputed cost of producing Y_1 exactly equals returns from Y_1. In other words, the value of the resources is determined by the revenues from the sale of the output. The farmer's objectives are met when the objective function is minimized. By doing so, he is ensuring that returns from Y_1 will equal returns elsewhere in his business.

The dual can now be stated in general notation. The original problem may be written as:

$$\begin{aligned} &\text{maximize} && P_{Y_1}Y_1 + P_{Y_2}Y_2 \\ &\text{subject to} && a_{11}Y_1 + a_{12}Y_2 \leqq b_1 \\ & && a_{21}Y_1 + a_{22}Y_2 \leqq b_2 \\ &\text{and} && Y_1 \geqq 0,\ Y_2 \geqq 0 \end{aligned}$$

where all variables are defined as before. The dual of this problem may be written as:

$$\begin{aligned} &\text{minimize} && b_1\lambda_1 + b_2\lambda_2 \\ &\text{subject to} && a_{11}\lambda_1 + a_{21}\lambda_2 \geqq P_{Y_1} \\ &&& a_{12}\lambda_1 + a_{22}\lambda_2 \geqq P_{Y_2} \\ &\text{and} && \lambda \geqq 0, \lambda_2 \geqq 0 \end{aligned}$$

where again all variables are as defined before. This notation is similar to that used earlier except that the general notation "a_{ij}" is used to denote production coefficients.

The dual solution can be illustrated using a simple example. Consider the following problem:

$$\begin{aligned} &\text{maximize} && \$3Y_1 + \$2Y_2 \\ &\text{subject to} && Y_1 + 0.5Y_2 \leqq 4 \\ &&& Y_1 + Y_2 \leqq 5 \\ &\text{and} && Y_1 \geqq 0, Y_2 \geqq 0 \end{aligned}$$

The solution is shown in Figure 9–5A. The isorevenue line has a slope of $-3/2$ and profit is maximized when $(Y_1, Y_2) = (3, 2)$ and total profit is \$13. The isorevenue line for \$13 touches the feasible set at (3, 2).

The dual for this problem would be written

$$\begin{aligned} &\text{minimize} && 4\lambda_1 + 5\lambda_2 \\ &\text{subject to} && \lambda_1 + \lambda_2 \geq 3 \\ &&& 0.5\lambda_1 + \lambda_2 \geqq 2 \\ &\text{and} && \lambda_1 \geqq 0, \lambda_2 \geqq 0 \end{aligned}$$

The solution to the dual problem is shown in Figure 9–5B. For this problem the isorevenue line has a slope of $-4/5$ where 4 and 5 are the total amounts of X_1 and X_2, respectively. The shadow price for X_1 at the point of optimal solution is $\lambda_1 = \$2$ and for X_2 is $\lambda_2 = \$1$. Total profit would be $4 \cdot \$2 + 5 \cdot \$1 = \$13$. As mentioned, the shadow prices measure the increase in profit that would result from the use of one more unit of input. Thus, an additional unit of X_1 would add \$2 to profit and an additional unit of X_2 would add \$1 to profit. The inputs are being used in the most efficient manner; any other allocation would reduce their earnings in the firm.

It is of interest to examine the nature of the dual for different output prices. Suppose that $P_{Y_1} = \$1$ and $P_{Y_2} = \$2$ so that the total

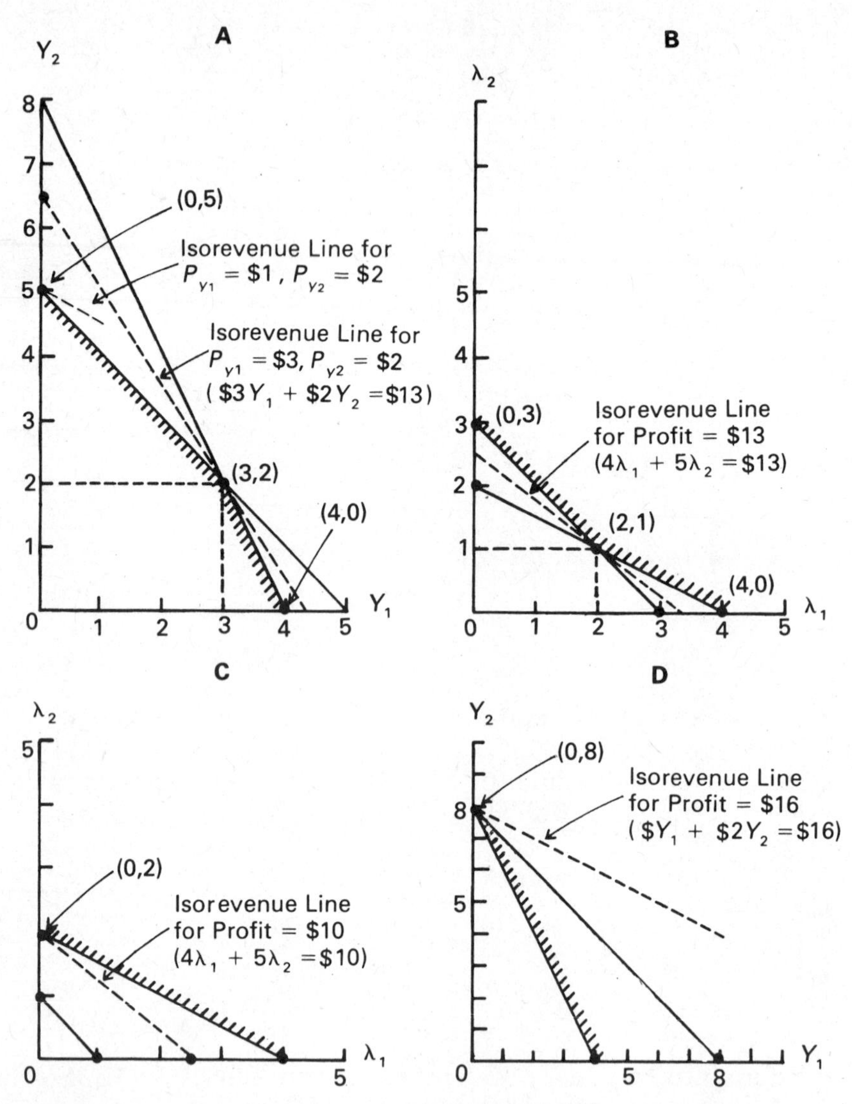

Figure 9–5. A linear programming problem and its dual.

profit is \$1Y_1 + \$2Y_2. A check of the feasible points in Figure 9–5A shows that the point (0, 5) is optimal for these prices and that total profit will be \$10. On the graph (Figure 9–5A), the isorevenue curve starts at (0, 5) and has a slope of $-1/2$. In this case, enterprise Y_2 dominates Y_1. A unit of Y_2 uses less X_1 than a unit of Y_1, the same amount of X_2 and, under the new prices, sells for twice as much. Because it uses fewer resources and earns a larger profit per unit, it has to dominate. The dual solution for this problem is shown in Figure 9–5C.

In Figure 9–5C, the constraint for Y_2 lies completely above the constraint for Y_1, demonstrating dominance graphically. (For later use, note that the equation for the Y_2 constraint is given by $\lambda_2 = 2 - 1/2\lambda_1$.) The isorevenue line has a slope of $-4/5$ and touches the Y_2 constraint at (0,2), where profit is $10. The shadow prices are zero for X_1 and 2 for X_2. X_2 is the limiting factor; the optimal solution, 5 units of Y_2, results in a surplus of 1.5 units of X_1. Therefore, an increase of X_1 would add nothing to profit. (Recall the isoquants for the fixed-proportion production function in Figure 9–1.) A one-unit increase of X_2 will increase profit by $2. The reason is that a unit of Y_2 requires one unit of X_2 and returns $2 profit.

As a final example, assume $P_{Y_1} = \$1$ and $P_{Y_2} = \$2$ as before, but now the supply of X_2 is increased to 8 units. X_1 remains at 4 units. The solution to the maximization problem in terms of Y_1 and Y_2 is presented in Figure 9–5D. Because Y_2 dominates Y_1, it is a corner solution, occurring at (0, 8) with total profit equal to $\$1 \cdot 0 + \$2 \cdot 8 = \$16$. However, both X_1 and X_2 are limiting. An increase in the output of Y_2 would require an increase in both inputs. The objective function for the dual solution in this case is $4\lambda_1 + 8\lambda_2 = \$16$. When this is simplified, it can be expressed as $\lambda_2 = 2 - 1/2\lambda_1$. But this objective function is coincident with the dual constraint for Y_2. If graphed on Figure 9–5C, it would be indistinguishable from the constraint that defines the lower boundary of the feasible set. Thus, the shadow prices are not unique in this case! Any pair of values for λ_1 and λ_2 that satisfy $4\lambda_1 + 8\lambda_2 = \$16$ with $\lambda_1 \geqq 0$ and $\lambda_2 \geqq 0$ will suffice.

When both inputs are limiting in this particular example, increases in output can be achieved only by moving out the expansion path associated with the fixed-proportion production function. The quantities of both inputs must be increased. But the two inputs are technical complements along the expansion path, and shadow prices therefore cannot be imputed to them individually. When one input is available in surplus amounts, the shadow price for the limiting input suggests the amount that profit would be increased, per unit of input added, when that input is increased until the input combination available is again on the expansion path, but at a higher level of output. When X_2 is increased from 5 units to 8 units, output expands and profit increases from $10 to $16, or $2 per unit of X_2 added, and neither input is in surplus. Thus, the first solution, point (0, 2) on Figure 9–5C, was similar to point B on the isoquant map for the fixed-proportion production function in Figure 9–1. Input X_2 was limiting output; increasing X_2 along the horizontal dashed line increases output until a higher level of production is achieved at point *D*. At point *D*, both inputs become limiting; this solution is similar to point (0, 8) in Figure 9–5D. A like interpretation could be given for point *C*, with the roles of the inputs reversed.

Shadow prices are sometimes called the opportunity costs of the inputs. Opportunity cost is usually defined as the return a resource could earn when put to its best alternative use. Because shadow prices determine the resource earnings within the farm business, they represent opportunity costs only when the farmer is considering using his inputs elsewhere.

A set of general principles or theorems applies to a linear programming problem, often called the *primal*, and its dual. These are:

1. The optimal values of the objective functions of the primal and the dual are always equal.
2. The optimal value of a variable in one system (primal or dual) is zero when the corresponding constraint in the other system is a strict inequality and is nonnegative when the corresponding constraint is a strict equality.

Study of the above examples will illustrate these general principles. For any maximization problem (primal) there will be a dual that is a minimization problem. The reverse is also true. For any minimization problem (primal) there will be a dual that is a maximization problem. Thus, arguing which is the primal or the dual is tantamount to the chicken-and-egg question. The researcher can formulate a given problem either way. Before the advent of electronic computers, one method of selection was to choose the form that was simplest computationally.

Assumptions and Selection of Activities

In the above sections, we defined the fixed-proportion production function and demonstrated its use to develop the basic maximizing and minimizing models of linear programming. With this background, we will now consider production processes in more detail. This section has two purposes. The first is to demonstrate how the linear programming model can be interpreted to represent solutions similar to those obtained from the classical theory presented in earlier chapters. The second is to present and explain the basic mathematical assumptions inherent in linear programming procedures. By doing so, we will show that, even with the assumption of linearity, linear programming models can be quite general—that is, how many of the concepts of the theory of the firm can be embodied in a linear model within the confines of the fixed-proportion production functions.

The fixed-proportion production function results from the manager's decision to produce a given output using a particular production process. After the method of production is selected, a budget is developed to determine the resource requirements of the production process. Finally, the resource requirements are converted to production coefficients, the a_{ij} defined above, by expressing the resource requirements per unit of output.

But a given production can usually be produced in a number of ways. Many systems are available for beef production, ranging from intensive drylot feeding to extensive pasture grazing. Similarly, hogs can be produced in confinement or by utilizing one of a number of less intensive systems. Farrowings can be planned for different times of the year. Similarly, crops such as corn and soybeans can be produced using different combinations of plant population, fertilizations, tillage and spraying, etc. Crop rotations can be varied. For a particular product, each method of production involves the use of a different set of technologies and requires a different mix of resources.

Each production method or process that is possible for the production of a given product defines a different fixed-proportion production function. In linear programming terminology, these alternative methods of production are called *processes* or *activities*. Because they are based on the production process, each activity necessarily embodies the technology and resource requirements of the production process. In Chapter 4, a crucial assumption was that the manager always selects the best technology, that is, the one that produces the most output per unit of input. In linear programming, the manager may select a number of alternative technologies—each the most efficient for its particular ratio of resource requirements—and include each in the model as a separate activity. As a result of the search for maximum profits, the linear programming solution will contain only those most suitable for the farm. The only restriction is that there can only be a *finite* number of such activities.

As an example, consider the production of an output, Y, using two inputs, X_1 and X_2, using a number of different production methods. Because the production functions are linear and emanate from the origin, they display constant returns to scale. Each activity, or production process, can be visualized as an arrow extending from the origin into the positive orthant; because there must be a finite number, all possible activities would appear as a cluster of arrows pointed into the positive space (Figure 9–6A). The arrowheads indicate constant returns; production is at a constant rate and diminishing returns never occur.

When the cluster of arrows representing activities is viewed directly from above, the resulting picture is not an isoquant map, as presented in Chapter 4, but rather will be a set of rays passing through the origin and extending outward indefinitely as in Figure 9–6B. Each ray, of course, represents a different production process.

All of the rays in Figure 9–6B represent activities to be considered in the linear programming model. The only ones that might occur in Figure 9–6A but not in Figure 9–6B will be those that are clearly dominanted. That is, if two activities use the same input ratio, their rays will be coincident in Figure 9–6B, and the most efficient one would be selected for inclusion in the model.

As an example, suppose three activities from Figure 9–6B are selected for further examination. All produce the same output, Y, but

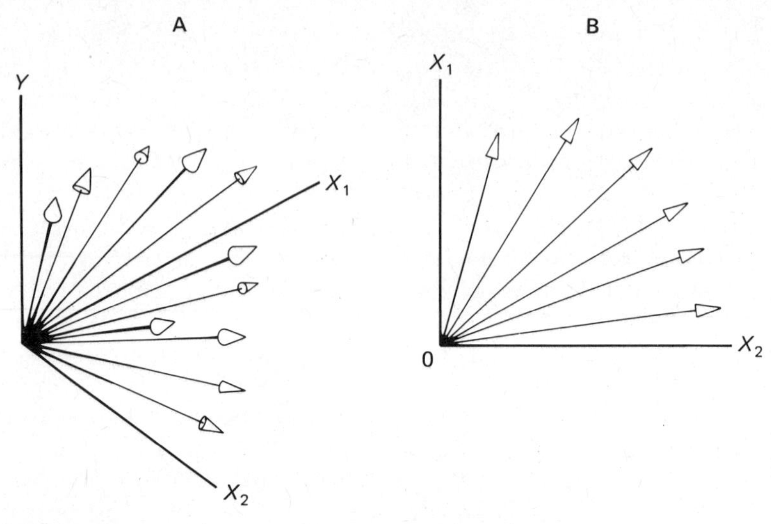

Figure 9–6. Linear programming activities.

each requires a different input ratio to do so. Activity one uses 5 units of X_1 to 3 of X_2; activity two uses 3 to 4, respectively, and activity three uses 1 to 4, respectively. The production coefficients, or the input requirements per unit of output, for the activities are

	A_1	A_2	A_3
X_1	0.555	0.375	0.250
X_2	0.333	0.500	1.000

These activities are shown as rays in Figure 9–7A. Output of Y resulting from various combinations of input are also shown in Figure 9–7A. These can be determined using the production coefficients. For A_3, 2 units of X_1 and 8 units of X_2 will produce 8 units of output; 6 units of X_1 and 24 units of X_2 will produce 24 units of output, etc.

Two additional properties of the linear programming activities can now be demonstrated. First, an activity can be operated at any rate desired to produce a given output level. Activity A_1 can produce, 8, 12, 24, or 36 units of output, but it can also produce 1.0, 3.14, 13, or 1000.01 units. Any non-negative value of output can be produced; this is known as the *divisibility* assumption. Second, the output of activities can be added. If A_1 is operated to produce 10 units and A_2 is operated to produce 15 units, the resulting output will be 25 units of Y. This is known as the *additivity* assumption; the combined output never exceeds the sum of the output of each activity.

Divisibility and additivity permit us to derive a concept somewhat similar to the isoquant of Chapter 4, although, in fact, input substitution is not permitted. Consider the production of 24 units of output.

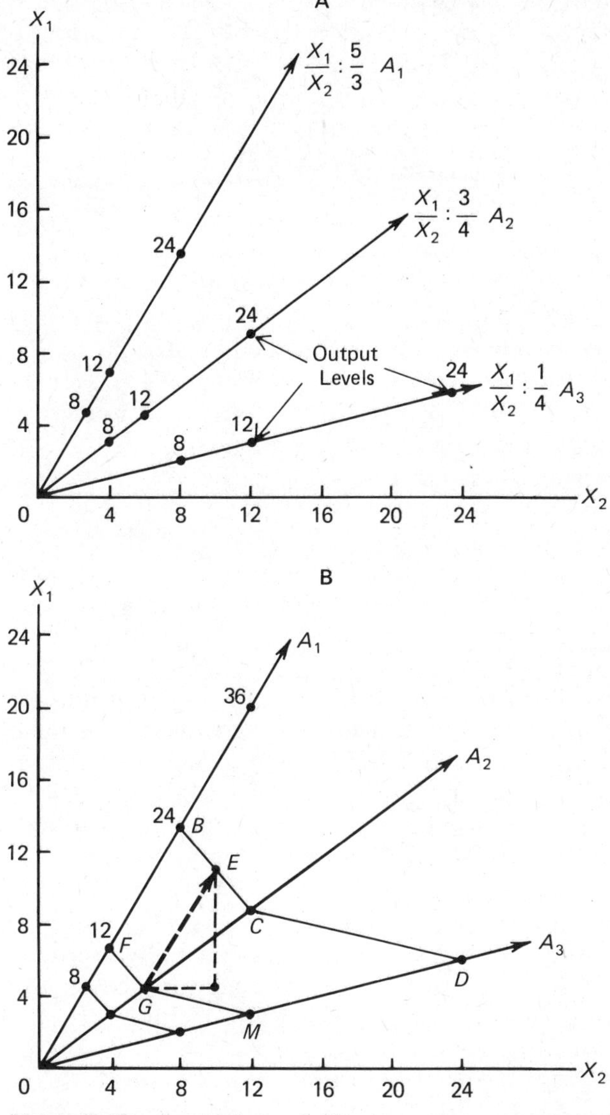

Figure 9–7. Combining activities to produce one product.

A_1 could be used to produce 24 units or A_2 could be used to produce 24 units (Figure 9–7A). Or A_1 could produce 23 units while A_2 produces one unit; A_1 could be used to produce 22 units while A_2 is increased to 2 units, etc. As production using A_1 is reduced while production using A_2 is increased, the line *BC* in Figure 9–7B is traced out. By combining A_2 and A_3 to produce 24 units of *Y*, line *CD* could also be determined. Thus, the connected line segments *BCD* represent input combinations required to produce 24 units of output most efficiently using combinations of pairs of the three activities. (Why are combinations of A_1 and A_3 not considered?) Similar lines are presented in Figure 9–7B for 8 and 12 units of output. In general, then, it can be seen that any input combination contained in the cone $A_1 0 A_3$ can be utilized to produce Y by combining two of the three activities.

A degression might help to clarify how the line BC is visualized. Suppose 12 units of Y are produced with A_2; this point, G, lies on ray A_2 where $X_2 = 6$ and $X_1 = 4$. But production of 12 units of Y with A_1 requires 4 units of X_1 and 6.67 units of X_2 (see point F on Figure 9–7B). In terms of the activities, A_2 is operating at G while A_1 is operating at F. In terms of total resource use, the requirements of the two must be added. Thus, starting at point G, adding 4 units of X_2 and 6.67 of X_1 results in total resource use of 10.67 for X_1 and 10 for X_2, and a total output of 24, all represented by point E. The right triangle with hypotenuse OF is identical to the right triangle with hypotenuse GE. This characterization is not unique; production for A_2 could be added to production from A_1, starting at point F.

We have seen how activities can be combined to utilize input combinations not possible from utilizing either activity alone. A second important point is that linear programming activities can also be designed to display diminishing returns to one input when the others are held constant. For example, assume X_2 is constant at 12 in Figure 9–7B. If so, production of 12 units of output (at point M) requires 3 units of X_1; an increase of output by 12 units to 24 (at point C) will require an additional 6 units of X_1. But if an additional 12 units of Y is desired, by increasing output from 24 to 36 (at point N), then 11 additional units of X_1 will be required. In this case, diminishing returns occurs because, with X_2 fixed, production can be increased only by utilizing activities that are successively less efficient in the use of X_1.

It is now time to consider the effects of prices on our analysis. For a given set of input prices, one activity will usually dominate the others and thus will produce at least cost (the exception being when two have identical costs). For example, if X_1 costs \$2 while X_2 costs \$1 per unit, then A_1 will produce a unit of output for \$1.44, A_2 for \$1.25, and A_3 for \$1.50. When the inputs can both be purchased, the manager will use A_2 and expand out the ray for A_2 until all capital is expended. He would not produce at a point such as E; production at E would cost \$2(10.67) + \$1(10) = \$31.34, and this amount of capital would permit the production of 25.1 units of Y using A_2 alone. However, linear programming models are often used to determine the optimal use of resources, at least some of which are fixed in quantity for the farm or, like machinery and labor, available in one month of a year but not another. When one or more inputs are fixed, activities may be combined to produce more output than one could alone. That is, when X_2 is fixed at 10, then the manager can purchase 10.67 of X_1 and produce 24 units of Y (at point E) and this represents the maximum possible production from the 10 units.

To summarize the assumptions, linear programming activities represent linear fixed-proportion production functions (*linear*); there must be a definite number of such activities (*finite*); an activity can produce any non-negative output (*divisible*); and the output of all activities sums to the total output for the farm (*additivity*). A review of the examples in the above sections will determine that these assumptions were implicitly used. We have now made them explicit.

The generality of the linear programming model can now be seen. A properly specified model will determine optimal resource allocation among enterprises and also will determine the most profitable combination of enterprises. Maximum profits for the firm will be determined and the shadow prices will place a value on limited resources. Therefore, although the assumptions appear rather restrictive, many useful linear programming models have been developed by imaginative researchers.

Supply Response

One important application of linear programming in agriculture has been its use to determine supply curves for agricultural products. Supply curves can be determined for individual firms and then aggregated over groups of "typical" firms to determine industry response.

The conceptual basis for determining supply response using linear programming is quite straightforward. The maximization model presented above will be used as an illustration. The production possibility curve for the two products, Y_1 and Y_2, first presented in Figure 9–2B is reproduced in Figure 9–8A. The isoprofit line for the two products, not shown in the figure, has the slope $(-P_{Y_1}/P_{Y_2})$, where, as always in this chapter, these symbols represent profit per unit of output from each enterprise. Supply response for a particular product, say, Y_2, is determined by holding the price and profit from Y_1 constant while varying the price and profit from Y_2. We will first show how output will respond to profit changes. The conversion from profit to price will then be discussed.

We start by assuming the firm is producing at point F in Figure 9–8A. Because the line segment EF has a slope of -3 and the profit from Y_1 is \$1, F can be the firm's profit-maximizing position only if profit from Y_2, P_{Y_2}, is less than \$0.33 and the isoprofit line has a larger (absolute) slope than EF. A zero amount of Y_2 will be produced. Next, let the profit from Y_2 become exactly \$0.33; the isoprofit line will have a slope of -3 and the producer may chose to produce any combination of Y_1 and Y_2 on EF, including either E or F. Therefore, at a profit of \$0.33, production of Y_2 can vary between and include 0 and 6. As profit increases above \$0.33, production of Y_2 will remain at 6 until the slope of the isoprofit line becomes $-3/2$, or when P_{Y_2} = \$0.66. At this price, the isoprofit line becomes coincident with line DE and any amount of Y_2 between and including 6 and 13.5 will be produced. As profits on Y_2 are increased, this adjustment process continues until profits from Y_2 exceed \$2.00. For all profit levels above \$2.00, 15 units of Y_2 will be supplied. The response in the supply of Y_2 to profit can be summarized in the following seven cases representing the four points C, D, E, F and the three line segments connecting them:

If $0 \leqq P_{Y_2} < \$0.33$, then $Y_2 = 0$.

If $P_{Y_2} = \$0.33$, then $0 \leqq Y_2 \leqq 6$.

If $\$0.33 < P_{Y_2} < \$0.66$, then $Y_2 = 6$.

If $P_{Y_2} = \$0.66$, then $6 \leqq Y_2 \leqq 13.5$.

If $\$0.66 < P_{Y_2} < \$2.00$, then $Y_2 = 13.5$.

If $P_{Y_2} = \$2.00$, then $13.5 \leqq Y_2 \leqq 15$.

If $P_{Y_2} > \$2.00$, then $Y_2 = 15$.

Because of the assumption of constant returns to scale, each unit of Y_2 is produced at the same (constant) variable cost; profit can be converted to price by adding the variable cost per unit to profit. For example, if Y_2 costs \$1.00 per unit to produce, then at any price of Y_2 below \$1.33 only Y_1 will be produced. When the price of Y_2 increases to exactly \$1.33, then profit is \$0.33 and $0 \leqq Y_2 \leqq 6$, etc. Assuming a variable cost of \$1.00 for Y_2, a supply curve is presented in Figure 9–8B. A change in the cost of production in Y_2 would shift the supply curve vertically.

The supply curve resulting from linear programming is called a *step* supply function. Note that it has a different interpretation than the supply curves discussed in Chapters 3 and 4. The supply function in Figure 9–8B is determined by substitution of Y_2 for Y_1 as the relative profitability of the two shifts in favor of Y_2. In the earlier chapters, output was increased along the marginal cost curve by buying and using additional variable resources as the output price increased. No additional resources must be utilized to obtain the supply response function derived in Figure 9–8.

The traditional supply curve can be obtained from the linear programming model. Referring to Figure 9–7B, if X_2 is fixed at 12 units, then output can be increased from 12 to 36 units by using additional units of X_1. In this case, as the profit potential is increased, additional quantities of X_1 are purchased and production is gradually shifted away from A_3, which is most efficient in the use of X_2, the fixed input. Supply response analyses of this type are made possible in linear programming by increasing the capital constraint, i.e., the capital available to purchase variable inputs. As the supply response causes shifts away from activities most efficient in using X_1, costs will rise and an upward-sloping supply function for Y will be obtained. As an exercise, the student should study Figure 9–7B to determine the shape of the supply curve for Y as output increases from 0 to 36, with X_2 fixed at 12 units.

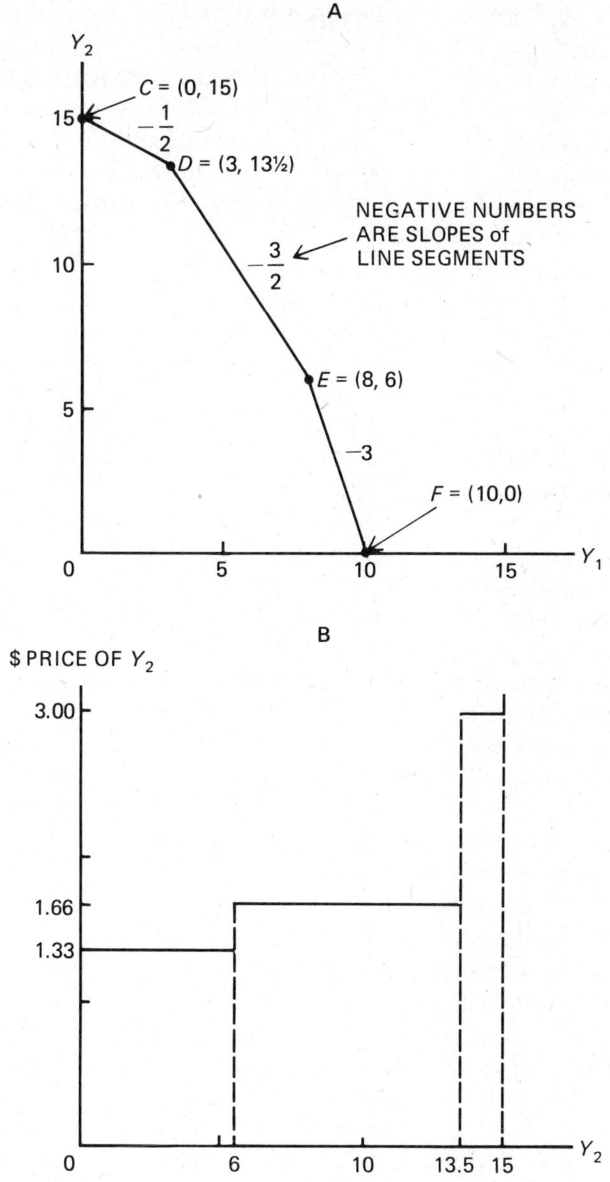

Figure 9–8. Step supply functions from linear programming.

THE SIMPLEX SOLUTION

The economic principles involved in linear programming have been presented using graphs and simple equations. However, a linear programming model of a farm, if it is to be at all realistic, will usually include several feasible enterprises using many different inputs. Graphic solutions for such problems become impossible; a systematic computational technique is required.

The simplex method is a computational procedure that starts with the budget data and proceeds step by step to a systematic solution of a programming problem. The method is a sequential set of arithmetic operations and as such is called an algorithm. In this section, a simple problem will be solved using the simplex algorithm. (The derivation of the algorithm will not be presented; it is mathematical in nature and beyond the scope and purpose of this text.)

The solution to the simplex method will determine (1) the combination of enterprises that maximize profit for the farm, (2) the amount of the maximum profit, and (3) shadow prices for the farm inputs. Thus, it includes information about the dual as well as the primal solution. While essentially a mathematical technique, detailed study of the algorithm will enhance the student's understanding of and appreciation for linear programming.

The Simplex Problem Matrix

Table 9–3 presents crop budgets for three crop enterprises.[2] The goal of the analysis will be to maximize gross returns minus variable costs. The objective function is often specified in this manner for farm planning purposes because only variable costs affect short-run decisions on the farm. Net farm income can be determined by subtracting fixed costs, such as taxes and depreciation, from the value of the final farm plan. The plan that maximizes income before fixed costs are subtracted will also maximize net farm income. Therefore, the budgets in Table 9–3 determine gross income minus variable costs.

The resources available on the farm are assumed to be 400 acres of land, 2200 hours of labor per year, and $5000 of operating capital. To solve the problem, the land, labor, and capital requirements must be estimated for each of the enterprises in Table 9–3. The budget information is based on units of one acre; therefore, one unit of each crop activity will require one acre of land. One acre of sorghum is estimated to require 4 man-hours of labor; one acre of wheat will require 2.5 man-hours of labor; and one acre of soybeans will require 3.5 man-hours of labor. For convenience, the operating capital requirements are rounded off to $15, $12, and $12 per acre for grain sorghum, wheat and soybeans, respectively.

The resource requirements per acre are called *production coefficients*. They are the "a_{ij}" defined in the previous section. Also called technical coefficients, they state the amount of input required per unit

[2]The problem presented here was developed by Professor Orlan Buller of Kansas State University. More detail on the problem is contained in his classroom manual, *Farm Management and Laboratory Manual and Supplemental Text,* Department of Economics, Kansas State University, Manhattan, Kan., 1975. We are indebted to him for permission to use this example.

Table 9–3 Budgets for Three Crop Enterprises

	Grain Sorghum	*Wheat*	*Soybeans*
Yield per acre	48 bu.	30 bu.	15.7 bu.
Price	$ 1.05/bu.	$ 1.50/bu.	$ 3.00/bu.
Gross income	50.40	45.00	47.10
Variable costs			
Seed	$ 0.80	$ 2.00	$ 2.50
Fertilizer	5.75	5.00	
Pesticides	3.00		4.50
Fuel and oil	3.00	3.00	3.00
Repairs	2.50	2.50	2.50
Total variable cost	$15.05	$12.50	$12.50
Income less variable costs	$35.35	$32.50	$34.60

of output for each activity. In this case there are three resources, and each enterprise has three production coefficients. The coefficient must always exist, even if it is zero.

The basic information needed for the simplex solution is now available. The next step is to formulate the problem in an orderly fashion to begin computations. The initial simplex tableau is presented in Table 9–4. At first glance this tableau may appear rather formidable, but we shall proceed to explain the concepts needed to start the solution. As each step is developed, the concepts involved should be clarified.

The mathematical technique underlying the simplex algorithm can be explained as follows: The inequalities that represent the resource constraints are converted to equalities. The programming model then consists of a set of linear equations with a linear objective function and non-negativity restrictions. A solution to the set of linear equations is found that is called "feasible"; that is, it satisfies both the linear restrictions and the non-negativity requirements. By using the principle of opportunity cost within the firm, the solution is examined to determine whether a change would increase profits. The technique is iterative—it can be repeated. If a change is profitable, a new solution is determined.

Disposal activities are used to convert the inequalities to equalities. In Table 9–4, three new activities are added to the real activities. The real activities are based on a one-acre unit, and their coefficients for land, labor, and capital are presented in columns labeled Y_1, Y_2, and Y_3. The disposal activities are labeled Y_4, Y_5, and Y_6. The disposal activity for land is also based on one acre; it requires one acre of land, uses no labor or capital, and earns nothing. When one unit of this "slack" activity is used, an acre of land is left idle. The disposal activities for labor and capital have a similar interpretation. Nonuse of a unit of a

Table 9–4 Initial Tableau for the Simplex Method

Activities in the Solution Set			*Real Activities*			*Disposal Activities*			*Ratio*
Net Price C_j	*Name*	*Amount* B	*Grain Sorghum* Y_1	*Wheat* Y_2	*Soybeans* Y_3	*Land* Y_4	*Labor* Y_5	Capital Y_6	R
0	Land-Y_4	400	1	1	1	1	0	0	400
0	Labor-Y_5	2200	4	2.5	3.5	0	1	0	550
0	Capital-Y_6	5000	15	12	12	0	0	1	333.33
Net Price of activities: C_j			35.35	32.50	34.60	0	0	0	
Opportunity cost of activities: Z_j			0	0	0	0	0	0	
Shadow prices: $Z_j - C_j$			−35.35	−32.50	−34.60	0	0	0	
Value of program = $0									

resource requires one unit of that resource, none of the others, and returns no profit. One unit of the labor disposal activity is defined to be one hour of labor, and one unit of the capital disposal activity is defined to be one dollar. In practical terms, the "production" of a unit of a disposal activity simply means that the resource does not have to be used to produce sorghum, wheat, or soybeans.

The coefficients for the real and disposal activities are contained in the upper center and right-hand portions of Table 9–4. The upper left-hand section of the table, "Activities in the Solution Set," contains a listing of the initial solution to the linear programming problem. This section contains the net price, the name, and the amount of the activities selected to be in the initial solution. The initial feasible solution always assumed for the simplex method is not exciting but it is easy to find. Initially, all resources are assumed to be used to produce disposal activities. Because these activities earn no income, the initial income (value of the program) is zero.

The initial feasible solution can be expressed in equations as follows: The constraint equations are

$$1Y_1 + 1Y_2 + 1Y_3 + 1Y_4 + 0Y_5 + 0Y_6 = 400$$
$$4Y_1 + 2.5Y_2 + 3.5Y_3 + 0Y_4 + 1Y_5 + 0Y_6 = 2200$$
$$15Y_1 + 12Y_2 + 12Y_3 + 0Y_4 + 0Y_5 + 1Y_6 = 5000$$

and the objective function is

$$\textit{Value of the Program} = \$35.35Y_1 + \$32.50Y_2 + \$34.60Y_3 + 0Y_4 + 0Y_5 + 0Y_6$$

The initial solution is $Y_1 = 0$, $Y_2 = 0$, $Y_3 = 0$, $Y_4 = 400$, $Y_5 = 2200$, and $Y_6 = 5000$. Thus, the constraints are fulfilled

$$1(0) + 1(0) + 1(0) + 1(400) + 0(2200) + 0(5000) = 400$$
$$4(0) + 2.5(0) + 3.5(0) + 0(400) + 1(2200) + 0(5000) = 2200$$
$$15(0) + 12(0) + 12(0) + 0(400) + 0(2200) + 1(5000) = 5000$$

and the objective function equals zero

$$0 = \$35.35(0) + \$32.50(0) + \$34.60(0) + 0(400) + 0(2200) + 0(5000)$$

A farm plan that leaves all resources idle and earns no income will not be regarded with favor by most farmers. A better plan must be found; the purpose of the three rows in the lower section of Table 9–4 is to provide the information needed to determine a better solution—if one exists. The first row of this section, "Net Price of Activities," lists the profit earned by a unit of each enterprise to be considered for the farm plan, including the disposal activities. The next row, "Opportunity Cost of Activities," lists the revenue within the farm that would have to be foregone to produce a unit of each activity. In the initial plan nothing of value is being produced; therefore, producing a unit of any real activity would cause no sacrifice of income. The disposal activities are being produced but have net prices of zero; therefore, reducing the production of the disposal activities does not reduce income.

The last row of the lower section presents the shadow prices for all activities. The shadow prices for the disposal activities give the imputed value of the inputs for this farm plan. Because nothing of value is being produced, the inputs have no imputed value. The shadow prices for the real activities are nonzero and assume an important role in the simplex solution. The shadow prices for the real activities suggest the apparent most profitable change that can be made in the initial solution. Grain sorghum returns \$35.35 per unit and has an opportunity cost of zero. Its shadow price, $Z_1 - C_1 = -\$35.35$, means that the objective function will be increased by \$35.35 for each unit of grain sorghum produced. Similar interpretations hold for the shadow prices of wheat and soybeans. Because a unit of sorghum would increase the objective function by the largest amount per unit of the activities (sorghum has the largest absolute shadow price), it appears that profits can be increased the most by producing grain sorghum.

All real activities have a negative shadow price, suggesting that profits could be increased by producing any of the three—producing something appears better than producing nothing. But sorghum is the one to start with because it appears to produce the most revenue; the word "appears" is intentional because the solution is iterative and the optimal farm plan could include an activity that produces less profit per unit of output, but more profit per unit of input. Nonetheless, the simplex method will always bring the activity with the highest net profit into the program first.

The final bit of data in the initial tableau is the value of the program. This has been shown to be zero and can be computed directly from the table by multiplying the amount of the activity in the solution set (column B) by the appropriate price of that activity (column C_j). Thus, value of the program $= \$0 \cdot 400 + \$0 \cdot 2200 + \$0 \cdot 5000 = \$0$.

Our explanation of the initial tableau and the initial feasible solution is now complete. To select a more profitable solution, we examine the $Z_j - C_j$ row of Table 9–4 and select the smallest negative (largest absolute) shadow price. As described above, that will be grain sorghum. We therefore decide to produce all the grain sorghum possible given the available resources. The maximum attainable production can be determined by dividing the sorghum production coefficients into the resource amounts. Therefore, the amounts of sorghum production each resource would allow are

Land	400/1 = 400 acres
Labor	2200/4 = 550 acres
Capital	5000/15 = 333.33 acres

These are included in Table 9–4 in the column labeled "*R*." Capital is the most limiting; therefore, the initial solution will be changed by adding 333.33 acres of sorghum. But this change will also change the number of units of the disposal activities that can be produced. To show these computations, a second tableau is needed.

The first iteration of the simplex solution is shown in Table 9–5. The new feasible solution is 333.33 acres of sorghum, 866.68 hours of unused labor, and 66.67 unused acres. The value of the program is now $11,783.22. This is an obvious improvement. All the calculations that follow are in essence needed to determine if a new plan will increase profit even more. The structure of Table 9–5 is exactly the same as Table 9–4. We now will derive the new numbers in Table 9–5.

The sorghum row: Because capital was the limiting resource for grain sorghum production, sorghum is said to "replace" capital in the feasible solution. This is intuitively reasonable: Sorghum is using all the capital. Another capital-using activity can be produced only by taking capital from sorghum (reducing sorghum production) and reallocating that "freed" capital to the new enterprise. Thus, in a real sense, sorghum *is* capital in the new solution. The grain sorghum row "comes in on" the capital row.

The values in the grain sorghum row of Table 9–5 are found by dividing each value in the capital row of Table 9–4 by the sorghum capital requirement. Thus, in order from left to right, the new coefficients are

$$\frac{5000}{15} = 333.33;\ \frac{15}{15} = 1;\ \frac{12}{15} = 0.8;\ \frac{12}{15} = 0.8;$$
$$\frac{0}{15} = 0;\ \frac{0}{15} = 0;\ \frac{1}{15} = 0.067$$

These values have a significant economic interpretation. As noted, 333.33 is the maximum amount of grain sorghum that can be produced in the new farm plan. The other six numbers are the marginal rate of product substitution between sorghum and the six productive activities. Thus, to grow one additional acre of grain sorghum, one acre of grain sorghum must be given up. That is rather trivial, but to grow one acre of wheat or one acre of soybeans, 0.8 of an acre of sorghum must be

Table 9–5 The First Iteration for the Simplex Solution

Activities in the Solution Set			*Real Activities*			*Disposal Activities*			*Ratio*
Net Price C_j	*Name*	*Amount* B	*Grain Sorghum* Y_1	*Wheat* Y_2	*Soybeans* Y_3	*Land* Y_4	*Labor* Y_5	*Capital* Y_6	R
0	Land-Y_4	66.67	0	0.2	0.2	1	0	−0.067	333.35
0	Labor-Y_5	866.68	0	−0.7	0.3	0	1	−0.268	2888.93
35.35	Sorghum-Y_1	333.33	1	0.8	0.8	0	0	0.067	416.66
Net price of activities: C_j			35.35	32.50	34.60	0	0	0	
Opportunity cost of activities: Z_j			35.35	28.28	28.28	0	0	2.36	
Shadow prices: $Z_j - C_j$			0	−4.22	−6.32	0	0	2.36	
Value of program = $11,783.22									

given up. Soybeans and wheat require \$12 of operating capital per acre while sorghum requires \$15. To add an acre of wheat or soybeans, sorghum must be reduced 0.8 of an acre because (0.8) (\$15) = \$12. The disposal activities require no capital, so the entries in those columns are zero. Finally, to allow a dollar of capital to go unused, sorghum production would have to be reduced by 1/15 = 0.067 acres.

Calculating column B: Production of 333.33 acres of grain sorghum will reduce the amount of unused labor and land. The amounts not used are assumed to remain in the disposal activities and are listed in column *B*. These are calculated as follows:

Land: $400 - (333.33 \times 1) = 66.67$

Labor: $2200 - (333.33 \times 4) = 866.68$

or, in general,

The amount of resource remaining in the disposal activity = (*the amount of the resource available*) – (*number of units of the new enterprise brought into the feasible solution multiplied by the resource requirement per unit of the new enterprise*).

Thus, 333.33 acres of sorghum require one acre of land each and four hours of labor each. When subtracted from the total available, 66.67 acres of land and 866.68 hours of labor remain.

The column coefficients: The remaining coefficients in the columns of the upper section of Table 9–5 will be computed. We will then discuss their interpretation. The general computational formula for any of the six activities (columns) is:

New coefficient in position (i, j) *in Table 9–5* = [*old coefficient in position* (i, j) *in Table 9–4*] – [*the marginal rate of substitution of activity* j *for sorghum that is found in the grain sorghum row* (*row three*) *multiplied by the coefficient of grain sorghum activity found in the row* i *of Table 9–4*].

The remaining column coefficients in Table 9–5 are given by

Grain Sorghum column:	$1 - (1 \times 1) = 0$ $4 - (1 \times 4) = 0$
Wheat column:	$1 - (0.8 \times 1) = 0.2$ $2.5 - (0.8 \times 4) = -0.7$
Soybeans column:	$1 - (0.8 \times 1) = 0.2$ $3.5 - (0.8 \times 4) = 0.3$
Land Disposal column:	$1 - (0 \times 1) = 1$ $0 - (0 \times 4) = 0$
Labor Disposal column:	$0 - (0 \times 1) = 0$ $1 - (0 \times 4) = 1$
Capital Disposal column:	$0 - (0.067 \times 1) = -0.067$ $0 - (0.067 \times 4) = -0.268$

The logic underlying these computations is as follows: Suppose one unit of the wheat activity is added to the feasible plan, using one of the 66.67 idle acres. This acre of wheat will require \$12 of capital which must be obtained by reducing sorghum by \$12/\$15 = 0.8 of an acre. Therefore, adding a unit of wheat actually requires only an additional 0.2 of an acre from the 66.67 idle acres. This coefficient in the wheat column is computed as [1 − (0.8 · 1) = 0.2] or (land requirement for wheat − land made available for use on wheat when capital is shifted from sorghum to wheat). The land available due to the reallocation of capital is equal to the number of units the sorghum activity is reduced per unit of wheat multiplied by the land requirement per unit of sorghum.

The labor requirement for wheat in Table 9–5 underscores this meaning. Introducing a unit of wheat into the plan will require 2.5 hours of labor. But a unit of sorghum uses 4 hours of labor. As shown, a unit of the wheat activity will displace 0.8 of a unit of sorghum, freeing 0.8 · 4 = 3.2 units of labor. Adding a unit of wheat thus frees 0.7 hours of labor to be used elsewhere on the farm; i.e., 2.5 − 3.2 = −0.7. Hence the negative sign. When sorghum production is limited by capital and wheat is introduced into the program, the wheat activity actually *supplies* labor to the farm—relative, of course, to sorghum!

Opportunity cost and shadow prices: The opportunity cost row measures the income sacrificed when an activity is increased by one unit. Sorghum is using all the available capital; therefore, an increase in any activity requires a reduction in sorghum. The amount that income is reduced by this shift is found by multiplying the marginal rates of substitution between each activity and sorghum by the income from one unit of sorghum. To produce an additional unit of sorghum, one unit would have to be sacrificed; therefore, the reduction in income would be \$35.35 · 1 = \$35.35. This is again trivial. To produce one unit of wheat, 0.8 of a unit of sorghum would be sacrificed; therefore, income would be reduced \$35.35 · 0.8 = \$28.28. The complete computations for this row are

Grain Sorghum:	\$35.35 (1) = \$35.35
Wheat:	\$35.35 (0.8) = \$28.28
Soybeans:	\$35.35 (0.8) = \$28.28
Land Disposal:	\$35.35 (0) = 0
Labor Disposal:	\$35.35 (0) = 0
Capital Disposal:	\$35.35 (0.067) = \$2.36

If one unit of capital is left unused, \$2.36 of income is lost. This could be computed directly from Table 9–4 as \$35.35/\$15 = \$2.36, the amount of income sorghum produces per dollar.

The shadow prices are computed by subtracting the net price of each activity from its opportunity cost. The economic logic of this is clear. Thus, the last row of Table 9–5 is given in each case by $Z_j - C_j$, or

Grain Sorghum:	\$35.35 − \$35.35 = 0
Wheat:	\$28.28 − \$32.50 = −\$4.22
Soybeans:	\$28.28 − \$34.60 = −\$6.32
Land Disposal:	\$0 − 0 = \$0
Labor Disposal:	\$0 − 0 = \$0
Capital Disposal:	\$2.36 − 0 = \$2.36

Examination of the shadow prices reveals two useful facts. First, two of the real activities have negative shadow prices, suggesting that income can be increased by introducing another real activity into the feasible plan. Of the two possibilities, soybeans has the smallest negative (largest absolute) shadow price; hence it will be introduced next. Second, of the three inputs only capital is limiting and only capital has a positive shadow price. By multiplying the shadow price of capital, \$2.36, by the amount of capital, \$5000, the value of the program is determined to be \$11,800, approximately the same as before (the difference is due to rounding).

Soybeans will be introduced into the feasible plan next. The limiting amount will again be found by dividing the coefficients in the soybeans column into the amounts of "resources" in column B. These are

Land:	66.67/0.2 = 333.35
Labor:	866.68/0.3 = 2888.93
Sorghum:	333.33/0.8 = 416.66

The number in the sorghum row, \$416.66, is actually the capital constraint for soybeans; it could have been computed directly from Table 9–4 as \$5000/12 = \$416.67. It determines the number of units of soybeans that could be produced if *all* capital is diverted to soybeans.

The solution of a linear programming problem using the simplex method is an iterative procedure. The soybean activity will replace the land disposal activity in the next tableau. All the computations will proceed in a manner similar to those just described. After each iteration, the shadow prices are examined to determine the need to introduce further real activities into the program. When all shadow prices are positive, the most profitable farm plan has been attained.

The most profitable plan is obtained for the present example by computing one additional iteration. The final tableau is presented in

Table 9–6 The Final Solution for the Simplex Method

Activities in the Solution Set			*Real Activities*			*Disposal Activities*			
Net Price C_j	*Name*	*Amount* B	*Grain Sorghum* Y_1	*Wheat* Y_2	*Soybeans* Y_3	*Land* Y_4	*Labor* Y_5	*Capital* Y_6	*Ratio* R
34.60	Soybeans-Y_3	333.33	0	1	1	5	0	−0.335	
0	Labor-Y_5	800.01	0	−1	0	−1.5	1	−0.158	
35.35	Sorghum-Y_1	66.67	1	0	0	−4	0	0.335	
Net price of activities: C_j			35.35	32.50	34.60	0	0	0	
Opportunity cost of activities: Z_j			35.35	34.60	34.60	31.60	0	0.25	
Shadow prices: $Z_j - C_j$			0	2.10	0	31.60	0	0.25	
Value of program = $13,890.00									

Table 9–6. The final solution includes 333.33 acres of soybeans, 66.67 acres of grain sorghum, and leaves 800.01 hours of labor idle. This plan will return $13,890 before fixed costs. Land and capital are both limiting; an additional acre of land would add $31.60 to profits while an additional dollar of operating capital would add $0.25 to profits. The imputed value of labor is zero. This does not suggest that labor produces nothing of value on the average, only on the margin. (Again, recall the properties of the fixed-proportion production function as shown in Figure 9–1.)

The value of the program can be computed in either of two ways, by using units of the real activities

$$\$13{,}890 = \$34.60\,(333.33) + \$35.35\,(66.67) + 0\,\$(800.01)$$

or using shadow prices,

$$\$13{,}890 = \$31.60\,(400) + \$0.25\,(5000) + \$0\,(2200)$$

Lastly, the final solution also can be shown to satisfy the constraint equalities. The final solution is $Y_1 = 66.67$, $Y_2 = 0$, $Y_3 = 333.33$, $Y_4 = 0$, $Y_5 = 800.01$, and $Y_6 = 0$. When substituted into the constraint equations, these values give

$$1(66.67) + 1(0) + 1(333.33) + 1(0) + 0(800.01) + 0(0) = 400$$
$$4(66.67) + 2.5(0) + 3.5(333.33) + 0(0) + 1(800.01) + 0(0) = 2200$$
$$15(66.67) + 12(0) + 12(333.33) + 0(0) + 0(800.01) + 1(0) = 5000$$

and the objective function is

$$\$13{,}890 = \$35.35\,(66.67) + \$32.50\,(0) + \$34.60\,(333.33) + 0(0) + 0(800.01) + 0(0)$$

This completes our discussion of setting up and solving maximization problems using the simplex algorithm. More detail can be found in textbooks devoted entirely to linear programming, such as the books by Beneke and Winterboer or Heady and Candler listed in the references at the end of the chapter. Fortunately, electronic computers can be used to solve linear programming problems. Solutions to large programs can be obtained at very little cost. A study of the simplex method adds to an understanding of linear programming—but interpretation of results does not depend on knowing the mathematical details. Comprehension of the economic logic underlying linear programming will enable the student to understand and use linear programming intelligently.

Interpreting the Results. Once the final plan is obtained, it must be evaluated from a practical point of view. Following are questions that Professor Buller suggests for consideration by a Kansas farmer:

1. Is it possible to get 333.33 acres of soybeans and 66.67 acres of grain sorghum planted in the spring and harvested, given the size of equipment implied by the labor coefficients?
2. Are 333.33 acres of soybeans and 66.67 acres of grain sorghum agronomically reasonable or does my farm require a rotation?
3. If no wheat is planted, what is the danger of losing my wheat allotment and feed grain base?
4. Are all 400 acres equally suited to either soybeans or grain sorghum or is there some difference in land quality that should be considered?
5. Do I want this sort of plan, or would I prefer one with more wheat and less soybeans?
6. Is an average yield of 15.7 bushels per acre reasonable if 333.33 acres are planted?
7. Should the plan include some livestock alternatives which have not been considered?

A review of the tableau in Table 9–6 suggests some of the changes that should be considered. For example, because 800 hours of labor are unused while capital is limiting, the manager should consider new activities that use more labor and less capital, that is, consider possibilities for substituting labor for capital. Or, because farmland has a shadow price of $31.60 and capital a shadow price of $0.25, perhaps the farmer should consider leasing land and borrowing additional capital, at rates below the shadow prices, to utilize the 800 hours of surplus labor. Thus, when the final farm plan includes large shadow prices and substantial amounts of unused resources, then the manager must reexamine the production activities included in the original plan. Such imbalances are difficult to remove from textbook examples but should not occur in realistic farm plans. Addressing this problem, Professor Buller very appropriately states:

If any or all of the considerations are valid, the original matrix, or problem, can be changed to include these. Just because you have set up a matrix and the computer has solved it doesn't mean the solution is a desirable one. It is up to the manager, if using linear programming, to make sure the problem has been defined adequately and properly.

The Simplex Solution for Minimization

The simplex algorithm is equally adaptable to problems of cost minimization. Suppose that a manager wishes to formulate a least-cost ration for an 800-pound steer in a finishing pen and decides to seek an average daily gain of 2.4 pounds.[3] To do so, the following three requirements must be met each day:

Dry matter, at most 18 pounds.

Digestible protein, at least 1.34 pounds.

Energy, at least 11.92 meg. cal.

In an actual feeding situation, other requirements would also be needed, but three will suffice to illustrate the method while keeping the example small. The nutritional content and costs of five available ingredients are listed in Table 9–7. Sufficient information is now available to formulate the linear programming problem.

The initial simplex tableau for the least-cost feed mix problem is contained in Table 9–8. It contains three new features: (1) the disposal activities for digestible protein and energy have negative coefficients; (2) two new columns, representing *artificial* activities for these two ingredients, are added to the right of the tableau; and (3) three new rows, called the M rows, are added to the bottom of the table.

The dry matter constraint represents a maximum and therefore has a disposal activity as described above for the maximization model. The negative signs on the two other disposal activities follow logically from the nature of the constraints; they now represent minimum amounts rather than maximum amounts. For example, the basic constraint for digestible protein would be

$$7.5Y_1 + 6.7Y_2 + 42.0Y_3 + 11.2Y_4 + 1.3Y_5 \geqq 1.34$$

Table 9–7 Nutritional Content and Cost of Five Feeds

	Dry Matter (lbs.)	*Digestible Protein (lbs.)*	*Energy (Meg. Cal.)*	*Cost (CWT)*
Corn (100 lbs.)	85.0	7.5	84.9	$2.65
Grain Sorghum (100 lbs.)	85.0	6.7	77.8	2.50
Soybean Oil Meal (100 lbs.)	90.6	42.0	79.6	6.00
Alfalfa (100 lbs.)	90.5	11.2	41.5	1.75
Sorghum Silage (100 lbs.)	25.3	1.3	15.0	0.50

[3]This example is also from Buller, *op. cit.*

Table 9–8 Initial Simplex Tableau for Cost-Minimizing Problem

Activities in Solution Set			*Real Activities*					*Disposal Activities*			*Artificial Activities*	
Net Price	*Name*	*Amount*	*Corn*	*G. Sorghum*	*S. Oil*	*Alfalfa*	*Silage*	*Dry Matter*	*Digestible Protein*	*Energy*	*Digestible Protein*	*Energy*
C_j		B	Y_1	Y_2	Y_3	Y_4	Y_5	Y_6	Y_7	Y_8	Y_9	Y_{10}
\$ *M*												
0 0	Dry Matter Y_6	18.00	85.0	85.0	90.6	90.5	25.3	1	0	0	0	0
0 1	Digestible Protein Y_9	1.34	7.5	6.7	42.0	11.2	1.3	0	−1	0	1	0
0 1	Energy Y_{10}	11.92	84.9	77.8	79.6	41.5	15.0	0	0	−1	0	1
\$ Cost of activities: C_j			2.65	2.50	6.00	1.75	0.50	0	0	0	0	0
\$ Opportunity cost of activities: Z_j			0	0	0	0	0	0	0	0	0	0
\$ Shadow prices: $Z_j - C_j$			−2.65	−2.50	−6.00	−1.75	−0.50	0	0	0	0	0
M Cost of activities: C_j			0	0	0	0	0	0	0	0	*M*	*M*
M Opportunity cost of activities: Z_j			92.4*M*	84.5*M*	121.6*M*	52.4*M*	16.3*M*	0	−*M*	−*M*	*M*	*M*
M Shadow prices: $Z_j - C_j$			92.4*M*	84.5*M*	121.6*M*	52.4*M*	16.3*M*	0	−*M*	−*M*	0	0
Value of program = \$13.26*M*												

To convert this to an equality and still ensure that the minimum requirement will be met requires that the disposal activity for digestible protein be given a negative sign. Therefore, the equality constraint is

$$7.5Y_1 + 6.7Y_2 + 42.0Y_3 + 11.2Y_4 + 1.3Y_5 - Y_7 = 1.34$$

The variable Y_7 measures the amount by which the quantities of the five ingredients in the solution exceed the minimum requirement for digestible protein. If the sum of the first five products in the above equation ($7.5Y_1 + \ldots + 1.3Y_5$) exceeds 1.34, the remainder is "deposited" in the disposal activity. Thus, the digestible protein in the amounts of the ingredients selected *must* equal 1.34 but *may* exceed 1.34. (Or, for any feasible solution, $7.5Y_1 + \ldots + 1.3Y_5 = 1.34 + Y_7$.) The disposal activity is still necessary but has a negative sign because its role is reversed from the function it performed in the maximization problem.

Because of the new role of the disposal activities, they cannot be used to provide the initial feasible solution. In the above constraint for digestible protein, when Y_1 through Y_5 are set equal to zero, then $-Y_7 = 1.34$. But if Y_7 is assigned the value of 1.34 in the initial feasible solution the resulting equation would be $-1.34 \neq 1.34$. Assigning Y_7 the value of -1.34 would satisfy the digestible protein constraint but violate the non-negativity requirement of linear programming. Therefore, the disposal activities cannot be used to form the initial solution for the simplex method.

This dilemma is solved in the simplex method by introducing two new activities called *artificial activities*. These activities have positive coefficients and thus can be used to provide the initial solution. For example, the addition of the artificial activity for digestible protein, Y_9, leads to a new linear equality

$$7.5Y_1 + 6.7Y_2 + 42.0Y_3 + 11.2Y_4 + 1.3Y_5 - Y_7 + Y_9 = 1.34$$

The initial feasible solution can be found by setting $Y_9 = 1.34$. $Y_{10} = 11.92$ and $Y_6 = 18.00$ to complete the initial solution, with all other activities set equal to zero.

The artificial activities solve one problem but create another. They cannot remain in the least-cost solution because they would permit the minimum requirements to remain unfulfilled. When 1.34 units of activity Y_9 is included, the dry matter requirement is met without utilization of the actual ingredients. Y_9 and Y_{10} must be forced out of the solution set; this is done by assuming that they are priced prohibitively high—so high, in fact, that they will never enter the final solution. This high cost of the artificial activities is denoted by M, where M could be a million, billion, or trillion dollars, just so that it is higher than the cost of any of the real activities. Given this assumption, the artificial activities will never appear in the final program.

If M were asigned a value, such as a million dollars, then the C_j, Z_j and $Z_j - C_j$ rows could be computed as usual. However, the programmer is never sure what value to pick for M; to resolve this, any M number is assumed to be larger than any number without M. This is accomplished by computing new MC_j, MZ_j, and $MZ_j - MC_j$ rows in exactly the same manner that $\$C_j$, $\$Z_j$, and $\$Z_j - \$C_j$ were computed in the maximizing problem. These calculations may be found in the last three rows of Table 9–8. The dollar values for Z_j, C_j, and $Z_j - C_j$ are also computed as before.

An initial feasible solution has been found. To minimize the cost function, the $\$Z_j - \$C_j$ and $MZ_j - MC_j$ rows are examined to determine the activity with the largest value for $\$Z_j - \$C_j$ or $MZ_j - MC_j$. Selecting the largest value will lead to largest decrease in the objective function—the cost of the ration. Because the M values always dominate, soybean oil has the largest $MZ_j - MC_j$, and it will enter the feasible solution first. Following the simplex procedure, 0.03 units of soybean oil will be brought in the solution in place of the digestible-protein artificial activity.

Because M costs are always larger than \$ costs, activities with positive $MZ_j - MC_j$ costs will be brought into the solution until all coefficients in $MZ_j - MC_j$ row are zero or negative. At that time, all artificial activities are eliminated from the solution, and the first realistic feasible solution is attained. The next step would be to select the activity that has the largest positive $\$Z_j - \$C_j$, choosing from those columns that have zeroes in the $MZ_j - MC_j$ row. When no positive values are found in the $MZ_j - MC_j$ row and no column with a zero in $MZ_j - MC_j$ has a positive value in the $\$Z_j - \$C_j$ row, the least-cost combination of feeds has been determined. The final solution for the problem in Table 9–8 is 12 pounds of corn and 3.8 pounds of alfalfa per day. Dry matter intake is 13.8 pounds. The least-cost mix costs \$0.39 per 2.4 pounds of weight gain or \$0.16 per pound of weight gain. The shadow prices are \$0.06 per pound of digestible protein and \$0.026 per megacalorie of energy. The impact on costs of ingredients not found in the least-cost mix are: forcing the mix to include one pound of grain sorghum increases the cost by \$0.08; forcing one pound of soybean oil meal into the mix would increase cost by \$1.41; and forcing one pound of silage into the mix would increase the cost of the mix by \$0.03. Further details on minimization problems can be found in Heady and Candler or Beneke and Winterboer.

Comments on Least-cost Feed Analyses. Professor Buller has noted several practical aspects related to this least-cost feed mix problem:

1. The answer to this problem will specify the least-cost feed mix to meet *minimum* daily requirements.

2. Instead of considering daily requirements, weekly or monthly feed requirements could be considered by making corresponding changes in minimum nutrients required. Changes in the minimum nutrient requirements as the steer grows can be taken into consideration by changing the appropriate values and rerunning the problem.
3. Many nutrients, some probably very important, are not considered for convenience purposes. Such requirements as vitamins, minerals, or ratios between several nutrients, if important, must be included for satisfactory results.
4. If the problem is to mix 100 pounds of feed, then nutrient requirements should be specified as needed for a 100-pound mix. Also, the requirement of an exact mix of 100 pounds must be added to the problem.
5. The nutrient content was specified for 100 pounds of each feed. Other units such as bushel or ton could also have been used. However, coefficients expressing quantities of nutrients must always correspond to the unit used. For example, 100 pounds of corn contain 85 pounds of dry matter. If the corn activity was defined in units of bushels, then the coefficient for dry matter in corn would be 47.6 pounds. All coefficients should be correct for the units used.
6. For any feed, all nutrients must be expressed in same units. That is, do not specify cost of corn in terms of bushels, dry matter in terms of hundred weight, and digestible protein in terms of pounds. However different feeds can be expressed in different units. Corn is listed in terms of 100 pounds, grain sorghum could have been listed in terms of a bushel, and alfalfa expressed in terms of a ton.
7. Each nutrient must be expressed in same units. All values in digestible protein row refer to *pounds* of digestible protein per unit of feed. Other units can be used as meg. cal. for energy. But be consistent in units row-wise and column-wise.

   85 pounds dry matter per 100 pounds corn.
   7.5 pounds digestible protein per 100 pounds corn.
   84.9 meg. cal. energy per 100 pounds corn.

   Column-wise, the denominators *must* be the same unit.

   85 pounds dry matter per 100 pounds corn.
   85 pounds dry matter per 100 pounds grain sorghum.
   90.6 pounds dry matter per 100 pounds soybean oil.
   90.5 pounds dry matter per 100 pounds alfalfa.
   25.3 pounds dry matter per 100 pounds silage.

   Row-wise the numerator *must* be in same unit whereas the denominator could have varied; for example, x *pounds* dry matter per 1 ton alfalfa.

The Dual of a Maximization Problem

The simplex tableau for the dual solution of the maximization problem in Table 9–4 is contained in Table 9–9. The dual solution uses the general format of the minimization problem in Table 9–8. The real activities in the dual are the three resources—land, labor and capital. Instead of representing net prices, the C_j are now resource amounts. The initial solution in the B column consists of the net prices of the three products. The M is needed as before to insure that the artificial activities do not appear in the final solution. The M is interpreted to

Table 9–9 Dual of the Maximization Problem in Table 9–4

Activities in Solution Set				*Real Activities*			*Disposal Activities*			*Artificial Activities*		
Resources C_j ***Units***	**M**	***Name***	***Amount*** **B**	***Land*** λ_1	***Labor*** λ_2	***Capital*** λ_3	***Sorghum*** λ_4	***Wheat*** λ_5	***Oats*** λ_6	***Sorghum*** λ_7	***Wheat*** λ_8	***Oats*** λ_9
0	*M*	Sorghum-λ_7	35.35	1	4	15	−1	0	0	1	0	0
0	*M*	Wheat-λ_8	32.50	1	2.5	12	0	−1	0	0	1	0
0	*M*	Sorghum-λ_9	34.60	1	3.5	12	0	0	−1	0	0	1
Resources: C_j			400	2200	5000	0	0	0	0	0	0	
Opportunity cost: Z_j			0	0	0	0	0	0	0	0	0	
Shadow prices: $Z_j - C_j$			−400	−220	−5000	0	0	0	0	0	0	
M Resources: C_j			0	0	0	0	0	0	*M*	*M*	*M*	
M Opportunity cost: Z_j			3*M*	10*M*	39*M*	−*M*	−*M*	−*M*	*M*	*M*	*M*	
M Shadow prices: $Z_j - C_j$			3*M*	10*M*	39*M*	−*M*	−*M*	−*M*	0	0	0	
Value of program = $102.45*M*												

be a resource amount larger than all other resource quantities. The simplex solution to the dual follows the same steps outlined for all simplex minimization problems.

Equality Constraints

Constraints that impose minimum or maximum limits on resource use have been discussed. The equality constraints described earlier in this chapter can be used in either maximizing or minimizing models. Such constraints are used to ensure that an exact amount of a resource is used. For example, if the farmer wanted to insure that labor was completely utilized in the maximizing model, the labor constraint would be written as

$$4Y_1 + 2.5Y_2 + 2.5Y_3 = 2200$$

The usual disposal activity could not be used to provide an initial solution because it would imply an inequality. Therefore, an artificial activity with the M price tag is used to ensure that the equality will be enforced in the final solution. Equalities must be used with care. Why would it not be reasonable to include this equality for labor in the maximizing solution shown in Table 9–6? Hint: equalities for land or capital would not change the solution in Table 9–6.

GENERAL CONSIDERATIONS

Linear programming is a systematic method of selecting the most profitable farm plan from a vast number of possible solutions. The validity of the final solution to a linear programming model will depend on the skill with which the activities in the model are defined, and the quality of the budgeting and technical data used to estimate the production coefficients.[4]

Defining Activities and Selecting Data

The real activities in a linear programming model are based on the production options available to the manager. Activities are usually defined for producing crops and growing livestock. But careful formulation of activities will enhance the realism of the model. For example, an activity for a corn enterprise could include all resource requirements and resulting net profit from growing and selling corn. Or, the harvesting and selling functions could be separated. Harvesting might be separated to allow for the possibility of an activity permitting custom harvesting; an activity permitting cash sales might be included as an option to be compared to feeding the corn in a livestock enterprise.

[4]This discussion will emphasize maximizing models, but similar comments would apply to minimizing models.

Real activities therefore also include selling products, harvesting crops, buying inputs, hiring labor and machinery services, hiring capital (obtaining an operating loan), transferring corn (feed) from one enterprise (corn production) to another (livestock production), buying feed, etc. The definition of a set of real activities depends on the nature of the types of management decisions the manager seeks to evaluate. Activities that provide the option to sell corn or feed it to livestock would not be of interest to the manager who prefers not to raise livestock.

Activities must be carefully defined. Enterprises that use different resources, use the same resource at different times of the production period, use different technologies, or produce outputs that are sold at different times must be defined as different activities. For example, a separate activity should be defined for each month that a hog enterprise would normally farrow. Hogs born in March will require labor at different times and sell for a different price than hogs born in June.

Real activities can be designed to reflect diminishing returns for a particular enterprise. Corn produced with low, medium, and high rates of two commercial fertilizers could be defined as three activities. All three would be included in the model, and the solution will select the corn activity that is most profitable for the farm. The three activities would represent three points along an expansion path. In like manner, different activities could be defined for points along an isoquant. This method of circumventing the constant returns inherent in a linear programming activity will not work for increasing returns, that is, when resource use falls and production coefficients become smaller as the size of an enterprises increases. Because of the assumptions of linear programming, the activities defined for the larger sizes will dominate those defined for smaller sizes. If the solution selects that enterprise at all, it will select the one that uses the least resources but may bring it in at a level that would in practice be too small to achieve the economies implied by the coefficients. In such a case, the model can be solved separately for each level of efficiency for the enterprise. Modern computer programs will permit this to be done with ease.

Resource constraints must be defined in detail comparable to enterprises. In addition to the types of input restrictions described in this chapter, constraints may also be defined to represent institutional factors such as acreage allotments and farm credit restrictions. The manager's personal preferences can be reflected using appropriate constraints.

The exact definition of constraints will again depend on the detail and realism needed to answer the manager's questions. Land may be classified by productivity class, upland, or bottomland, etc. Physical facilities such as barn space may be included. Labor may be specified by weeks or months during critical periods. Cash flow and family living needs can be built into the model.

Accurate specification of prices and production coefficients is a necessity if useful farm plans are to be obtained with programming. The managerial ability of the farmer must be reflected in the production coefficients. If a manager does very poorly with a particular enterprise, the coefficients of that enterprise should be adjusted accordingly. Prices of inputs and outputs should be representative for the planning period. When a product is sold throughout the year, seasonal price effects should be estimated. Prices may trend up or down in the future periods, but, as emphasized earlier in this text, relative prices are often as important as actual prices. If input and output prices double, the optimal plan is unaffected. Therefore, although the programmer should always obtain the best possible information on actual future prices, the use of accurate relative prices for activities and inputs will ensure that a plan approximating the optimal plan will be selected even when estimated profit levels are inaccurate.

Sensitivity Analysis of Solutions

The simplex method leads iteratively to the farm plan that maximizes farm income before fixed costs. After examining the plan, the manager may wish to know how the optimal plan and resulting profits might change when a coefficient or set of coefficients are changed, when a price or a set of relative prices change, or when the resource constraints are changed. Changes in the optimal plan and income that result from such changes in the basic data fall under the general heading of sensitivity analysis.

Sensitivity analysis may be used for a variety of reasons. For example, suppose a manager is not sure of the accuracy of the production coefficients used for a particular activity that plays an important role in the final farm plan. She may wish to raise and lower the coefficients of that activity to test the stability of the final plan. If small changes in the coefficients cause the activity to enter or leave the final plan, she may seek more reliable production data before actually implementing the plan. On the other hand, if the final plan is invariant to large changes in the coefficients, then the manager may proceed to implement the plan with the knowledge that small data errors are not influencing her decisions.

Sensitivity analyses may be used for reasons other than to test the effects of unreliable data. The manager may anticipate a possible large change in absolute or relative prices in the future and consequently want to know the appropriate farm plan and resulting income for such a change. She would then adopt the plan (or not) based on her subjective evaluation of the probability that the price change will occur (see Chapter 8). Or, she may want to know which categories of resources to expand. In general, many different situations might arise wherein sensitivity analyses would be useful.

The shadow prices of the final plan provide some information about the effects of changes. Shadow prices on real activities that are not in the final plan suggest the amount by which one unit of the activity would reduce income if it were introduced into the program. By examining these, the manager can determine the effects on income of small changes in the final plan. The same is true of shadow prices on disposal activities for resources. These prices indicate the contribution to income of the last unit of input used and can be used to suggest which resources should be increased.

The shadow prices represent changes that result from changes of one unit—for either a resource or activity. Thus, their interpretation is necessarily limited. For the sensitivity analysis to be useful, the manager needs to know the types of changes that will occur as a result of larger changes. The question asked can take two forms: (1) What will happen as a result of a given change in a price, coefficient, or resource amount? (2) How much must a price, coefficient or resource amount be changed to cause a change in the final plan?

At one time, sensitivity analyses of the types described above could be solved only by long, tedious computational methods. Heady and Candler describe these methods in detail. Now, extremely efficient linear programming algorithms are available for use on electronic computers. Using such programs, many types of sensitivity questions can be obtained along with basic solutions simply by exercising the options available in the algorithms. Beneke and Winterboer have explained these options in detail; we will summarize them in the next paragraph.

Range analysis can be used to estimate the range over which a particular shadow price will remain unchanged. *Parametric programs* can be used to estimate the effects of changing the *B* column or the *C* row. Using a parametric routine, for example, Beneke and Winterboer point out that the effects of changing hog prices from $16 to $22 in intervals of $0.25 can be done with one computer "run," as compared to 24 runs using a standard simplex solution. Most computer routines also permit the use of *multiple objective functions*. Finally, options are available to change coefficients in the model. Often known as *revise* options, these modifications permit the final plan to be obtained, the coefficient revisions to be entered, and the new final plan to be obtained. The flexibility added to linear programming techniques by modern electronic computers has greatly enhanced its use in farm planning and research. In passing, we might also note that modern computer algorithms routinely insert disposal and artificial activities in models, freeing the researcher of the need to even specify such activities.

Uses of Linear Programming

The uses of linear programming have been so profuse and varied that we cannot begin to attempt a complete survey. Fortunately, one is available elsewhere. Professor Harald Jensen has prepared a comprehensive listing of the many uses of linear programming in farm management in the chapter entitled "Farm Management and Production Economics," Lee R. Martin, ed., A *Survey of Agricultural Literature,* Volume I, Minneapolis University of Minnesota Press, 1977. We strongly recommend that Professor Jensen's chapter be used as a supplemental reading for any course utilizing this textbook.

Linear programming has been used to derive optimal farm plans and least-cost feed mixes. These examples have been discussed; variations on these uses have proven extremely valuable. Models have been designed to develop long-run capital investment plans, evaluate the effects of field work time available from machinery, study cash flow problems, simulate the farm business, plan forage activities, develop irrigation practices, study allocation and development problems over a period of years (among time periods), and integrate the goals of family living into the farm business.

Transportation models have been used to minimize the costs of moving products among several locations. Other, similar models have been used to estimate the optimal location of entire industries, for example, finishing beef cattle. Transshipment problems also have been addressed.

Sensitivity analyses have provided many interesting research applications. Heady and Candler present the usefulness of techniques such as price mapping and variable capital programming. Price mapping depicts the changes in optimal plans caused by changes in one or more prices and has been widely used to derive normative supply functions for agricultural products.

Linear programming models have been adapted to include considerations of risk and uncertainty. Integer programming enabled researchers and managers to consider indivisibilities directly in the optimal programs.

All of this is not to suggest that linear programming does not have limitations. Some of the limitations are theoretical and have been explained earlier; others are pointed out by Jensen. For readers interested in other types of programming, a more general but rather simplified introduction is contained in Appendix II.

Problems and Exercises

9–1. Solve the following linear programming problem using the graphical method:

Objective:

Maximize $3Y_1 + 4Y_2$

subject to:

$$5Y_1 + 2Y_2 \leq 10$$

$$2Y_1 + 6Y_2 \leq 12$$

$$Y_1 \geq 0, Y_2 \geq 0$$

9–2. Solve the following linear programming problem using the graphical method:

Brand *XYZ* hog feed is made from a combination of wheat and corn. Wheat contains 10 percent protein, 40 percent starch, and 50 percent fiber by weight. Corn contains 15 percent protein, 50 percent starch, and 35 percent fiber by weight. The contents for Brand *XYZ* hog feed calls for a minimum of 45 percent starch and 40 percent fiber; there is no restriction on protein because it is supplemented with another type of feed.

If a 1000-pound batch of Brand *XYZ* hog feed is produced, how many pounds of corn and how many pounds of wheat should be used to minimize costs? Corn costs 5 cents per pound and wheat costs 3 cents per pound.

9–3. Solve the following linear programming problem using the graphical method and/or simplex procedure:

A processor has the ability to manufacture two products. Each product requires a blend of three ingredients. Two of the three ingredients in each product are used in both products.

The available ingredients, X_1, X_2, X_3, and X_4 in pounds (lbs.) amount to 6000, 4000, 350, and 800 pounds, respectively. One hundred pounds of product Y_1 requires 70 lbs. of ingredient X_1, 25 lbs. of X_2, and 5 lbs. of X_3; one hundred pounds of product Y_2 requires 35 lbs. of ingredient X_1, 25 lbs. of X_2, and 13 lbs. of X_4.

How much of Y_1 and Y_2 should be made to result in maximum net returns when the net price of product Y_1 is \$3.20 per unit and that of product Y_2 is \$2.40 per unit?

Suggested Readings

Baumol, W. J., "Activity Analysis in One Lesson." *American Economic Review,* Volume 48, December 1958, pp. 837–73.

Baumol, W. J. *Economic Theory and Operations Analysis.* Englewood Cliffs, N.J.: Prentice-Hall, Inc., 1961, Chapters 5–8 and 12.

Beneke, Raymond, R. and Winterboer, Ronald. *Linear Programming Applications to Agriculture.* Ames: Iowa State University Press, 1973.

Bennion, E. G. *Elementary Mathematics of Linear Programming and Game Theory.* East Lansing, Mich.: Bureau of Business and Economic Research, Michigan State University, 1960.

Buller, Orlan. *Farm Management Laboratory Manual and Supplementary Text.* Manhattan, Kan.: Department of Economics, Kansas State University, 1975.

Charnes, A., Cooper, W. W., and Henderson, A. *An Introduction to Linear Programming.* New York: John Wiley and Son, Inc., 1953.

Chiang, A. C. *Fundamental Methods of Mathematical Economics,* Second Edition. New York: McGraw-Hill Book Company, 1974, Chapters 18 and 19.

Dorfman, R., Samuelson, P., and Solow, R. *Linear Programming and Economic Analysis.* New York: McGraw-Hill Book Company, 1958.

Hadley, G. *Linear Programming.* Reading, Mass.: Addison-Wesley Publishing Company, Inc., 1962.

Heady, Earl O. and Candler, Wilfred. *Linear Programming Methods.* Ames: Iowa State College Press, 1958.

Longworth, John W. and Menz, Kenneth M. "Bridging the Gap between Production Economics Theory and Practical Farm Management Procedures." *Review of Marketing and Agricultural Economics,* Volume 48, April 1980, pp. 7–20.

FARM ADJUSTMENTS IN A CHANGING ECONOMY

CHAPTER 10

The purpose of this chapter is to present some factual information about agriculture and illustrate the use of some of the concepts presented in previous chapters. Only selected topics are included, and the discussions are brief; obviously, a complete review of all the many social and economic forces affecting farms in the nation and within regions of the nation cannot be contained in one chapter. The topics were selected with the twofold purpose of providing subject matter of general interest and of illustrating the application of economic analyses to varied situations. While foregoing chapters present concepts which agricultrual economists need to know, this chapter gives a glimpse of what agricultural economists do.

ROLES OF AGRICULTURE IN ECONOMIC DEVELOPMENT

In primitive times the human race proceeded slowly on the road of development but nevertheless made some progress. Human society has never remained in a totally static condition. Tools were slowly but surely perfected. With the domestication of cattle, cattle raising and agriculture began, which increased agriculture production and made possible pursuits in other endeavors. Increased production not only means more food for direct consumption but also more raw materials used for manufacturing nonagricultural products needed by society. Surplus food products and raw materials, for example, not only support the populace of nonagricultural sectors but in many instances are used to earn important foreign exchange that finances imports of capital goods for industrial development. Canada built its industry by exporting timber; Japan, by exporting silk; and the United States, in the early years of development, paid off loans with shipments of grain and tobacco.

In the early stages of economic development, then, it is important to have a strong and prosperous agriculture. Industrial development can accompany agricultural development but it seldom can precede it.

The economic development of a country or region can be measured by adjustments in three sections of employment: (1) primary employment—people employed in extractive industries such as farming, fishing, forestry, and mining; (2) secondary employment—people having jobs in the manufacturing industries, processing materials into semifinished and finished goods; and (3) tertiary employment—people who serve the needs of the other two employment sectors in such areas as education, financing, transportation, construction, and government.

The three phases of economic growth have been observed historically. The relative magnitude of employment in the three sectors varies from country to country as the level and rate of economic growth varies. In large areas of the world such as Asia and Africa the economy still depends almost solely on primary employment (Table 10–1). Productivity per farm worker is low, so the vast majority of the population in these areas must engage in farming for survival.

Increases in farm productivity permit a movement of some workers from employment on farms to various types of nonfarm work. In so doing, secondary employment is given impetus. The first migrants out of farming generally manufacture materials and utensils for use on farms. As secondary employment increases, the improved quality of agricultural implements represent new technology and enables continued movement of people from primary to secondary employment.

As the number of people in secondary employment grows over time, there will be a more than proportionate increase in the number of people employed to service the needs of the primary and secondary sectors. Fewer and fewer people are needed in farming because new technologies enable more food to be raised by fewer people. These changes have important implications. People shift from food production to the production of other goods and services desired by society.

The process of economic development never stops. It can be demonstrated by examining changes in the employment patterns in the United States economy (Table 10–2). The primary industries have

Table 10–1 Percentage of the Working Force in Agricultural Sector in Selected Countries and Years

	Years		
Country	*1890*	*1950*	*1982*
Australia	43.1	11.6	14.0
Belgium	18.2	10.5	3.4
China (mainland)	—	—	85.0
Germany	35.0	11.8	6.0
Great Britain	10.4	4.5	1.5
India	75.0	77.4	74.0
Japan	52.0	32.6	12.0
Norway	45.2	24.0	7.4
Sudan	—	—	86.0
Uganda	—	—	90.0
United States	43.1	11.6	3.1
U.S.S.R.	80.0	48.0	20.0

Source: The World Almanac & Book of Facts, New York: Newspaper Enterprise Association, Inc., 1951 and 1983.

Table 10–2 Percentage Distribution of Employment by Major Groups in the United States, Indicated Years

	Primary (Farming & Mining)	*Secondary (Contract Construction and Manufacturing)*	*Tertiary (Transportation, Trade, Finance, Selected Services, and Government)*
1920	28.9	28.6	42.5
1925	26.8	27.7	45.5
1930	25.8	26.6	47.6
1935	26.6	25.5	47.9
1940	22.6	27.6	49.8
1945	17.8	32.4	49.8
1950	14.8	32.5	52.7
1955	12.0	32.9	55.1
1960	10.3	32.0	60.0
1970	5.2	32.5	62.3
1980	4.5	28.4	67.1

Source: Statistical Abstract of the United States (various issues), Washington, D.C.: U.S. Department of Commerce; and *Empolyment and Earnings* (various issues), Washington, D.C.: U.S. Department of Labor.

declined from over one-fourth of all employment in the United States in 1920 to less than 5 percent of the total in 1980. Both secondary and tertiary employment have increased. Their growth came from two sources: the migration from mining and farming, and the new entrants to the labor force provided by natural population growth.

ROLE AND IMPORTANCE OF FARMING IN THE U.S. ECONOMY

Farming has gradually evolved from a way of life to a business in its own right during the past century. Even though the number of farms has dropped steadily during this period, farming plays a vital role in the U.S. economy. More than 2 million farm units are closely interwoven with all other sectors of the nation's economic activity.

As farms grew larger and became more mechanized, goods produced off the farm were substituted for these formerly produced by the farmer. At first, tractors replaced horses and tractor fuel was substituted for horse feed. Chemicals made manure obsolete as fertilizer. Purchased seeds were used instead of seeds grown on the farm. Groceries purchased in supermarkets all but eliminated the farm garden as an important source of food.

As they increased production of livestock and crops for sale and increased purchases of supplies produced off the farm, farm families developed a new dependence on the total marketing system. It is true that the farm share of the gross national product is declining, but because agriculture is becoming increasingly interdependent with other industries, farm income still is an important influence on the entire economy.

This interdependence has come about because of the responsiveness of different sectors, including farming, to demands of our growing economy. It did not come about easily. At the root of an increased interdependence and growth among different sectors lies change that is at times painful. It is painful because of human nature and the conflict between Economic Man (homo-economicus) rationally administering scarce resources and Anti-Economic Man emotionally clinging to the past, demanding a continuity of life that is at times contrary to the demands of economic calculation.

Nowhere has this conflict been more apparent than in agriculture. As one author put it, "Not much harmony is possible in a split-minded society which demands simultaneously the fruits of dynamic mobility and the privilege of immobilism."[1]

Many are concerned about the changes that are taking place in today's agriculture. These concerns are not new. Back in 1891, U.S. Senator W. A. Peffer was very much opposed to the changing rural economy. He spoke of the changes taking place during the 50 years prior to 1891, commenting that "the farmer sells his hogs, and buys bacon and pork; sells his cattle and buys beef—fresh, canned, or corned; sells his fruit and buys it back in cans." He went on to state that the farmer now (1891) buys nearly everything that he produced at one time and that it all costs money.

Senator Peffer continued:

> In earlier times not one home in one thousand was mortgaged and the little money needed could be borrowed from farmers. Now (1891), where 10 times as much is needed, little or no money can be obtained. Nearly half the farms are mortgaged for as much as they are worth and interest rates are exhorbitant.

He went on to say that farmers were maintaining an army of middlemen, loan agents, and bankers, whose services ought to be dispensed with.

Think what the situation would be today if farmers were still operating as in earlier times. Nearly all people would be farmers. The national wealth would not have increased, and farmers would be dependent on primitive means of production and marketing. Farmers would be doing for themselves many things others can do for them cheaper and better. There would be no modern-day tractors, chemicals, or electric power lines to farms. People who were released from farm work to work elsewhere in our economy would not number 50 or more per farmer; as is now the case. Thus, there would be no labor

[1]Love, Harry M., "Some Farm Trends and Implications," *Virginia Farm Economics,* Jan–Feb. 1967.

to produce the vast array of supplies and services needed to support modern-day farming. If the Peffers had succeeded in keeping the city industries from developing so that farmers could have their wagons, shoes, and clothing made at home, neither the people nor capital would be available to provide the level of living and economic strength we enjoy today.

The comparative advantages of large-scale farm production has concentrated farm production so that our country has become dependent on a few regions to supply farm product needs. In such areas a major portion of the demand for farm products is determined by forces outside the area. For example, export sales are now an important component of farm income. Farm sectors in agriculturally oriented economies use a high percentage of such income for local capital, production, and consumption expenditures; foreign sales therefore create a multiplier effect within the economy and increase the impact of farm income on nonfarm income and subsequently the total income of the economy.

In a rapidly growing industrial economy like that of the United States, the farm sector's importance and contribution to the total economy is often overlooked. The increasing complexity of highly mechanized farming has created greater and stronger linkages between the farm sector and the rest of the economy. On-farm adjustments and developments have brought about commercialization of the farm sector as farm production has become more dependent on purchased off-farm inputs, which have important influences on local, state, and regional economies.

The volume of purchased inputs has increased five-fold since 1960, from $26 billion to over $130 billion in 1980. While part of this increase is due to inflation, the data do suggest a growing interdependence and reliance of the farm sector on nonfarm sectors for farm inputs. In 1960, for each dollar of gross farm income, the farmers spent less than 70 cents for purchased inputs.

The purchased farm inputs include fertilizer and lime, fuel, lubricants, and farm machinery such as tractors, combines, and other equipment as well as their maintenance, farm buildings, seed, and many others. These expenditures have resulted in the development of a thriving agri-business industry. This industry, in turn, provides income and employment opportunities in other economic sectors. Thus, the sectors of the state's economy are fastened together like links of a chain. The strength of the chain is dependent on the strength of the different sectors of the economy.

Farming is one of the important links. Year in and year out, farming helps keep the feed mills grinding, gasoline stations pumping, meat-packing plants operating, and machinery dealers and repair shops working, etc. All of these operations and their respective monetary transactions are reflected in the farm income multiplier. Its effect increases the importance of farm income, as do the multipliers of other economic sectors, in determining nonfarm income and, subsequently, the total national income.

The farm income multiplier measures the impact of an increase in farm income on total income. It shows the number of dollars of total income generated by each dollar of farm income. A multiplier of 3, for example, indicates that $1 of farm income generates $3 of income in the total economy. A 10 percent increase in farm income with a multiplier of 3 would add 30 percent of farm income, not just 10 percent, to the total economy. Despite a lack of precision in measurement, we know that farm income has a definite impact not only on the local economy but on the state and nation's economy as well.

An increase in farm income benefits all other sectors of the economy. This is particularly true for the agri-business industries such as farm machinery, fertilizer, and agricultural chemical dealers. It was true decades ago, when farmers were less dependent on purchased goods, and it is even more true today, when a major portion of the farmer's income is spent for production expenses.

Decreases in farm income, on the other hand, have adverse effects not only on farming but on other sectors as well. When farm purchases decline, so does employment in the industries supplying farmers' production needs. These industries in turn reduce expenditures and cause unemployment in other industries. This chain reaction eventually reaches everyone in the society. All these operations and their respective monetary transactions are reflected in the farm income multiplier.

These interdependencies and linkages of different sectors in our economy cannot be overlooked. Farmers, as other segments in the economy over the years, have been challenged to contribute to economic progress and growth. They have been meeting these challenges by making adjustments that affect not only them but all of us.

Number and Size of Farms

The number of U.S. farms declined from 4 million in 1960 to 2.4 million in 1980, a drop of 1.6 million or more than 1 of every 3 farms. In 1980, there was one farm for every 75 people living in the United States; in 1960, the figure was one farm for every 48 persons. Average farm size increased from 297 acres in 1960 to more than 400 acres in 1980.

The decline in farm numbers has not been spread evenly across the land. The largest decreases occurred in the South, where many share-tenant farms were absorbed by larger farms. Similarly, the South Atlantic and New England areas lost a large number of farming units because of inadequate size of operations and returns received from farming. This is particularly true for farming areas where, in addition

Table 10–3 U.S. Farm Numbers, Output and Income by Economic Class, 1970 and 1980

		Economic class, by farm sales							
	Year	*Ia $100,000 and over*	*Ib $40,000 to $99,999*	*II $20,000 to $39,999*	*III $10,000 to $19,999*	*IV $5,000 to $9,999*	*V $2,500 to $4,999*	*Other Less than $2,500*	*Total*
No. of farms (1000)	1960	23	90	227	497	660	617	1,849	3,963
	1970	55	178	343	390	397	435	1,156	2,934
	1980	282	383	282	288	352	332	509	2,428
(Percent of all farms)	1960	0.6	2.3	5.7	12.5	16.7	15.6	46.6	100.0
	1970	1.9	6.0	11.6	13.2	13.5	14.6	39.1	100.0
	1980	11.6	15.8	11.6	11.8	13.7	14.5	21.0	100.0
Cash receipts from farming (Average per farm)	1960	265,261	60,634	28,714	14,966	7,818	4,016	1,104	8,881
	1970	328,946	67,046	32,379	17,142	8,744	4,294	1,422	18,551
	1980	342,695	68,875	28,691	14,445	7,717	4,272	1,175	57,634
(Percent of all receipts)	1960	17.3	15.5	18.5	21.1	14.7	7.1	5.8	100.0
	1970	33.0	21.8	20.3	12.2	6.3	3.4	3.0	100.0
	1980	69.1	18.8	5.8	3.0	1.8	1.1	0.4	100.0
Realized net farm income (Average per farm)	1960	30,826	13,812	8,084	5,095	3,211	1,931	806	2,806
	1970	38,600	16,664	10,120	5,707	3,160	1,634	893	4,665
	1980	33,972	16,614	8,280	4,299	2,512	1,582	1,821	9,002
Off-farm income (Average per farm)	1960	N.A.	N.A.	1,678	1,258	1,574	1,848	2,732	2,140
	1970	7,618	3,949	3,359	4,187	5,448	6,184	7,433	5,874
	1980	12,922	7,922	9,358	12,847	16,768	20,156	20,242	14,820
Total net income per farm	1960	N.A.	N.A.	9,762	6,353	4,785	3,779	3,538	4,946
	1970	46,218	20,613	13,612	9,894	8,608	7,818	8,326	10,539
	1980	46,894	24,596	17,638	17,146	19,280	21,738	22,063	23,822

Source: USDA *Economic Indicators of the Farm Sector,* Economic Research Service, ECIFS1-1, Washington, D.C. Aug. 1982.

to poor soil, the location is mountainous and hilly. The rate of decline in number of farms in the Corn Belt, Northern Plains, and the Lake States, on the other hand, has been slower.

Between 1960 and 1980 the number of farms that sold less than $10,000 worth of farm products was reduced by over one and a half million, from 3.126 million in 1960 to 1.193 million in 1980. This statistic represents a continuation of the trend that began in the mid-Thirties and has been accelerating since World War II. Since then, the number of farms in the United States has decreased by 60 percent, from 5.9 million in 1947 to 2.4 million in 1980.

Most farms selling under $10,000 worth of farm products could no longer provide an adequate family income. These farms generally use family labor and capital on units that are two small to yield adequate incomes and returns to resources. Mechanical cotton and corn pickers, for example, are doing work that once required much unskilled labor. These technological advances usually also require large-size operations and thus make small farms less able to compete. The resulting trend has been toward larger and larger farms. Farms selling $40,000 or more of farm products, for example, have increased nearly six-fold in number in the last 20 years, from 113,000 in 1960 to 665,000 in 1980 (Table 10–3). Increasingly, managers of smaller-size farms depend on off-farm income as a source of livelihood.

Changes in Farm Population and Productivity of Farm Labor

Urbanization has been dramatic in the United States, beginning a century and a half ago. The period 1960–1981 was no exception. During this 21-year period the farm population declined by nearly 10 million, or about 60 percent (Table 10–4 and Figure 10–1). The average annual migration rate in the Sixties, approximately 6 percent, was higher than in any other period in history. Since 1960, the number of people engaged in farming has been reduced by 3.35 million (Table 10–5 and Figure 10–2). Yet those who remained on farms produced more output on fewer acres than the labor force of 1960.

When the war and post-war boom increased employment opportunities in other sectors of the economy, the rush of farm people for jobs in industry suggested that the lack of job opportunities in industry may have been the only force holding them on the farms. The movement from farm to city will undoubtedly continue in the future, but at a slower pace.

The increased productivity of American agriculture has helped provide workers to strengthen the nonfarm economy and has enabled the United States economy to grow to its present status. If farmers produced today at the productivity level of 1910, we would have to add over 30 million people to the farm labor force and thus would have 30 million fewer employed in manufacturing, processing, and service industries.

Table 10–4 U.S. Total and Farm Population, 1960–1981

Year	Total Resident Population	Farm Population: Number	Farm Population: Share of Total Population
	Thousands	Thousands	Percent
1960	179,323	15,635	8.7
1963	187,837	13,367	7.1
1966	195,045	11,595	5.9
1969	200,887	10,307	5.1
1972	207,797	9,610	4.6
1975	212,542	8,864	4.2
1978	217,771	6,501[a]	3.0[a]
1981	224,064	5,790[a]	2.6[a]

Source: U.S. Department of Commerce, Bureau of the Census, and U.S. Department of Agriculture, Economic Research Service, *Farm Population of the United States: 1981,* Series P-27, No. 55, November 1982, p. 4.

[a]Current farm definition.

Labor and other inputs used in 1980 differ from those of 1960s. Changes indicated in the mix of inputs used in farming, such as those shown in Table 10–6 and in Figures 10–3 and 10–4, do not accurately reflect the qualitative changes in different resources, particularly for human resources. Farmers and farm workers are better educated and more knowledgeable than ever before. As a result, man-hour inputs in 1980 embody a relatively greater proportion of managerial input than in 1960. Because of these and other technological changes, farm productivity and farm output per unit of input is now considerably

Table 10–5 Farm Labor: Annual Average Number of Workers on Farms, United States, 1960–80

Year	Average number of persons: Total Workers[a]	Average number of persons: Family Workers	Average number of persons: Hired Workers
	Thousands	Thousands	Thousands
1960	7057	5172	1885
1963	6518	4738	1780
1966	5214	3854	1360
1969	4596	3420	1176
1972	4373	3227	1146
1975	4342	3025	1317
1978	3957[a]	2689	1268
1980	3705[a]	2402	1303

Source: USDA, *Agricultural Statistics,* U.S. Government Printing Office, Washington, D.C., 1981, p. 429.

[a]Hawaii included in U.S. averages beginning 1978.

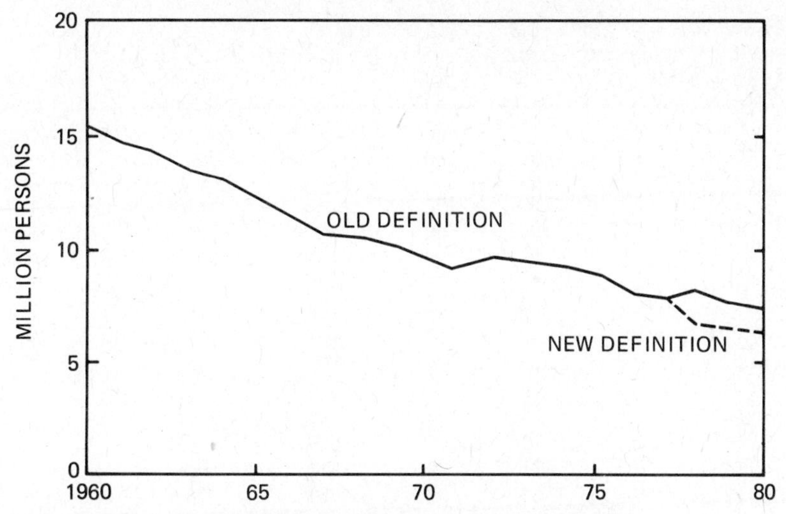

Figure 10–1. Farm population. Present definition includes those living on places with $1000 or more in agricultural product sales. Previous definition was those on places of 10 or more acres with at least $50 in sales and under 10 acres with at least $250 in sales. (*Source:* USDA, *1982 Handbook of Agricultural Charts,* Washington, D.C., November 1982.)

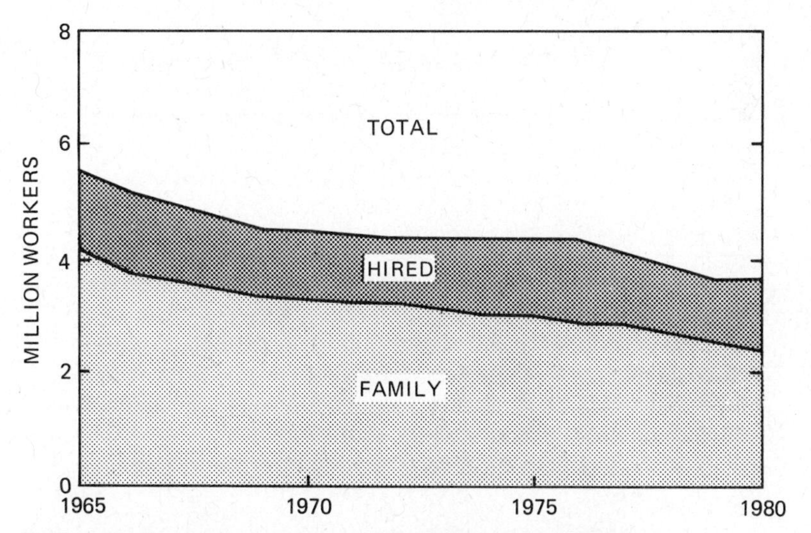

Figure 10–2. Annual average farm employment. Average number of persons employed in one survey week each month through 1974, the last full calendar week ending at least one day before the end of the month; beginning with 1975, estimates are quarterly and include the week of the 12th of January, April, July, and October. (*Source:* USDA, *1982 Handbook of Agricultural Charts,* Washington, D.C., November 1982.)

Table 10–6 Index of Farm Total Output and Input, Selected Farm Inputs and Output per Unit of Input, 1967–1980 (1967 = 100)

Year	*Total Output*	*Total Input*	*Labor*	*Real Estate*	*Parts and Machinery*	*Agricultural Chemicals*	*Output per Unit of Input*
1967	100	100	100	100	100	100	100
1968	102	100	97	99	101	105	102
1969	102	99	93	98	101	111	103
1970	101	100	89	101	100	115	101
1971	110	100	86	99	102	124	110
1972	110	100	82	98	101	131	110
1973	112	101	80	97	105	136	111
1974	106	100	78	95	109	140	106
1975	114	100	76	96	113	127	114
1976	117	103	73	97	116	145	104
1977	119	105	71	99	120	154	113
1978	122	105	67	97	126	161	116
1979[a]	129	108	66	96	130	184	119
1980[b]	122	106	65	96	128	174	115

Source: USDA *Agricultural Statistics,* U.S. Government Printing Office, Washington, D.C., 1981, pp. 438–439.

[a]Preliminary
[b]Estimated

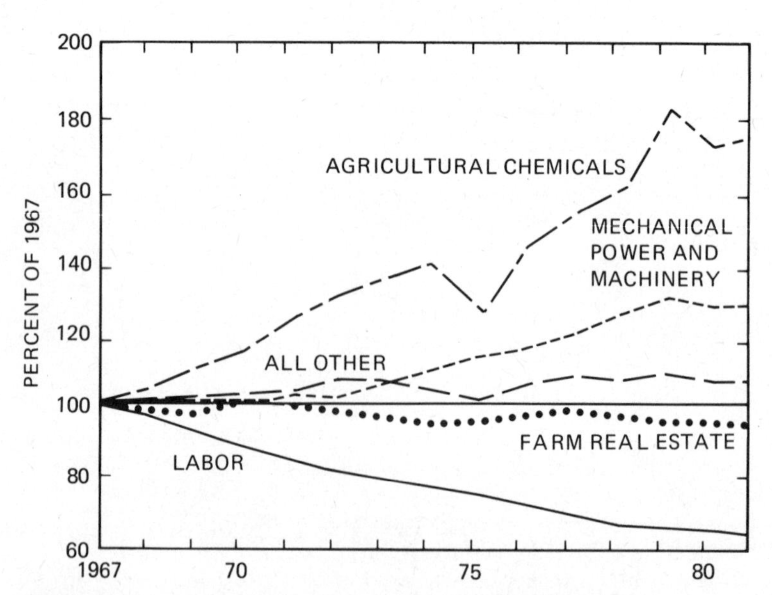

Figure 10–3. Use of selected farm inputs: 1980 preliminary, 1981 projected. (*Source:* USDA, *1982 Handbook of Agricultural Charts,* Washington, D.C., November 1982.)

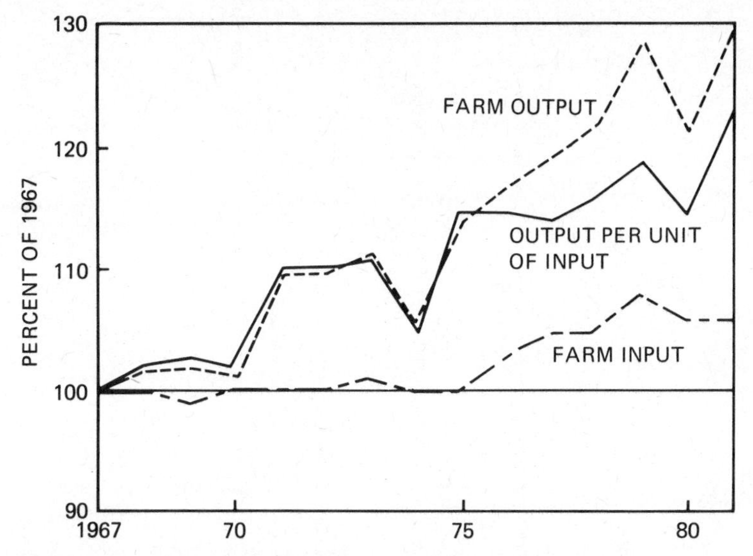

Figure 10–4. Farm productivity: 1980 preliminary, 1981 projected. (*Source:* USDA, *1982 Handbook of Agricultural Charts,* Washington, D.C., November 1982.)

higher than at the start of the 1960s (Figure 10–4). The average U.S. farm worker now produces enough food and fiber to supply 75 persons. In 1960 the average worker produced enough to supply only 27 persons.

The rapid advances in agricultural production increased the interdependencies between the farm sector, businesses that provide services and inputs to the farm sector, and businesses that process and market agricultural products. It is estimated that there are now more than two nonfarm employees in agriculturally related industries for every farm worker.

Increasing Importance of Purchased Inputs

Changes in American agriculture manifest themselves most dramatically in continuously rising productivity. In recent times, the most spectacular rates of productivity increases have not come from giant factories but were realized on the family farms. Applications of machine and electrical power, improved farming practices, better breeding and feeding of livestock, new crop varieties, chemicals, and fertilizers have contributed to these advances.

The volume of purchased inputs continues to increase. It is estimated that 75 percent or more of all inputs used on U.S. farms today are purchased. Purchases of feed, seed, and livestock, the use of commercial fertilizer, liming materials, and inputs of mechanical power and machinery have increased sharply (Tables 10–6 and 10–7).

Table 10–7 Farm Production Expenses, United States, Selected Years 1960–1980 (million dollars)

Year	*Feed*	*Livestock*	*Seed*	*Fertilizer*	*Hired Labor*	*Repairs Machinery*	*Taxes, Interest, Rent, & Other*	*Total*
1960	4,552	2,506	519	1,344	3,062	3,982	11,411	27,376
1965	5,674	2,912	720	1,994	3,604	3,943	14,803	33,650
1970	8,028	4,324	928	2,435	4,340	4,539	19,830	44,424
1975	12,805	4,954	2,138	6,660	6,586	7,806	34,812	75,863
1980[a]	18,618	10,539	3,351	9,922	10,411	16,096	61,719	130,656

Source: USDA, *Economic Indicators of the Farm Sector,* ECIFS1-1, Economic Research Service, Washington, D.C., August 1982, pp. 62–65, and USDA, *Agricultural Statistics,* U.S. Government Printing Office, Washington, D.C., 1981.

[a]Preliminary

Increases in purchased inputs are reflected in substantial changes in farm production expenditures. In 1960, for example, current farm operating expenses amounted to about $19 billion; in 1980 they were estimated at $92 billion. Total production expenses increased during the same period from $27 billion to $131 billion.[2] While some of the changes represent inflation, the data do suggest a growing interdependence and reliance of the farm sector on nonfarm sectors for farm inputs. In 1980, for example, farmers spent nearly $19 billion for feed, $11 billion for livestock, $10 billion for fertilizer, and over $16 billion for the use and repair of farm equipment and machinery (Table 10–7).

CAPITAL AND CREDIT REQUIREMENTS

Continued substitution of purchased for nonpurchased inputs, expanding farm size, and greater specialization in farming have increased the capital and credit requirements of today's farms. During the 1960–1982 period, the aggregate investment in farming increased more than five-fold (Table 10–8 and Figure 10–5). On a per farm basis, because of fewer farm units, it increased even more (Table 10–8).

In 1960 the assets in agriculture were $203.5 billion; by January 1, 1981, they were estimated at $1091.8 billion.[3] During the 1960s and mid-1970s the value of real estate assets (largely because of increasing land prices) increased more rapidly than nonreal estate assets. Real estate and nonreal estate debts increased even more dramatically during the period. While concern is often expressed about the weakening of the financial conditions of farmers, it should be remembered that farm operators are not the only people sharing in the total farm debt. The latest Census of Agriculture survey of farm debt indicates that landlords—those who own an interest in a farm but do not actually operate it—owe more than 20 percent of the total farm debt. As might be expected, landlord debt is predominantly in the form of real estate debt. Of the total farm real estate debt, more than one-fifth is held by people other than farm operators. This is significant since—unlike nonreal estate debt, which is generally self-liquidating—real estate debt represents an annual fixed commitment that can be especially burdensome when farm incomes are low. In 1982 total claims of $19.8 billion against farm assets still left an equity of $896.9 billion, or 87.2 percent, which may be considered high by nonfarm industry standards.

The sharp decline in the number of farms caused phenomenal gains in agricultural assets per farm during the 20-year period. Assets

[2]*USDA Agricultural Statistics,* U.S. Government Printing Office, Washington, D.C., Table 659, 1974, p. 468; and 1981, p. 462.

[3]In constant 1967 prices the comparable figures for total farm assets are 260.8 billion for 1960 and 395.6 billion for 1982; an increase of 52 percent over the 22-year period.

Table 10–8 Assets, Liabilities, and Equity in U.S. Agriculture, January 1, 1960, and 1982

Item	*1960*			*1982*		
	Total	***Per Farm***	***Per Worker***	***Total***	***Per Farm***	***Per Worker***
	($bil.)	($1000)	($1000)	($bil.)	($1000)	($1000)
Assets						
Real Estate	130.2	32.9	18.4	823.8	338.0	222.3
Nonreal Estate:						
Livestock	15.2	3.8	2.2	53.6	22.0	14.5
Machinery and motor vehicles	22.7	5.7	3.2	111.4	45.7	30.1
Crops stored on and off farms	7.7	2.0	1.1	36.5	15.0	9.9
Household furnishings	9.6	2.4	1.4	21.7	8.9	5.9
Financial assets	18.1	4.6	2.6	44.8	18.4	12.1
Total	$203.5	$51.4	$28.9	$1091.8	$448.0	$294.8
Liabilities						
Real Estate Debt	12.1	3.1	1.7	102.0	41.9	27.5
Nonreal Estate Debt:						
Commodity Credit Corporation	1.1	0.3	0.2	8.0	3.3	2.2
Other reporting institutions	11.6	2.9	1.6	84.8	34.8	22.9
Total liabilities	$ 24.8	$ 6.3	$ 3.5	$ 194.8	$ 80.0	$ 52.6
Proprietor's Equities	178.7	45.2	25.4	896.9	368.0	242.1
Percent equity	87.8	87.8		87.2	82.0	
Debt-to-asset ratio	12.2	12.2		17.8	17.9	

Source: USDA, *The Balance Sheet of Agriculture, 1974,* Ag. Infor. Bul. No. 376, Economic Research Service, Washington, D.C. September 1974; and *Economic Indicators of the Farm Sector,* ECISF 1-1, Economic Research Service, Washington, D.C. August 1982, p. 131.

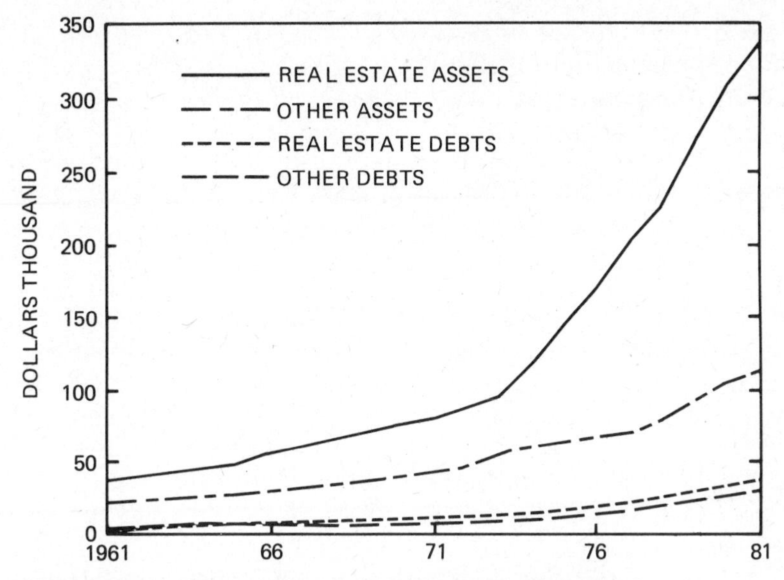

Figure 10–5. Farm assets and debts per farm. Data as of January 1; 1981 preliminary. (*Source:* USDA, *1982 Handbook of Agricultural Charts,* Washington, D.C., November 1982.)

per farm increased from $51,400 in 1960 to $448,000 in 1982 (Table 10–8). While capital and credit requirements vary considerably among different types of farm and regions, all types of farms have increased the use of capital. More and more commercial farms add heavier doses of capital and proportionately less labor and land in their production mix. Input-output coefficients are in continuous flux.

Credit Use In Farming

The vast array of supplies and services that support modern-day farming require large sums of capital. In this regard, farming is not different from other businesses depending on capital markets. Farmers need the capital to finance real estate and other purchases.

The use of credit—borrowing money—is similar in one respect to renting land. By renting, a farm operator is able to use productive assets otherwise unavailable. By combining his limited amount of capital with another's land, he is able to develop a farm organization and resulting income which would otherwise be impossible. In fact, he may attain a higher income level by renting an adequate-size unit than he would as a debt-ridden owner-operator of a small farm. In the same way, credit enables the operator to purchase land or other inputs needed for production. Because major inputs are complementary, use of credit may permit the operator to utilize superior production techniques, resulting in more income at lower risk.

At one time the use of credit was considered evil. This might have stemmed from the time when human rights and justice had not attained its present status (in the United States), and debtors could be thrown into prison. Mothers at that time more than likely taught children to stay out of debt with the ultimate aim of keeping them out of prison. At any rate, long after the threat of prison was lifted, the abhorrence of debt was incorporated into the minds of the people.

The use of credit in production (and consumption) has become more prevalent in recent decades. Attitudes toward credit have been changed, and the amount of money available for lending has increased. As society became more affluent and savings increased, for example, by means of investment in insurance programs and time deposits in banks, those who lend money found they had more to lend. The increase in credit use resulted from encouraging people to borrow, changing the image of lending institutions, offering credit on a wider variety of purchases, and developing more flexible and useful credit plans.

The use of credit in agriculture has increased substantially since 1960. Trends in nonreal estate and farm mortgage debt since 1960 are shown in Figure 10–6. Liabilities of the farm sector have been increasing steadily since the early 1940s. In 1960, for example, the farm debt stood at $24.8 billion dollars compared to $194.8 billion in 1982—an eight-fold increase.

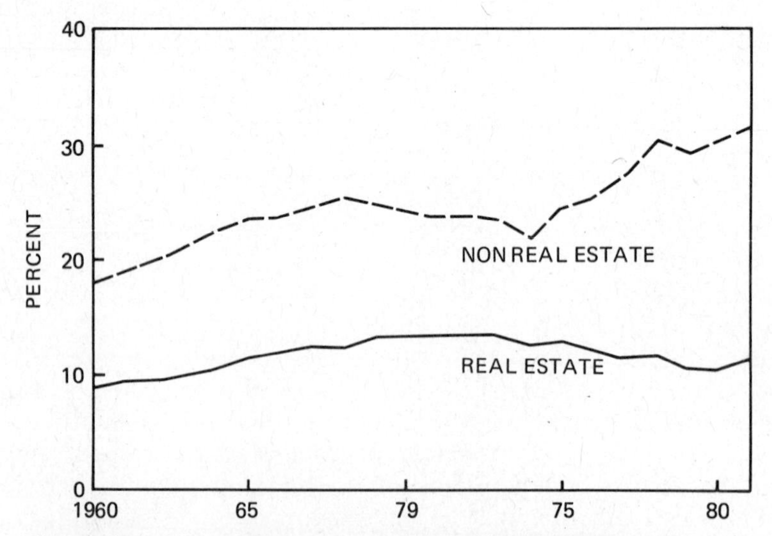

Figure 10–6. Farm debts as percentage of assets. Debt is shown as percentage of real estate debts to real estate assets and percentage of non-real estate debts to normal estate assets. Data as of January 1. (*Source:* USDA, *1982 Handbook of Agricultural Charts,* Washington, D.C., November 1982.)

While debt or, to put it another way, the use of credit has increased in absolute amounts, indebtedness relative to asset values has not spiralled. In 1940, real estate debt was 20 percent of assets. During the profitable war years, debts were retired. Equities of farmers and farm owners increased as a result of three separate influences. By far the most important was the increase in real estate values. A similar increase took place in crop, livestock, and machinery prices. A second influence was the increased physical inventories of livestock and, to a lesser extent, increased quantities of machinery and equipment on farms in 1945 as compared with 1940. The third influence was the reduction in debts owed by farmers and farm owners during this period.

In addition, reductions in world food production in 1945, resulting from unfavorable weather and economic disorganization in both Europe and Asia, put an unexpected strain on agriculture in the United States. Instead of falling prices after the war, as many expected, prices of grains and other farm products rose. Continued shortages and high prices of farm products, together with returning veteran's demands, stimulated both land prices and sales. Thus, in 1950 farm debt was 7 percent of real estate assets. In spite of sharp increases in farm debt, the debt-to-asset ratio has increased from 12.2 percent of asset value in 1960 to only 17.8 percent in 1982 (Table 10–8).

These ratios are, of course, only averages for the farm economy as a whole. Not all farm operators are in debt, and the farm debt is not evenly distributed among borrowers. Farm debt tends to be concentrated in the hands of the largest farmers. Even though smaller farmers may have a much lower debt-to-income ratio, a much larger proportion of their income must be used to meet family expenses. Although larger farmers tend to have higher debt-to-income ratios, because of their higher incomes a greater proportion of their income is available for repaying debts and making additional investments.

In post-war years the increase in nonreal estate debt is striking. The proportion of nonreal estate loans continues to increase, suggesting the increasing importance of nonreal estate credit in agriculture (Figure 10–6). As farm size and value of the assets needed for modern methods of production have increased, farmers have increased credit use in their efforts to maintain income.

As farmers increase the size of operations and expand the volume of purchases needed for efficient production, many farmers find it profitable to use even larger amounts of borrowed capital. This should not be a cause of concern as long as the credit is used to increase productivity and earnings rather than family consumption.

Economics of Credit Use

Except for its role as a common denominator, a function that theoretically could be performed by any input, borrowed capital is not uniquely different from other purchased inputs. Each dollar added to the productive process will result in a change in total product. Thus, a production function for capital can be visualized for any enterprise or firm. The interest rate is the cost or price per unit of capital; the manager will purchase capital and add it to the production process as long as the added profit exceeds the interest rate. That is, a dollar invested must return, after all other costs, a dollar plus the interest on the dollar. A limited amount of capital is allocated among enterprises in such a way that the marginal earnings of a dollar are equal in each enterprise.

Two important concepts often encountered in the literature of agricultural economics are *internal* and *external* capital rationing. Capital is available to the farmer from several sources. Real estate mortgages usually represent the lowest-cost loans. Interest rates are then higher for bank loans secured by productive assets, production credit loans, second mortgages, and usually even higher for personal loans or loans from finance companies. Thus, the capital supply curve faced by the farmer would appear as a "step" function (Figure 10–7A). The cost of a dollar is $\$(1 + i)$, where i is the interest rate. In Figure 10–7A, *OA* amount of money is available at the lowest interest rate, and *CD* at the highest rate. After total amount *OD*, no more capital can be borrowed.

External credit rationing is depicted in Figure 10–7B. Here the operator desires to equate the *VMP* of capital to its cost. He finds, however, that he can only borrow *OA* capital at the lower interest rate and *AB* at the higher rate, a total of *OB*. He would like to borrow more (at any interest rate which would intersect the *VMP* within the arc labeled *C*), but lending agencies will not extend the credit.

The farmer, when deciding to use credit or any other input, is faced with all the technical, institutional, weather, and other uncertainties described in Chapter 8. The lending firm making the loan must weigh the same uncertainties plus additional uncertainties associated with the farmer himself, including the farmer's ability to use the added capital effectively, and his general reputation. The lending firm wants to stay in business and to do so must be sure it can reclaim its money. Thus, in the absence of sufficient security (mortgagable assets) or knowledge of the operator, the firm will not supply all the farmer's needs and external credit rationing exists.

Internal credit rationing occurs as a result of expectations or attitudes on the part of the farmer. In Figure 10–7C, the farmer's supply curve for capital intersects the *VMP* of capital and amount *OB* should be borrowed. But in this case the farmer subjectively "discounts" the *VMP* of capital and borrows only amount *OA*. What accounts for this discounting? First, the farmer may have the aversion to debt mentioned earlier. As the debt grows, his aversion and the discount increase. Second, he may also believe that uncertainty increases with the size of debt and may often be correct in this belief. Third, due to the principle

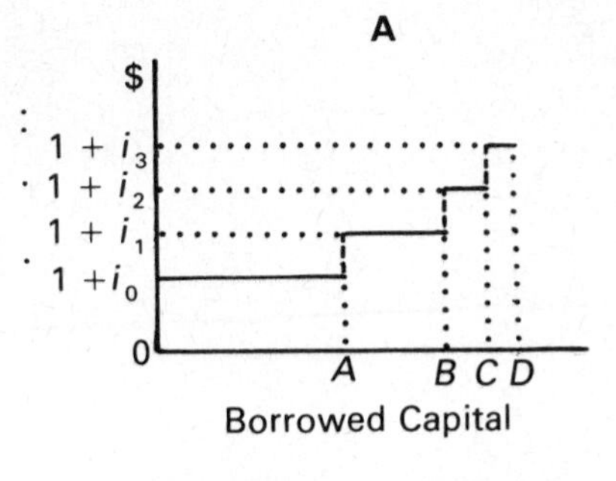

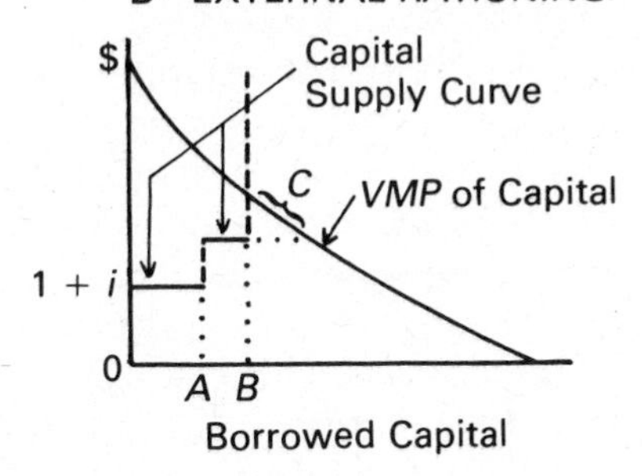

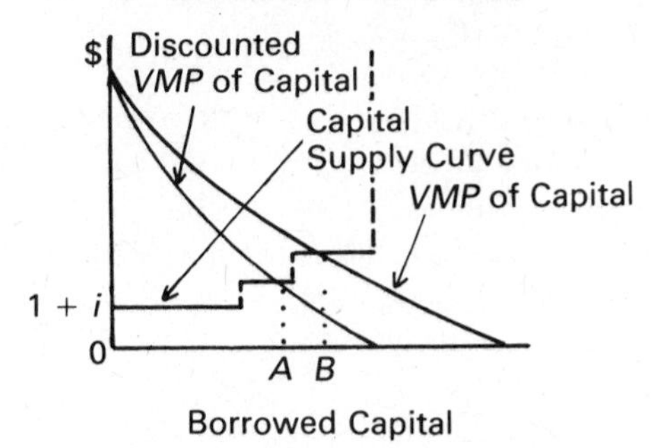

Figure 10–7. Internal and external capital rationing.

of increasing risk, capital losses on the farmer's owned capital (equity) increase as debt and size of operation increase. For example, if a farmer possessed $10,000 and borrowed another $10,000, a 10 percent loss would leave him with $8,000 after repayment of the loan. If he borrowed $90,000, a 10 percent loss would wipe out his equity. This can occur even though the uncertainty, as measured by the percentage loss, does not change as borrowing increases.

In general, given the goal of efficiency in resource use, credit should be extended on a basis that would maximize efficiency. A lending institution with a given amount of capital should therefore follow a set of lending criteria designed to allow capital to flow where it is most limiting—where its *VMP* is the highest. While the old saw that "in order to get a loan from a banker, you must first prove you don't need it" is perhaps unfair, it is probably true that loans to users with high *VMP*s are considered more risky by the banker than loans to firms more amply endowed with capital. Or, to restate it, if a farmer has good security in the form of assets or capital, he can easily obtain loans and consequently the *VMP* of capital in his business is approximately equal to cost per dollar.

Care should be taken to distinguish between capital "needs" and capital earnings. A small or poorly organized farm may "need" capital badly if it is to survive, but the earnings of that farm may never be high; on the other hand, a loan to a well-organized, well-financed farm may result in large earnings. Right or wrong, the efficiency criterion would allocate capital where its earnings are the highest.

Interest rates on loans vary only within narrow ranges around prevailing rates determined by financial and governmental institutions. Moreover, maximum rates are established by law in many states. Lenders are thus not able to vary interest rates with the uncertainty of the loan. In the absence of restrictions, the lender would presumably regard interest rates and security as substitutes. A loan thought to involve a given amount of uncertainty might be made at a low interest rate if sufficient security (mortgagable assets) is provided or at a higher interest rate when less security is available. A very high rate might be charged when security is completely lacking; the farmer would then have to decide if the capital would earn more than it costs. Present lending practices do not allow this type of flexibility. Therefore, when the amount of security available drops below the amount the lender believes to be needed for the man and investment involved, he simply will not extend the credit.

It is often argued that, because of the uncertainty involved in agricultural production, the repayment of farm loans should be scheduled to depend on farm earnings. Large payments would be made in good years and small payments in poor years. In this way, a portion of the uncertainty is transferred to the lender, who, in turn, would be reluctant to accept such uncertainty except for increased earnings—higher interest rates—or increased security through mortgage of assets. The farmer would then have to decide if the increased costs to such loans were justified.

FARM INCOME AND FARM RETURNS

As noted earlier, farm production expenses have increased sharply since the 1960s; production costs for U.S. agriculture went up fivefold (Table 10–7). Figure 10–8 shows relatively steady increases in production expenses and sharp fluctuations in net farm income, even though cash receipts from marketing have shown modest but steady increases throughout the 20-year period. The major reason for the

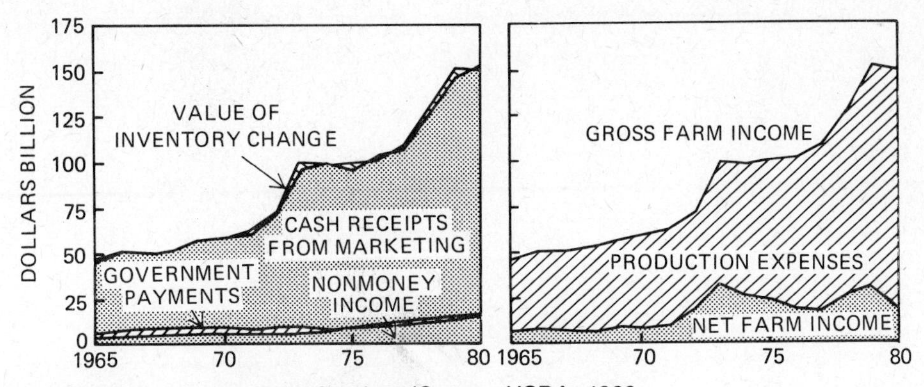

Figure 10–8. Income from farming. (*Source:* USDA, *1982 Handbook of Agricultural Charts,* Washington, D.C., November 1982.)

fluctuations in net farm increase is the relationship between the prices farmers received for their products and the prices they paid for the production inputs (Figure 10–9). In recent years the price increases for farm products have not kept pace with the prices paid for the inputs and thus net farm income has been negatively affected (Figure 10–10). The net income decline has been most pronounced for farms with annual sales of $100,000 or more (Figure 10–11). These are the farms that depend heavily on purchased inputs. In 1980, for example,

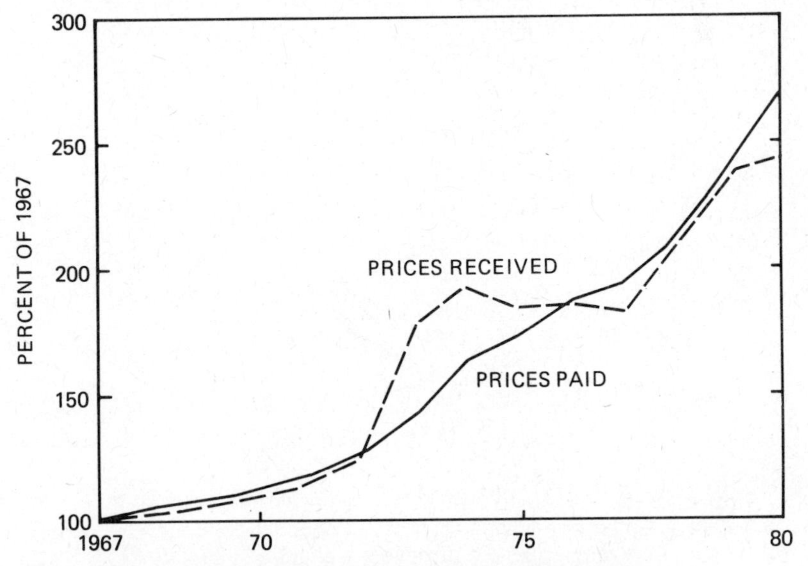

Figure 10–9. Prices received and paid by farmers. Prices paid include commodities and services, interest, taxes, and wage rates. (*Source:* USDA, *1982 Handbook of Agricultural Charts,* Washington, D.C., November 1982.)

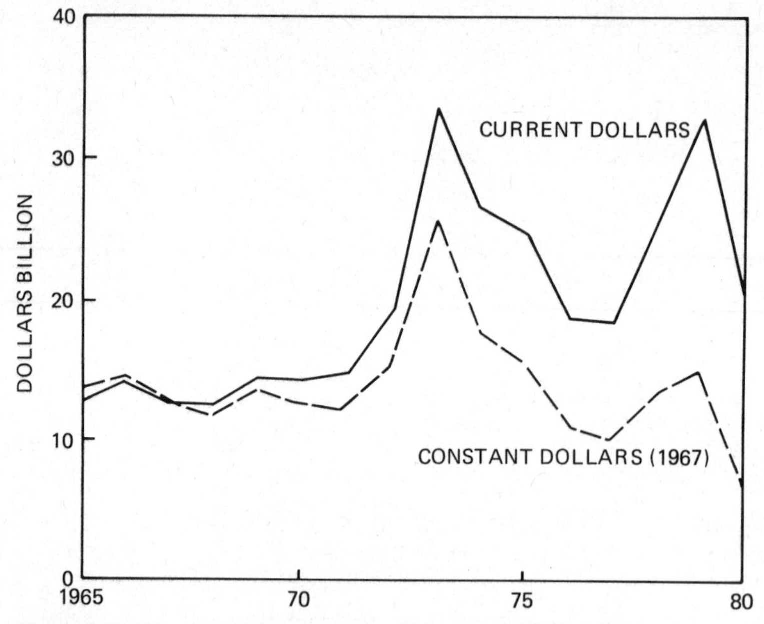

Figure 10–10. Net farm income. (*Source:* USDA, *1982 Handbook of Agricultural Charts,* Washington, D.C., November 1982.)

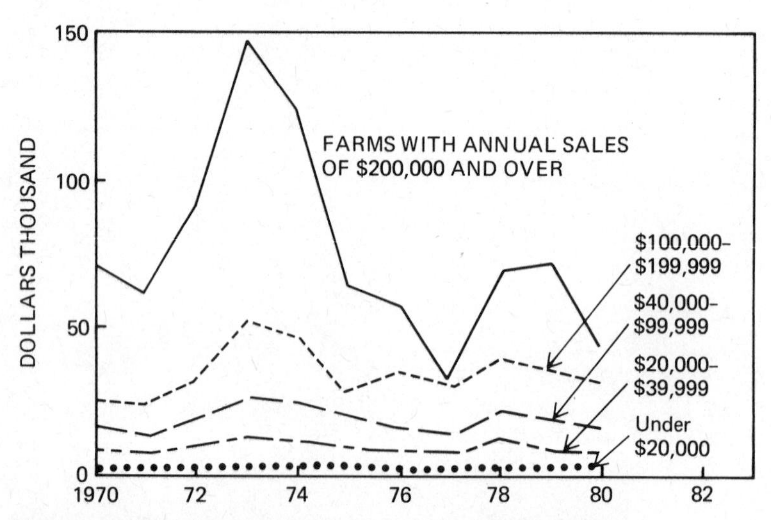

Figure 10–11. Average net farm income by sales class. Net income before adjustment for inventory change. Beginning in 1977, farm income data reflect a change in the farm definition from a place of 10 or more acres with $50 in agricultural product sales and under 10 acres with $250 in sales to a place with $1000 in sales. (*Source:* USDA, *1982 Handbook of Agricultural Charts,* Washington, D.C., November 1982.)

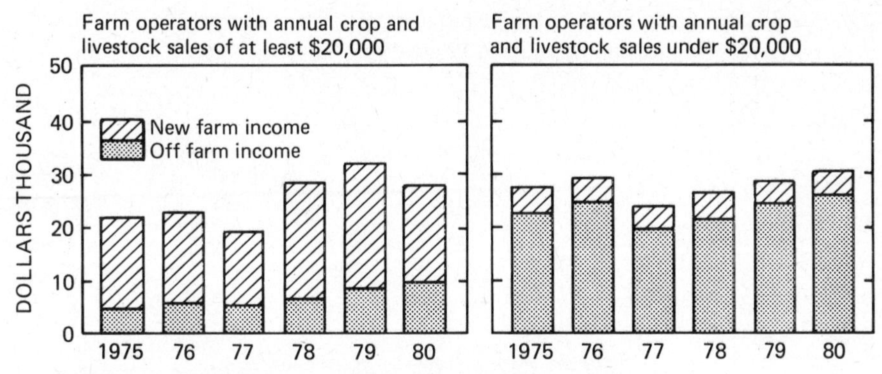

Figure 10–12. Average farm family income all sources. Net income before adjustment for inventory change. Beginning in 1977, farm income data reflect a change in the farm definition from a place of 10 or more acres with $50 in agricultural product sales and under 10 acres with $250 in sales to a place with $1000 in sales. (*Source:* USDA, *1982 Handbook of Agricultural Charts,* Washington, D.C., November 1982.)

they spent 90 cents for each dollar of gross income. This compares with 80 cents or less for farms with lower sales class categories (Table 10–3).

Net farm income does not include the farmer's income from work off the farm or other nonfarm income of the farmer or his family. Farm income from nonfarm sources has been making steady gains and now represents the major income component for farms selling less than $20,000 worth of farm products. (Table 10–3 and Figure 10–12).

The 1960s and 1970s saw a marked improvement in the average disposable personal income of farm people. While incomes of farm people have been increasing over time, their progress relative to nonfarm people has been slow. In 1955, for example, the per capita money income of farm families was only 47.8 percent of the per capita income of the nonfarm population. By 1974, the per capita disposable income of the farm population was equivalent to that of the nonfarm population (Table 10–9 and Figure 10–13).

The relatively striking improvement in per capita disposable income of the farm population during that period was due in part to a high level of national prosperity that increased job opportunities in rural areas. While the location of jobs and employment opportunities still falls short of an ideal distribution throughout the country, the problem has become a matter of national concern. The Appalachia Act and the Public Works and Economic Development Act represented attempts to improve the economic status of low-income rural people found in the underdeveloped regions of the country.

The decade of the 1960s and most of the 1970s can be considered favorable from the viewpoint of resource adjustments and gains in

Table 10–9 Disposable Personal Income per Capita, Farm and Nonfarm Population, Selected Years 1955–80

	Per Capita Income From All Sources of the			
Year	*Farm Population (dollars)*	*Nonfarm Population (dollars)*	*Total Population (dollars)*	*Farm as Percentage of Nonfarm (percent)*
1955	847	1773	1666	47.8
1960	1071	2020	1937	53.0
1965	1688	2486	2436	67.9
1970	2482	3421	3376	72.6
1975	4520	5099	5075	88.6
1980	6553	8042	8002	81.5

Source: Data from USDA, Economic Research Service, *Farm Income Situation,* Washington, D.C., July 1975, and, USDA, *Agricultural Statistics,* U.S. Government Printing Office, Washington, D.C., 1981.

disposable per capita incomes by farm people. Averages per farm, however, may not provide meaningful evaluations of the economic well-being of farm people because of the large differences among U.S. farms. Averages include the 282,000 farms that in 1980 sold $342,695 worth of farm products per farm; they also include 509,000 farms that in 1980 sold $1175 worth of farm products peer farm (Table 10–3).

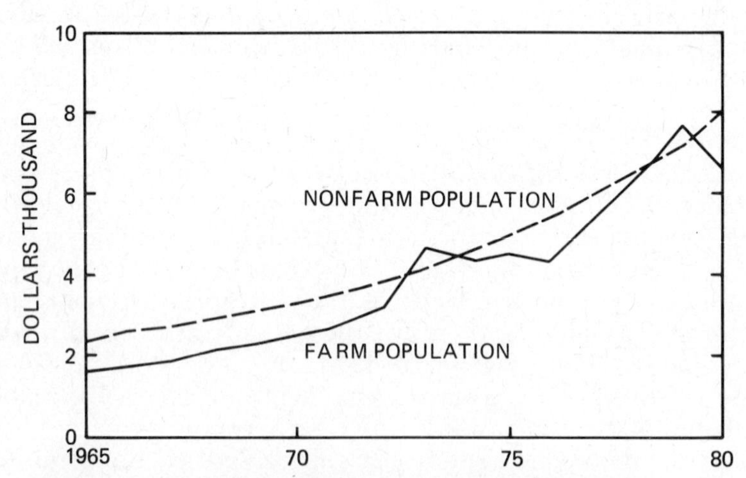

Figure 10–13. Disposable income per capita. Income from all sources less personal contributions for social insurance, personal tax, and nontax payments. beginning in 1977, farm income data reflect a change in the farm definition from a place of 10 or more acres with $50 in agricultural sales and under 10 acres with $250 in sales to a place with $1000 in sales. (*Source:* USDA, *1982 Handbook of Agricultural Charts,* Washington, D.C., November 1982.)

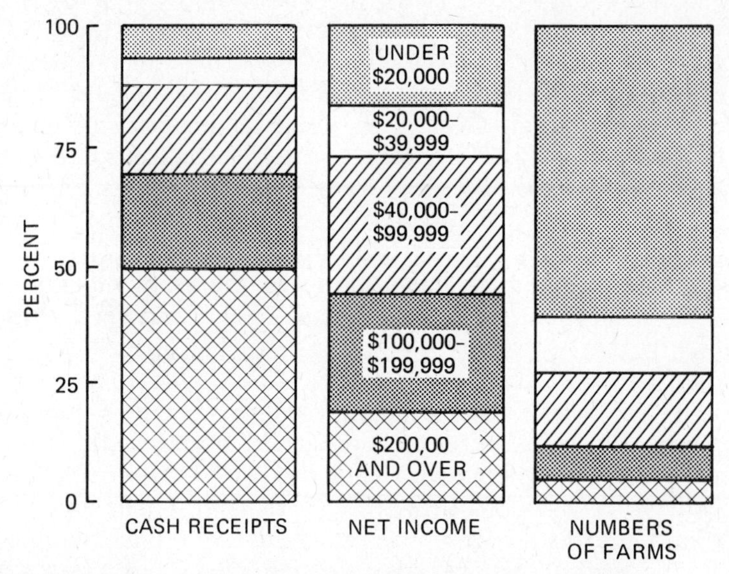

Figure 10–14. Cash receipts, net income, and farm by sales class. Net income before adjustment for inventory change, 1980 data. (*Source:* USDA, *1982 Handbook of Agricultural Charts,* Washington, D.C., November 1982.)

Heterogeneity among U.S. farms continues to increase. The gap between successful commercial farms and farms characterized by relatively low incomes has been widening. Evidence of the pressure toward dualism becomes more apparent from an examination of changes in farm output from 1960 to 1980 (Table 10–3). Rather sharp demarcations appear to aggregate growth trends based on farm numbers and cash receipts from marketings.

The proportion of farms in all-sales-size categories above $40,000 increased between 1960 and 1980. These are the farms that made sizable percentage gains in cash receipts from farming. There appears to be popular acceptance of the notion that one-fourth of our nation's farms are commercial producers affected by farm policy, technology, and development. In 1980 the top one-fourth (40,000 or more in terms of gross sales) produced 70 percent of the cash receipts from marketings (see Table 10–3 and Figure 10–14).

TECHNOLOGICAL CHANGE

One of the most striking features of the American economy has been its rapid rate of technological innovation. Agriculture has participated in this advance. Agricultural productivity, the ratio of farm output per unit of input, is presented as an indexed series in Figure 10–4. Since about 1935, productivity has been increasing at almost a straight-line rate and has doubled in 40-years.

This increase in productivity stems from two factors: (1) public and private investment in agricultural research, and (2) the farmer's willingness to adopt new techniques. A substantial amount of research has been devoted to improving agricultural production techniques. Much of this work has been done by the Agricultural Experiment Stations established within Land Grant Universities and by the United States Department of Agriculture, either directly or by providing funds to the experiment stations. Contributions have also been made by industrial firms supplying agricultural inputs, that is, by machinery companies, chemical and fertilizer companies, feed and seed companies, etc. These firms are continually attempting to develop and sell products better than those of their competitors. The ability to market a new product first, such as a self-propelled combine or a new feed additive, can mean the difference between success and failure in the farm supply market. By pursuing their own profit-motivated objectives, farm supply firms have contributed directly to agricultural efficiency.

Technological advances have touched every part of agriculture and include chemical fertilizers, sprays, better feeds and feed additives, improved and new types of machinery, high-yield crop varieties, improved strains of livestock, and, most important but perhaps immeasurable, a substantial improvement in managerial skills of farmers. All the types of technological improvements and the resulting impacts on farm production cannot be discussed here. Rather, the economic concepts presented in earlier chapters will be used to present the elements of an economic analysis of technical change.

Impact of Technology on the Farm

Technological change can cause old products to be replaced by new ones, can create new inputs or improve old ones, and can otherwise affect the production process. Some technological developments, chemical fertilizers for example, provide a new input while others, hybrid corn varieties as an example, represent an improvement in an established practice. When the technological development is a new input, this input would have a production function and a unit cost; the manager would thus equate the value of the marginal product of the input with unit cost. In this case the result of the technological change would be an increase in output and costs, revenue and profit.

If the new input were a replacement for an input previously used in the production process and if it has exactly the same production function as the old input, the new input would be utilized as a replacement for the old input only if it costs less. In such a case the input would be called a "cost-reducing" technology. If output remained constant, costs would be lowered and profits increased.

Output would not remain constant, however. As shown in Figure 10–15A, the new input, X_t, has a lower cost and the manager will use the added amount of *DE*, will increase profit by the area under the

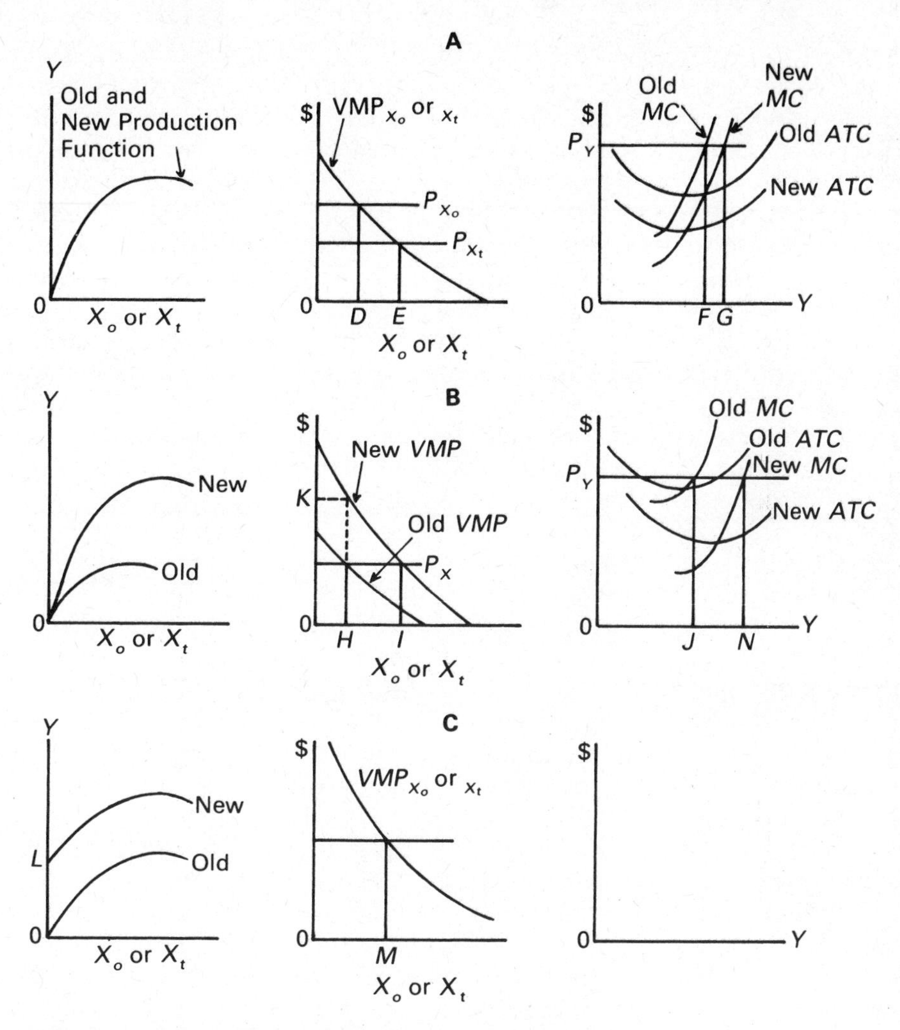

Figure 10–15. Technological change and the production function.

VMP curve and between the two price lines (center diagram), and will increase output by *FG* (right-hand diagram). Perhaps in a very short period, for example, when appropriate adjustments in durable resources cannot be made, the new inputs will reduce costs without increasing output. In general, however, the manager will adjust as quickly as possible to gain the added profit. Thus, cost-reducing technological improvements tend to increase output, given constant prices for output and other inputs.

Suppose the new input shifted the production function upward as shown in Figure 10–15B. For any given input amount, output is increased and production function slopes more steeply, indicating an increase in marginal productivity. In this case the new input, X_t, will be used to replace the old, X_0, even if it (X_t) costs more. The criteria is, of course, that profit using the new input must equal or exceed profit using the old input. If the inputs are identically priced, an additional amount, *HI*, will be used (center diagram), resulting in increased output, *JN* (right-hand diagram), and an increase in profits. In fact, output and profits will be increased for all prices of the new input equal to or less than *OK* (center diagram 10–15B) and even for some higher prices (why?). Thus, a cost-increasing technology can also result in increased outputs and profits.

A third effect of a technological change might be to increase output at each input amount while leaving marginal physical product unchanged. In this case the production function would increase a constant amount at each input level, but its slope (*MPP*) would remain unchanged. This situation is represented in Figure 10–15C; the production function shifts upward a constant amount and the *VMP* curve remains unchanged. As long as profits using the new input exceed those from using *OM* amount of the old, the new technology will be adopted and an amount more or less than *OM* (center diagram, Figure 10–15C) will be used depending on the unit cost of the new input.

Assuming for the moment that the new input costs as much or more than the old input, then an equal or smaller amount will be used. An increase in production has resulted without an increase in the amounts of other inputs. Additional resources are not attracted to the firm and some may even be freed for use elsewhere. Also, output has increased. Thus, a technology having the effect demonstrated in Figure 10–15C would appear to be ideal—output and profits increase while the price mechanism diverts resources to other uses. Such a conclusion is fallacious in the presence of fixed inputs and can be regarded as valid only in the short run. The technology increases profits accruing to the fixed inputs—which, of course, are not actually fixed—and the manager will thus seek to acquire more of the fixed asset and thereby increase output. Also, economic rent is increased and additional firms may enter production. Thus, only when the time period is so short that fixed assets cannot be expanded will this technological innovation result in an increase in profits along with a stable output. (The effect of this type of technological change on the average and marginal cost curves is left to the student as an exercise.)

Technology and Farms

If the assumptions of the preceding section are accepted, the conclusion is that technological change increases the farm's output. Because farmers are pure competitors and regard prices facing them as set by the market, the individual farmer can benefit from innovation. By adopting a new technique, a farmer can reduce costs and/or increase profits in the short run. As more farmers use the technique, total aggregate output of the commodity will increase and, without a comparable increase in aggregate demand, cause product prices to fall. Thus, because of the competitive nature of agriculture, the technique will be adopted and aggregate output will increase regardless of the action of an individual farmer. It is therefore to the benefit of the individual farmer to adopt new techniques as quickly as possible, thereby capturing the returns possible before others follow. Hence, good managers are innovators.

Technological changes such as hybrid seed corn require little added capital investment while changes such as those taking place in Grade A dairies (including milking parlors, milking machines, and bulk tanks) not only require substantial investment but also put a lower limit on enterprise size. A modern Grade A dairy cannot be established for a herd of two or three cows. With some exceptions, technological change has increased the capital needed for farming.

The same types of comments may be made about technology and risk. While new technologies at first represent an unknown to be mastered by the manager, ultimately they may result in a reduction of risk for the good manager. New technologies, because of the added capital investment and increased farm size, may tend to force the poor or average manager past his natural management abilities and thereby increase risk for him.

Land is a residual claimant of agricultural earnings. To the extent that technological change increases profits, it will also increase the economic rent accruing to land. According to the analysis above, this rent increase will increase the inherent value of land and tend to bring additional land into production.

Growth in agricultural productivity results in problems as well as benefits. Increased productivity benefits the consumers through savings in the resources necessary to produce a given output of agricultural products. In the short run, however, increases in productivity may cause a surplus of output that the market cannot absorb at acceptable prices, resulting in low returns to agricultural resources.

An important source of the chronically low earnings of resources in agriculture is the economic organization of agriculture itself. While certain features or perfect competition are foreign to all sectors of our economy (perfect knowledge, perfect product and resource divisibility, perfect mobility), certain aspects are found in agriculture—especially when compared to many other sectors. Farming is a unique industry. It is made up of millions of individual firms. Each is an individual business—generally with its own self-employed manager.

The lack of price policy in such an industry follows from its organization; no producer has control over a sufficiently large fraction of the market to affect price. Aggregate output thus cannot be planned as it is, say, in the automobile industry. As a result, prices and income in agriculture are often unstable and fluctuating.

Because of the large number of farms, the output of each producer contributes a small part of the total industry's output. The individual farm faces a perfectly elastic demand function (Figure 10–16A). If there were, for example, 1 million wheat producers each producing 2000 bushels, market supply would be 2 billion bushels. If one producer changed hit output by 50 percent, total supply would change by only 1000 bushels, or 0.000005 percent. This small change in total supply would not affect market price.

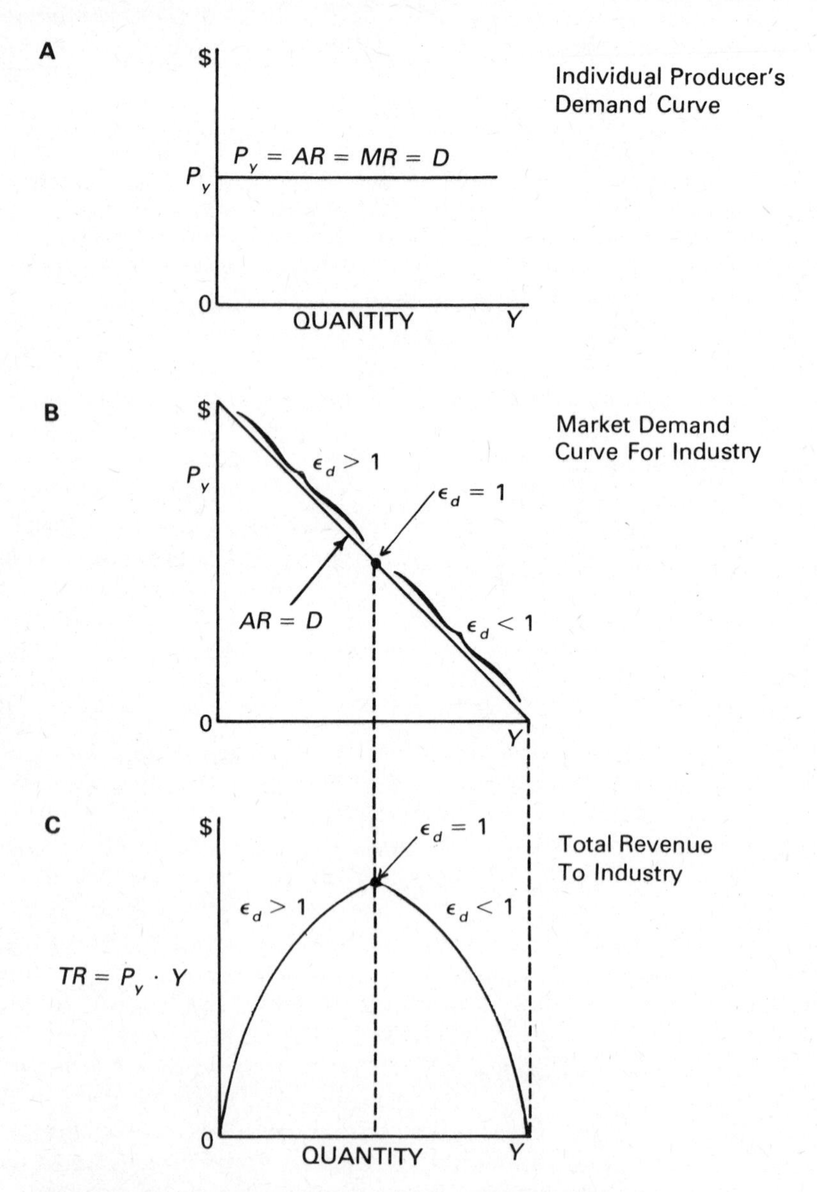

Figure 10–16. Hypothetical demand situation faced by an individual producer in price competition with industry demand and total revenue curves.

Wheat farmers as a group can cause market price to vary as shown by the industry's demand function, Figure 10–16B. If all the wheat producers were to change production by 50 percent, a large change in the price of wheat would result because total supplies would change by 50 percent. Thus, when a sector of an economy, organized as is United States agriculture, is provided with output-increasing techniques, the economic impacts that follow are largely predetermined. The producer cannot influence price, but he can influence his costs by adopting new techniques. He does not consider what his neighbors will do, although this is not irrelevant; he simply has no control over what his neighbors will do. When all or many have improved their methods of production, costs will fall but, generally, so will farm prices.

Another reason for low earnings in agriculture is the low elasticity of demand for farm products. Adam Smith's remarks in 1776 are still much to the point:

> The desire for food is limited in every man by the narrow capacity of the human stomach; but the desire of the conveniences and ornaments of building, dress, equipage and household furniture, seems to have no certain boundary.[4]

Figures 10–16B and C show the market demand function and the associated total revenue function. Total revenue is, of course, price times quantity and represents total revenue to the industry. The elasticity of demand relates the impact of a price change to the amount of total revenue received by an industry. In absolute terms, as long as elasticity of demand is greater than one ($\epsilon_D > 1$), an increase in quantity sold, Y, will increase the total revenue; when $\epsilon_D < 1$ (inelastic), an increase in quantity sold reduces the revenue to the industry. Table 10–10 shows the effects of changing outputs and costs on industry total returns under different demand elasticities situations. Figure 10–16C should be helpful in foreseeing the effect of new technologies (innovations) on the industry's profits. For example, in Case I, Table 10–10, if demand for a product is elastic $\epsilon_D > 1$, and an innovation is both output- and cost-increasing, net returns will increase as long as total revenue increases more than total cost. In the second case, Table 10–10, net returns will decrease because demand is inelastic, $\epsilon_D < 1$; the innovation is output-increasing (moving to the right and down on the total revenue curve Figure 10–16C) even though costs remain constant. (In Figure 10–16, AR = average revenue, MR = marginal revenue, and D = demand.)

In general, when a rich country grows richer, the demand for food increases relatively little. The demands for resources in industries producing products with high income elasticities will rise relative to the demands in industries producing products with low income elas-

[4]Smith, Adam, *An Inquiry into the Nature and Causes of the Wealth of Nations,* James E. Thorold, ed., London, England: MacMillan and Co., 1869, pp. 174–175.

Table 10–10 Effects of Innovations (New Technologies) on Industry's Net Returns Under Different Demand Elasticities

		Effect of Innovation on		
Case	*Demand*	*Output*	*Costs*	*Net Returns to Industry*
1	Elastic	Increasing	Increasing	Increase if increase in *TR* (total revenue) > increase in *TC* (total cost)
2	Inelastic	Increasing	Constant	Decrease
3	Elastic	Constant	Decreasing	Increase
4	Inelastic	Constant	Decreasing	Increase
5	Elastic	Increasing	Decreasing	Increase
6	Inelastic	Increasing	Decreasing	Increase if decrease in *TR* < decrease in *TC*
7	Elastic	Decreasing	Decreasing	Increase if decrease in *TR* < decrease in *TC*
8	Inelastic	Decreasing	Decreasing	Increase

ticities. In the competition for resources, the industries that sell products with low income elasticities will be outbid for resources by industries that sell products with high income elasticities. Because of the very low income elasticity for food, agriculture can be outbid by most other industries. This tends to depress prices of farm products relative to prices of resources used in agriculture. This is a partial explanation of the so-called cost-price squeeze in farming.

The demand for most agricultural products is inelastic with respect to price and income. This means that reductions in prices of farm products are accompanied by proportionately smaller changes in quantities sold; a 1 percent change in price will change quantity of product purchased by less than 1 percent, and the price elasticity is less than one—inelastic. Similarly a 1 percent change in real income changes the quantity of farm products demanded by less than 1 percent. Thus, the nature of demand for farm products can affect the net income of agriculture adversely.

COMPETITION AMONG REGIONS

Production in a region is influenced by the comparative advantage in the use of the resources. The quality of transportation services available also plays a critical role in the assembly of the raw product in farming areas and in the distribution of processed products from producing areas to consumer markets. With highly developed transportation facilities, food production does not have to take place near population centers. This results in substantial economies in production and dramatic reductions in the relative cost of food purchased by U.S. consumers.

Opportunities to ship from low-cost producing areas to large urban markets benefits both producers and consumers. On the other hand, high-cost producing areas are placed at a disadvantage relative to the new trading pattern, and shifts of resources out of the less competitive areas can cause temporary hardships until new adjustments are made.

The movement of goods from one region to another takes place in response to price differentials that make such movements profitable. For many commodities, and for most agricultural products, the United States can be viewed as one large market with prices in any one region closely tied to those that exist at the same point in time in other regions of the country. Regional specialization in farming reflects the differences in the adaptability of different regions' resources to the production of agricultural commodities.

The costs of transporting agricultural commodities from production sites to the processing centers or consumers, as noted earlier, is but one of many costs that are a part of the production process. Yet differences in transportation costs do have a unique bearing on the intensity in the use of resources and on net returns to producers.

Improved transportation systems have an effect on the cost of production similar to the development of a high-yielding variety that is adaptable only to a certain region. Both affect the costs of production. Interstate highways and air freight, for example, provide opportunities for marketing the products grown in areas that have not previously had access to major markets.

The example that follows illustrates the effect of different production costs, in this case transportation cost, on the firm's output, profit, and even the possible viability of the enterprises.

Assume that 50 producers in each of three different locations produce the same product, Y. The basic production functions are the same in all three locations; the product is homogeneous and therefore cannot be differentiated. The only difference to be considered in this problem is the cost of transporting the product to a common market place, where demand, $D = -10P_Y + 3900$.

The hypothetical total cost function is

$$TC = 0.5Y^2 \tag{10.1}$$

and the cost of transporting each unit of Y to market is \$3 for location 1, \$6 for location 2, and \$9 for location 3. Thus, the total cost for the three different locations is

$$
\begin{aligned}
TC_1 &= 0.5Y_1^2 + 3Y_1 \\
TC_2 &= 0.5Y_2^2 + 6Y_2 \\
TC_3 &= 0.5Y_3^2 + 9Y_3
\end{aligned}
\tag{10.2}
$$

As noted in Chapter 3 for perfectly competitive markets, the marginal cost curve above the average variable cost curve serves as a supply function for the individual producer. The individual supply function has a value of zero for all levels of market price below the minimum point on the average variable cost curve.

From total cost functions, (10.2) in our example, it can be seen that the marginal costs exceed the average variable costs at all positive values of output. Thus, marginal costs serve as individual supply functions for all values of output exceeding zero. In a competitive market, the producers maximize profits by producing a quantity of Y at the point where $MC = P_Y$.

The corresponding marginal cost equations for producers in each of the three locations are:

$$
\begin{aligned}
\frac{dTC_1}{dY_1} &= Y_1 + 3 \\
\frac{dTC_2}{dY_2} &= Y_2 + 6 \\
\frac{dTC_3}{dY_3} &= Y_3 + 9
\end{aligned}
\tag{10.3}
$$

The profit-maximizing quantities are:

$$
\begin{aligned}
Y_1 + 3 &= P_Y \\
Y_2 + 6 &= P_Y \\
Y_3 + 9 &= P_Y
\end{aligned}
\tag{10.4}
$$

And the individual producer's short-run supply functions are:

$$
\begin{aligned}
Y_1 &= P_Y - 3 \\
Y_2 &= P_Y - 6 \\
Y_3 &= P_Y - 9
\end{aligned}
\tag{10.5}
$$

The above equations, (10.5), of individual firms indicate the amounts individual producers would offer for sale as functions of the market price. If the market price is \$10, then producers at location 1 would supply 7 units each (10 − 3 = 7); producers from location 2 would supply 4 units each (10 − 6 = 4); and producers from location 3, 1 unit each (10 − 9 = 1). If the price were to fall to $5, producers at locations 1 and 3 would not offer any quantity for sale because marginal cost would be above price, while the producers in location 1 would reduce the quantity offered from 7 to 2 units (5 − 3 = 2).

The market supply curve is obtained in general by the addition of the individual firm's supply curves. As indicated in Figure 10–17, the market supply curve is upward-sloping because the individual producer's supply curves are also upward-sloping. The higher the price (measured on vertical axis), the larger the quantity offered for sale (measured on horizontal axis). The speed with which the quantity of supply changes in response to changing price is indicated by the elasticity of supply. The elasticity of supply is defined in a manner similar to the elasticity of production discussed in Chapter 2 or the elasticity of substitution discussed in Chapter 4. The elasticity of supply is the percentage change in quantity supplied, divided by the percentage change in price. A large elasticity of supply indicates that producers are more responsive to changes in price. A supply curve that is horizontal (parallel to the quantity axis) is said to be infinitely elastic. A supply curve that is vertical (parallel to the price axis) is completely inelastic. Most supply curves lie between these two extremes.

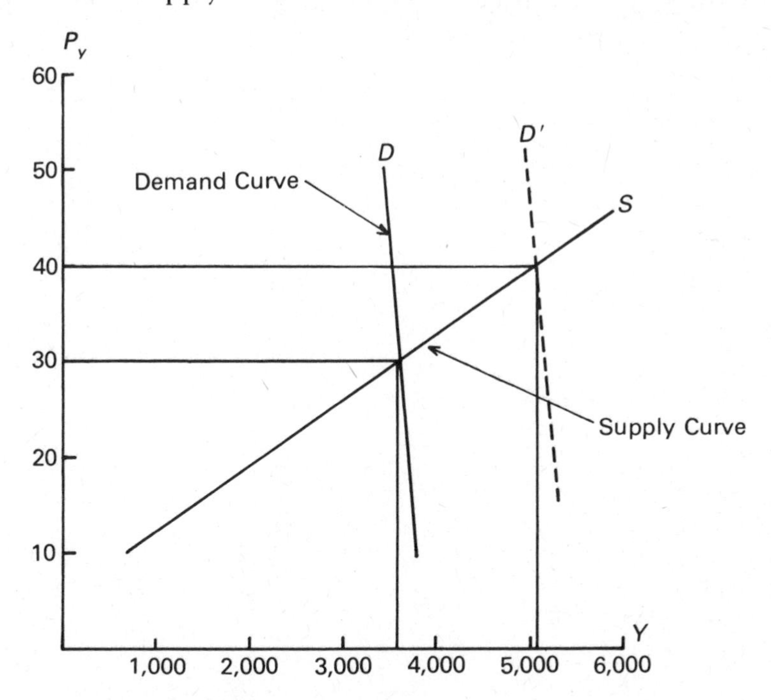

Figure 10–17. Aggregate market supply and demand curves for producers in three locations.

The market supply function is obtained for this example by adding the individual supply functions and, because all producers at a location have identical input-output and cost relationships, by multiplying the individual supply functions by the number of producers at each location. If each location has 50 producers, the market supply function would be

$$\begin{aligned} S_{MKT} &= (P_Y - 3)50 + (P_Y - 6)50 + (P_Y - 9)50 \\ S_{MKT} &= 150P_Y - 900 \end{aligned} \tag{10.6}$$

The market clearing price is determined at the point where market supply equals market demand:

$$\begin{aligned} S_{MKT} &= D_{MKT} \\ 150P_Y - 900 &= -10P_Y + 3900 \\ 160P_Y &= 4800 \\ P_Y &= 30 \end{aligned} \tag{10.7}$$

The total quantity of Y demanded and supplied at $P_Y = 30$ is 3600 units. This is derived by substituting the market clearing price in either the market supply or market demand equation; for example, $150 \cdot 30 - 900 = 3600$ or $-10 \cdot 30 + 3900 = 3600$.

Because each producer receives the same price regardless of the location of production, the producer's share of the market can be determined by substituting the market clearing price, 30, in the individual's supply equation, (10.5). Each producer in location 1 would supply 27 units of Y $(30 - 3)$, and producers from locations 2 and 3 would supply 24 and 21 units of Y, respectively.

Transportation cost is a variable cost and hence is a part of marginal cost, which in turn determines the intensity of resource use. The marginal cost for producers in location 1 intersects or equals the price line at higher level of output than the marginal unit costs for producers at locations 2 and 3 (see Figure 10–18).

Differences in transportation costs also affect net returns to producers. The net return (profit) per unit of output to producers is \$13.50 in location 1, \$12.00 to producers in location 2, and \$10.50 to producers in location 3. The total net return to each producer is \$364.50, \$288, and \$220.50 in locations 1, 2, and 3, respectively.

Similarly, the effects on producers in different locations due to differences or changes in either production functions or demand can be calculated. For example, if the demand for the product increases to the extent that the market clearing price $(S = D)$ is \$40, and quantity demanded and supplied is 5100 (Figure 10–17), each producer in locations 1, 2, and 3 would supply 37, 34, and 31 units, respectively. The net returns to producers would increase accordingly (Figure 10–18).

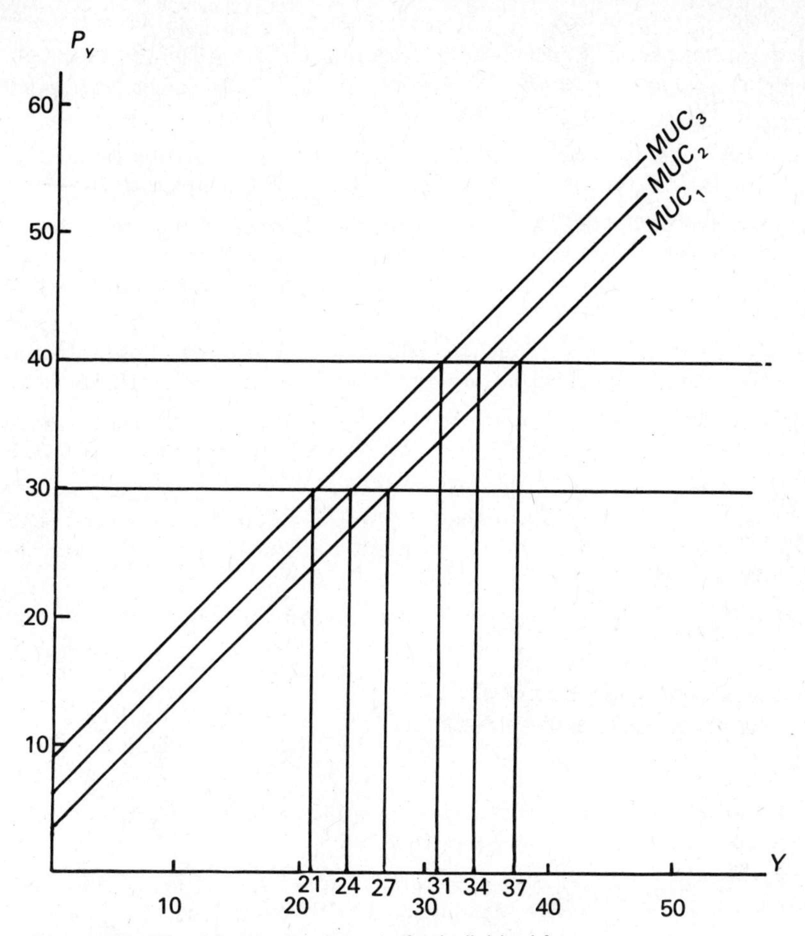

Figure 10–18. Marginal unit costs for individual farms producing product Y in three different locations.

Suggested Readings

Ball, Gordon A. and Heady, Earl O., eds. *Size, Structure, and Future of Farms.* Ames, Iowa: Iowa State University Press, 1972.

Benedict, M. R., Tolley, H. R., Elliott, F. F., and Conrad, Taeuber. "Need for a New Classification of Farms." *Journal of Farm Economics,* Volume 26, November 1944, pp. 694–708.

Bernard, Harvey R. *Technology and Social Change.* New York: Macmillan Company, 1972, Chapter 7.

Bieri, J., Schmitz, A., and Janvry, A. D. "Agricultural Technology and Distribution of Welfare Gains." *American Journal of Agricultural Economics.* Volume 54, December 1972, pp. 801–808.

Brandow, George E. *Interrelations Among Demands for Farm Products and Implications for Control of Market Supply,* Bulletin 680. Pennsylvania Agricultural Experiment Station, Pennsylvania State University, 1971.

Christensen, Laurits R. "Concepts and Measurement of Agricultural Productivity." *American Journal of Agricultural Economics,* Volume 57, December 1975, pp. 910–915.

Cramer, Gail L. and Jensen, Clarence W. *Agricultural Economics and Agribusiness,* Second Edition. New York: John Wiley and Sons, Inc., 1982, Chapters 2, 10–16.

Daly, Rex F. *Agricultural Growth Structural Change and Resource Organization, Food Goals, Future Structural Changes, and Agricultural Policy: A National Casebook.* Iowa State Center for Agricultural and Economic Development, Ames, Iowa: Iowa State University Press, 1969.

Davis, John H. and Goldberg, Ray A. *The Concept of Agribusiness.* Boston Division of Research, Graduate School of Business Administration, Harvard University, 1967.

Doll, John P. and Widdows, Richard. *The Value of Agricultural Land in the United States: Some Thoughts and Conclusions,* Paul Velde, ed. ERS Staff Report No. AGES820323, Washington, D.C., April 1982.

Doll, John P. and Widdows, Richard. *A Comparison of Cash Rents and Land Values for Selected U.S. Farming Regions,* Paul Velde, ed. ERS Staff Report No. AGES820415, Washington, D.C., April 1982.

Doll, John P. and Widdows, Richard. *Imputing Returns to Production Assets in Ten U.S. Farm Production Regions,* Paul Velde, ed. ERS Staff Report No. AGES820703, Washington, D.C., July 1982.

Doll, John P. and Widdows, Richard. *A Critique of the Literature on U.S. Farmland Values,* Paul Velde, ed. ERS Staff Report No. AGES830124, Washington, D.C., January 1983.

Economic Research Service, USDA. *Agricultural Markets in Change.* Agricultural Economic Report No. 95, Washington, D.C., 1966.

Economic Research Service, USDA. *Economic Indicators of the Farm Sector, Income and Balance Sheet Statistics, 1981.* Washington, D.C., August 1982.

Foster, George M. *Traditional Societies and Technological Change,* Second Edition. New York: Harper & Row, 1973, Chapter IX.

Griliches, Z. "Agriculture: Productivity and Technology." In *International Encyclopedia of the Social Sciences,* Volume 1. New York: The Macmillan Company and the Free Press, 1968.

Grove, Ernest W. "Farm Labor and the Structural Analysis of Agriculture." *Journal of Farm Economics,* Volume 49, December 1967, pp. 1245–1253.

Headley, J. C. "Agricultural Productivity, Technology, and Environmental Quality." *American Journal of Agricultural Economics,* Volume 54, December 1972, pp. 749–756.

Krause, Kenneth R. and Kyle, Leonard R. "Economic Factors Underlying the Incidence of Large Farming Units: The Current Situation and Probable Trends." *American Journal of Agricultural Economics,* Volume 52, December 1970, pp. 748–760.

National Academy of Sciences. *Agricultural Production Efficiency.* Washington, D.C., 1975.

Nikolitch, Radoje and McKee, Dean E. "The Contribution of the Economic Classification of Farms to the Understanding of American Agriculture." *Journal of Farm Economics,* Volume 47, December 1965, pp. 1546–1554.

Preusche, G. "How Much Longer Will Technological Revolution in Agriculture Last?" *Landtechnik,* Volume 28, No. 18, 1973.

Roy, Ewell P. *Exploring Agribusiness.* Danville, Ill.: The Interstate Printers and Publishers, Inc., 1965.

Ruttan, Vernon W. "Agricultural Policy in an Affluent Society." *Journal of Farm Economics,* Volume 48, December 1966, pp. 1100–1120.

Ruttan, Vernon W., Waldo, A. D. and Houck, J. P. *et al.,* eds. *Agricultural Policy in an Affluent Society.* New York: W. W. Norton & Company, 1969, Part One.

Schertz, Lyle P. *et al. Another Revolution in U.S. Farming?* Washington, D.C.: USDA, Agricultural Economic Report 441, 1979.

Thomas, Gerald W. *Progress and Change in the Agricultural Industry.* Dubuque, Iowa: William C. Brown Book Company, 1969.

Tweeten, Luther G. "Low Returns in a Growing Farm Economy." *American Journal of Agricultural Economics,* Volume 51, November 1969, pp. 810–811.

Welsch, Delane E. and Moore, Donald S. "Problems and Limitations Due to Criteria Used for Economic Classification of Farms." *Journal of Farm Economics,* Volume 47, December 1965, pp. 1555–1564.

Widdows, Richard and Doll, John P. *Four Econometric Models of the U.S. Farmland Market: An Updating with Comparisons,* Paul Velde, ed. ERS Staff Report No. AGES820702, Washington, D.C., July 1982.

NOTATION, GEOMETRIC CONSIDERATIONS, AND A NOTE ON CALCULUS

APPENDIX I

NOTATION

Functional Notation

The production function relates output, Y, to input, X. The definition of a *function* is: If a variable Y depends on another variable X so that when each value of X is known, a unique (single) value of Y is determined, then Y is called a function of X. A shorthand expression for a function is

$$Y = f(X)$$

This notation is called functional notation and is read "Y is a function of X." Y is usually called the dependent variable, and X, the independent variable.

Subscripts

Subscripts are useful when symbols are used. Their use adds some additional meaning to the symbol. Consider, for example, the notation for the production function

$$Y = f(X)$$

where X is the amount of input and Y is the resulting amount of output. Because there is only one input and one output, there can be no confusion about identification of input or output. In this case, subscripts can be used to denote *amounts:* X_1 is an amount of X, X_2 is a greater amount than X_1, X_3 is a greater amount than X_1 or X_2, etc. Also, subscripts on Y can be used to denote amounts of Y. Thus, amount Y_1 of Y results from the use of X_1 amount of X or

$$Y_1 = f(X_1)$$
$$Y_2 = f(X_2)$$
$$Y_3 = f(X_3)$$

When more than one input or output is included in a problem, subscripts can be used as a means of identification. For example, when output is a function of three inputs, the production function can be written

$$Y = f(X_1, X_2, X_3)$$

where X_1, X_2, and X_3 are distinct and different inputs. The subscripts identify the inputs. If amounts are to be denoted, additional subscripts must be used. X_{11} is an amount of X_1; X_{12} is a greater amount of X_1; X_{21} is an amount of X_2; X_{22} is a greater amount of X_2; etc. Subscripts can also be used to identify outputs. Thus, Y_1, Y_2, and Y_3 can be distinct outputs; amounts could be shown by adding another subscript.

The "Δ" Notation

When dealing with problems concerning the change in a variable, the following notation is common: The change in any variable is denoted by "Δ" (the Greek letter delta) placed before the variable. For example, the change in the variable X is denoted by ΔX, the change in Y is ΔY, the change in Z is ΔZ, etc. Note that ΔX does not mean a quantity Δ multiplied by a quantity X, but denotes a single quantity, the change in X. If ΔX is positive, the change in X is an increase. If ΔX is negative, the change in X is a decrease, and similarly for other variables. The change ΔX is called the increment of X.

Discrete and Continuous Data

The difference between discrete data and continuous data is in the divisibility of the inputs (or outputs). An example of a discrete input is a cow. A dairy herd may be composed of two, three, or more cows. However, one and a half, two and an eighth, three and a quarter cows, etc., will not be found in a dairy herd because, at least at the time of this writing, cows come only in units of one. Commercial fertilizer, on

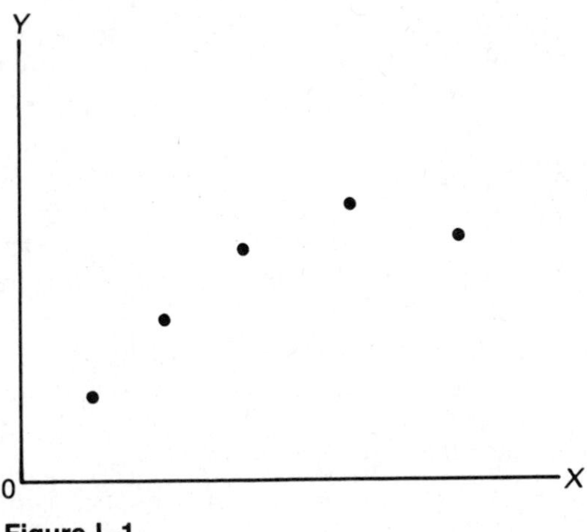

Figure I–1

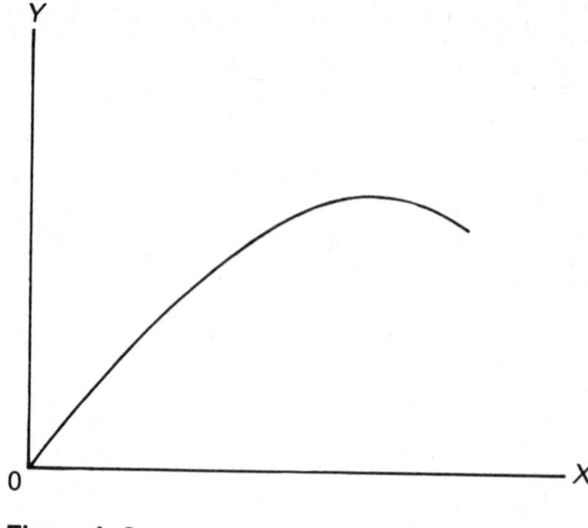

Figure I–2

the other hand, is an example of a continuous input. Fertilizer can be divided into any size unit, and for each size unit there is a resulting yield.

The production schedules used in this text are examples of discrete data. A discrete production function is depicted in Figure I–1. Here, no yield is forthcoming between the dots. One and a half cows give no more milk than one cow. A continuous production function is depicted in Figure I–2.

SOME GEOMETRIC CONSIDERATIONS

The Slope of Lines and Curves

Determining the slope of a curve is a problem in applied mathematics. Mathematicians are interested in the slope of a curve because it denotes the rate of change in a variable such as output, Y. Economists are interested in the slope of a curve (production function) because it represents a marginal concept, such as the marginal product of an input, X.

Determining the slope of a curve is easier said than done because, generally speaking, the slope differs for each point on the curve. In Figure I–3, the slope of the curve at A is different from the slope at B. Also, the slope at any point between A and B is different from the slope at A or B.

In contrast to a curve, the slope of a straight line is the same for all points on the line and is therefore easily determined. In Figure I–3, the slope of the straight line is the same at either C or D. It is also the same for all points other then C or D. If the slope of a straight line is easily determined but the slope of a curve is difficult to determine, the question might arise: Why not use the straight line to determine the slope of a curve? That is what we will do. First, we will determine

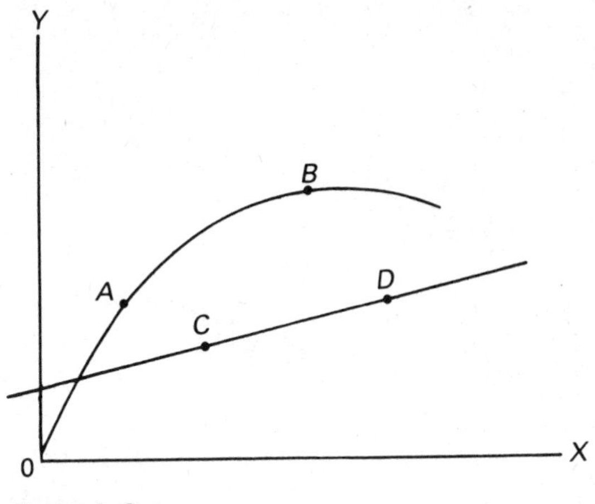

Figure I–3

how to measure the slope of a straight line, and then we will show how a straight line can be used to determine the slope of a curve.

The Slope of a Line. Through any point C on a straight line draw CD parallel to the X axis (Figure I–4). Next, draw line DE parallel to the Y axis. Then,

$$\text{the slope of line } AB = \frac{DE}{CD}$$

This is the "rise" over the "run" and measures the gradient or slope of the line AB. Notice that in Figure I–4, CED and $C_1E_1D_1$ are similar triangles. Therefore,

$$\frac{DE}{CD} = \frac{D_1E_1}{C_1D_1}$$

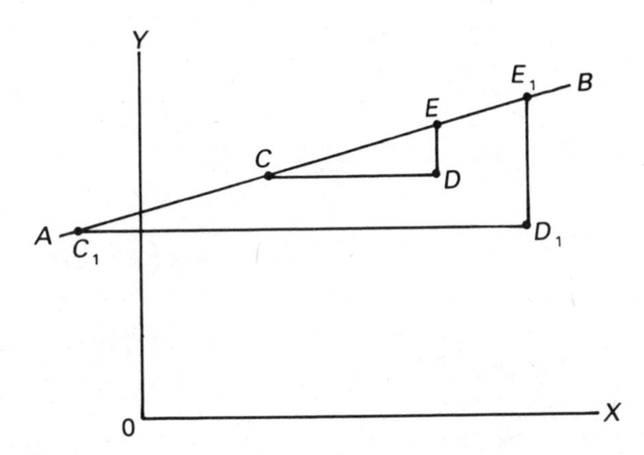

Figure I–4

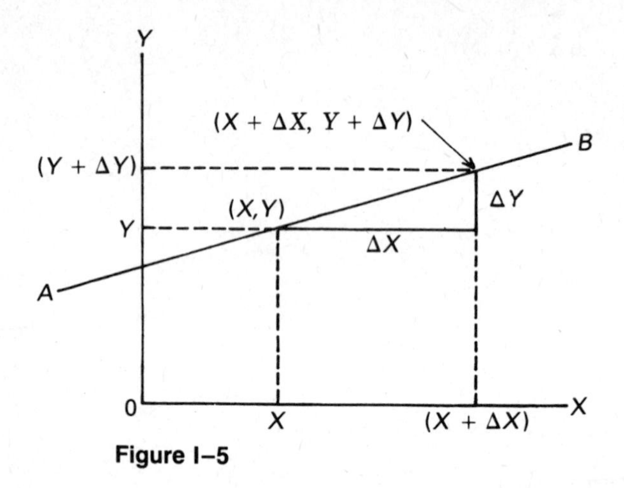

Figure I–5

The slope of the line *AB* is the same at any point on the line. On a straight line, the size of the triangle *CED* is immaterial because only the ratio of the sides of the triangle determines the slope.

The following can be said of the slope of a straight line:

1. The slope of a line parallel to the *X* axis is zero.
2. The slope is the number of units that *Y* increases (or decreases) when *X* increases one unit.
3. The slope of a line that falls to the right is negative.
4. The slope of a line that rises to the right is positive.
5. The slope of a line parallel to the *Y* axis is undefined.

Figure I–5 is essentially the same as Figure I–4. However, Figure I–5 is labeled differently. Point *C* is now denoted by the coordinates (X, Y). Point *E* is now $(X + \Delta X, Y + \Delta Y)$. ΔX is any change in *X*, and ΔY is the corresponding change in *Y*. Now

$$\text{the slope of line } AB = \frac{\Delta Y}{\Delta X}$$

But when *Y* is an output and *X* is an input

$$\frac{\Delta Y}{\Delta X} = \text{marginal physical product of } X$$

Because the slope is a measure of *MPP*, *MPP* is a constant for a straight-line (linear) production function.

Determining the Slope of a Curve. Figure I–6 illustrates one way a straight line may be used to determine the slope of a curve. Here,

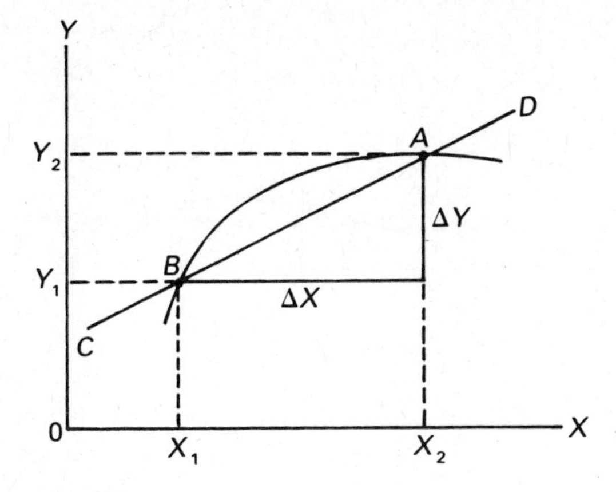

Figure I–6

points B (X_1, Y_1) and A (X_2, Y_2) are known and we want to measure the slope of the curve between B and A. (The subscripts are used to denote amounts). The slope is

$$\frac{Y_2 - Y_1}{X_2 - X_1} = \frac{\Delta Y}{\Delta X}$$

$\Delta Y/\Delta X$ does not measure the slope of the curve at A or the slope at B, but measures the slope of the line CD. However, the slope of CD can be regarded as an estimate of the slope of the curve between A and B. As the distance between A and B decreases, the slope of the curve is more closely approximated by the slope of the line.

Figure I–6 presents pictorially the situation that occurs when marginal physical products are computed from a production schedule containing only a few values of the continuous input. When used to make graphs, the marginal physical products computed in this manner are placed halfway between the input levels rather than on either level. That is, in Figure I–6, the *MPP* resulting from adding the input amount $(X_2 - X_1)$ would be graphed halfway between X_1 and X_2. (Line CD is called a secant.)

A straight line can also be used to determine the exact slope of a curve at any point. In Figure I–7, the slope of the curve and the line AB are equal at C. AB is tangent to the curve at C. By determining the slope of line AB, the exact slope of the curve at C can be determined. Thus,

$$\text{the slope of } AB = \frac{CD}{HD} = \text{slope of curve at } C$$

and

$$\frac{CD}{HD} = \frac{CD}{OE} = \text{the exact } MPP \text{ of } X \text{ at } C \text{ (for } OE \text{ amount of } X\text{)}$$

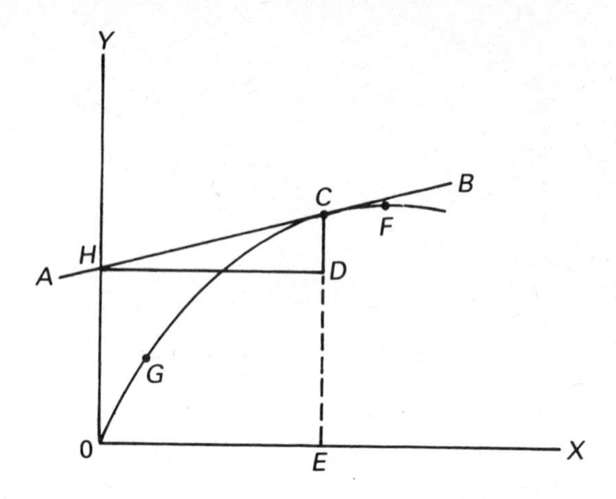

Figure I–7

when the curve represents a production function. Similarly, by drawing a straight line tangent to any other point on the curve in Figure I–7, such as *G* or *F*, the *MPP* at that point could be determined.

A Line passing Through the Origin. So far, the straight line has not necessarily gone through the origin (point where $X = 0$ and $Y = 0$). Consider now the special case of a straight line that passes through the origin. The slope of line *AB* in Figure I–8 is GF/OF. Also,

$$\frac{GF}{OF} = \frac{\Delta Y}{\Delta X} = MPP$$

for the line (or production function) *AB*. When the straight line passes through the origin,

$$GF = OH = Y = \text{an amount of } Y$$

and

$$OF = X = \text{an amount of } X$$

so

$$\frac{GF}{OF} = \frac{OH}{OF} = \frac{Y}{X} = \frac{\text{yield}}{X \text{ used to produce yield}} = APP$$

Therefore, the slope of a straight line (production function) passing through the origin is equal to *APP* at any point on the line. And, if the straight line is a production function, $APP = MPP$ for all amounts of *X*.

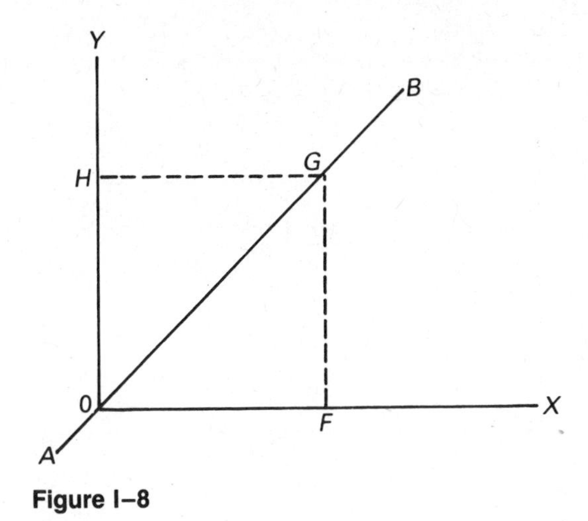

Figure I–8

In summary, the following points have been discussed:

1. Determining the slope of any straight line.
2. Use of a straight line (secant) to determine the average slope between two points on a curve.
3. Use of a straight line (tangent) to determine the exact slope at a point on a curve.
4. The special case of a straight line passing through the origin.

Application to the Production Function

Given a graph of the production function, straight lines can be drawn tangent to the production function at any point to measure the exact *MPP* at that point. Also, straight lines can be drawn through the origin to determine the *APP* at the point where the line intersects the production function; the slope of the straight line measures *APP* at the point(s) where the line intersects the production function. However, two uses of straight lines that are of particular interest are shown in Figure I–9.

In Figure I–9, the line *OA* passes through the origin and is tangent to the production function at *E*. At *E*

$$\frac{ED}{OD} = \frac{Y}{X} = APP = MPP$$

Therefore, the line *OA* is tangent to the production function at *E* where marginal and average physical products are equal. *OA* is used to determine the least amount of *X* that would be used, the boundary between Stages I and II. If any *X* were to be used, *OD* amount of *X* would be used. *APP* for this production function must be a maximum at *E*

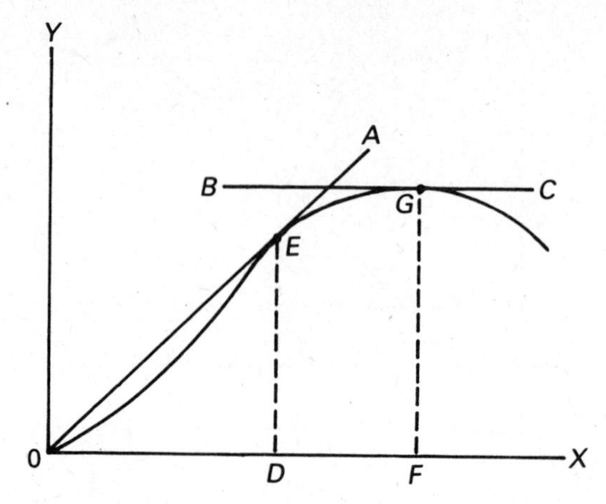

Figure I–9

(for *OD* amount of *X*) because any straight line passing through the origin and steeper than *OA* would not touch the production function.

Line *BC* in Figure I–9 is parallel to the *X* axis and tangent to the production function at *G*. Because the slope of any line tangent to the production function at a point determines the *MPP* at that point, the *MPP* at *G* (resulting from *OF* amount of *X*) is zero. *OF* is the most *X* that would be used even if *X* were a free good. For quantities of *X* exceeding *OF*, total product decreases. Thus, the line *BC* is used to determine the maximum amount of *X* that would be used, the boundary between Stages II and III.

The values of *X* between and including *D* and *F* make up Stage II of the production function, sometimes called the region of *economic relevance*. This latter term arises because input and output prices are needed to determine the most profitable amount of *X* to use *within* Stage II.

Application to Elasticity

When average measures of *MPP* are used (such as demonstrated in Figure I–6), an average measure of elasticity is used. This measure, called the arc elasticity formula, is

$$\epsilon_P = \frac{X_1 + X_2}{Y_1 + Y_2} \cdot \frac{Y_2 - Y_1}{X_2 - X_1}$$

or

$$\epsilon_P = \frac{\dfrac{X_1 + X_2}{2}}{\dfrac{Y_1 + Y_2}{2}} \cdot \frac{\Delta Y}{\Delta X}$$

When $\Delta Y/\Delta X$ is a measure of the average slope of the arc (Figure I–6), then $X_1 + X_2/Y_1 + Y_2$, the average value of X and Y between points B and A, is also used.

When the exact *MPP* is determined by the use of a tangent (Figure I–7), then an exact measure of elasticity is used. This measure, called the point elasticity formula, is

$$\epsilon_P = \frac{X}{Y} \cdot \frac{\Delta Y}{\Delta X} = \frac{MPP}{APP}$$

Here, $\Delta Y/\Delta X$ must be defined to be the exact *MPP* at C (Figure I–7), and X and Y are the exact amounts of input (OE) and output (EC) at point C.

The names of the two formulas suggest their use. The arc formula is used to measure the elasticity over an arc. The point formula is used to measure the elasticity at a point.

DIFFERENTIAL CALCULUS

We demonstrated how the slope of a tangent can be used to determine the slope of a curve at a particular point (Figure I–7). By measuring the slope of the tangent, the slope of the curve at the point of tangency was determined. This technique will work on graphs if the function is plotted with sufficient accuracy, but is of little value when the function is expressed as an equation. Of course, the equation could be graphed and the slope determined on the graph by drawing tangents, but that would be awkward at best. The question is: Is there a method to determine the slope of a function directly from the equation of the function? The answer is yes, under most conditions, and the equation that tells the slope of the function is called the derivative of the function. To restate, the derivative of a function is another function whose value at a point is the slope of the original function at the same point.

Conditions for the Existence of a Derivative

To have a derivative, a function must be a smooth curve that has a slope at all points of interest to the particular analysis; for example, for production functions or cost functions, Stages I and II define the area of relevance. Intuitively, a curve is continuous if you can draw it without lifting your pencil from the paper. But to have a derivative, a function must be more than continuous; it must be smooth.

Three examples are given in Figure I–10. A function is being considered over an interval from $X = a$ to $X = b$. In the left-hand diagram, the function obviously has no slope at X_0. Indeed, it is not even defined when X takes the value X_0 (no Y exists for the value X_0; see the definition of a function). Because it is not defined and has no slope at X_0, it has no derivative at X_0. It is also not continuous at X_0.

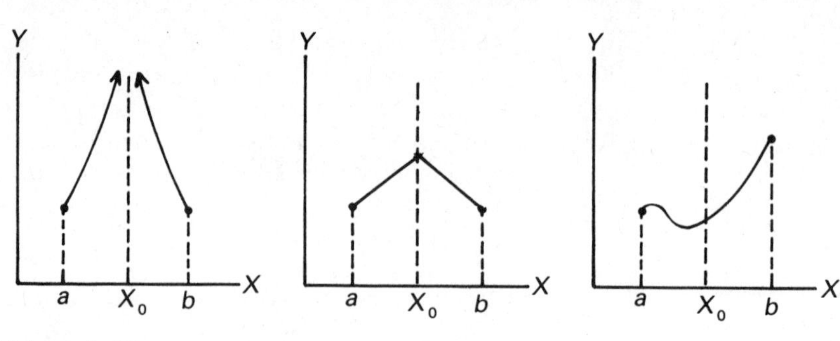

Figure I–10

The center diagram of Figure I–10 depicts a continuous function. This function is defined at X_0 (has a value of Y corresponding to X_0) but does not have a slope at X_0. The function comes to a point at X_0, much like the ridge of a house roof. If a tangent were to be used on this "razor's edge" to measure the slope at X_0, it would either take the slope of the straight line between a and X_0 or the straight line between X_0 and b. Thus, it cannot be said that there is one slope at X_0, that is, one tangent that could be used to determine the slope at X_0. The diagram on the right (Figure I–10) does depict a curve that is smooth at every point between a and b, including X_0, and thus has well-defined slopes and derivatives at all the points within the interval. This curve would have a tangent at all points in the interval and is typical of functions that have derivatives everywhere.

In summary, we want to know if a particular function has a derivative at a point (X_0, Y_0), where $Y_0 = f(X_0)$. If there is a straight line tangent to the graph at (X_0, Y_0) and if this line is not parallel to the Y axis, then the function has a derivative at X_0. The value of the derivative at X_0 will be the slope of the line tangent to the function at X_0.

Construction of a Derivative

Now that we know what a derivative is and the nature of functions that have derivatives, how can the derivative be found? Refer back to Figure I–6. Suppose we want to measure the slope of the curve at point B $(X = X_1)$. Then it would appear reasonable to reduce the size of ΔX, that is, move X_2 back toward X_1. Because the curve is smooth, X_2 can be moved as close to X_1 as we choose. For example, suppose $X_1 = 2$, then $X_2 = 3$ would be a good first estimate—but why not think small? X_2 could be 2.5 or 2.1 or 2.01 or 2.001 or even 2.0000001! In essence, then, we can get the most accurate estimate of the slope at B by letting the change in X become infinitesimally small. When we do, that should give us the slope of the tangent at B.

Consider a very simple function:

$$Y = X^2 \tag{1}$$

and let X be any value of the independent variable for which we want to evaluate the slope. For example, X could be X_1 in Figure I–6. Then we want to add ΔX to X; if we do, Y will be changed to a new value, $Y + \Delta Y$. Therefore, because $Y = X^2$, it must be true that

$$Y + \Delta Y = (X + \Delta X)^2 \tag{2}$$

But, as shown before, our measure of the slope is $\Delta Y/\Delta X$; therefore, we need to work toward an expression of that form. To do so, expand the right side of (2) to get

$$Y + \Delta Y = X^2 + 2X(\Delta X) + (\Delta X)^2 \tag{3}$$

where (ΔX) is manipulated just as any variable in algebra. But $Y = X^2$, so examination of (3) suggests that

$$\Delta Y = 2X(\Delta X) + (\Delta X)^2$$

so that

$$\frac{\Delta Y}{\Delta X} = \frac{2X(\Delta X) + (\Delta X)^2}{\Delta X} = 2X + \Delta X \tag{4}$$

Now we see that as ΔX becomes very small, the slope must approach $2X$. This can be written

$$\frac{dY}{dX} = \lim_{\Delta X \to 0} \frac{\Delta Y}{\Delta X} = 2X. \tag{5}$$

To get the limit value of $2X$ contained in (5), the value $\Delta X = 0$ is substituted directly into (4). This can be done because we are working with nicely behaved continuous functions. The notation $\Delta X \to 0$ is read "as ΔX approaches zero."

We now have another function, $2X$, which is called a derivative and which gives us the slope of our original function, $Y = X^2$, at any point X. For example, when $X = 3$, the slope of $Y = X^2$ will be $2 \cdot 3 = 6$.

From (4), the approximate slope of $Y = X^2$, $\Delta Y/\Delta X$, was shown to be $(2X + \Delta X)$. The term $2X$ we have shown to be the exact slope at a point; more specifically $2X_0$ is the slope of $Y = X^2$ at the point X_0, $2X_1$ is its slope at X_1, $2X_2$ its slope at X_2, etc. Thus, the term ΔX as found in (4) is the error resulting from measuring the slope using a secant (as CD in Figure I–6) rather than a tangent (as AB in Figure I–7). When $X_1 = 3$, for example, the exact slope of $Y = X^2$ was 6. If we were to estimate the slope using the formula $\Delta Y/\Delta X$ between $X_1 =$

3 and $X_2 = 4$, the error would be $X_2 - X_1 = \Delta X = 1$ and our estimate of slope would be $2X_1 + (X_2 - X_1) = 2 \cdot 3 + (4 - 3) = 7$. If we were to choose $X_2 = 3.0001$, then the slope estimate would be 6.0001 and the error becomes 0.0001. As X_2 approaches X_1, the error goes to zero. Note that the term (ΔX) represents the error for the equation $Y = X^2$; other equations will have a different "error term."

In general, the derivative of a function, $Y = f(X)$, is defined as

$$\frac{dY}{dX} = \lim_{\Delta X \to 0} \frac{f(X + \Delta X) - f(X)}{\Delta X}$$

This can be related directly back to Figure I–6, where $X = X_1$, $\Delta X = X_2 - X_1$ and $\Delta Y = Y_2 - Y_1$. Again, $\Delta X \to 0$ means that X_2 becomes infinitesimally close to X_1.

As an example of using this more general notation, consider the function

$$Y = \frac{1}{X^2}, \; X \neq 0$$

Then

$$f(X + \Delta X) = \frac{1}{(X + \Delta X)^2}$$

And

$$\frac{\Delta Y}{\Delta X} = \frac{-2X - \Delta X}{(X + \Delta X)^2 X^2} \tag{6}$$

so that the derivative is

$$\frac{dY}{dX} = \lim_{\Delta X \to 0} \frac{\Delta X}{\Delta Y} = -\frac{2}{X^3}$$

Note that in this case the error associated with $\Delta Y/\Delta X$ is not simply ΔX.

Derivatives are also often denoted as $df(X)/dX$ or $f'(X)$. When no particular value for X is to be implied, they are often written simply dY/dX, df/dX or f'. When a derivative is to be evaluated at a particular point, X_0, that intention is denoted by writing one of the following

$$\frac{df(X_0)}{dX} = f'(X_0) = \left.\frac{dY}{dX}\right|_{X=X_0}$$

Rules for Taking Derivatives

If you worked the last example, you may have decided that the algebra involved in finding derivatives is rather tedious. Moreover, the answers are always the same; the derivative of $Y = X^2$ is always $2X$ and the derivative of $Y = 1/X^2$ is always $-2/X^3$. As a result, you may anticipate the existence of simple rules for finding derivatives. If so, you are correct. We summarize the important rules without proof. They can be derived using the techniques described above. The functions $f(X)$ and $g(X)$ used below are any smooth, continuous functions of X:
Exponents of X:

$$\text{If } Y = X^n,$$
$$\text{then } \frac{dY}{dX} = nX^{n-1}.$$
$$\text{If } Y = X^6,$$
$$\text{then } \frac{dY}{dX} = 6X^5.$$

Constant times a function of X:

$$\text{If } Y = af(X) \text{ where } a \text{ is constant,}$$
$$\text{then } \frac{dY}{dX} = a\frac{df(X)}{dX}.$$
$$\text{If } Y = 3X^2,$$
$$\text{then } \frac{dY}{dX} = 3\frac{dX^2}{dX} = 6X.$$

Constant functions of X:

$$\text{If } Y = \text{C}$$
where C is constant,
then
$$\frac{dY}{dX} = 0.$$
$$\text{If } Y = 17{,}001,$$
$$\text{then } \frac{dY}{dX} = 0.$$

Sums or differences of functions of X:

If $Y = af(X) \pm bg(X)$
where a and b are constants,
then

$$\frac{dY}{dX} = a\frac{df(X)}{dx} \pm b\frac{g(X)}{dx}.$$

If $Y = 3X - 2X^2$,

then $\dfrac{dY}{dX} = 3 - 4X.$

Products of functions of X:

If $Y = f(X)\,g(X)$,

then $\dfrac{dY}{dX} = f(X)\dfrac{dg(X)}{dX} + g(X)\dfrac{df(X)}{dX}.$

If $Y = 3X^2 \cdot 4X$,

then $\dfrac{dY}{dX} = 3X^2 \cdot 4 + 4X \cdot 6X = 36X^2.$

Direct evaluation verifies this result.

If $Y = 3X^2 \cdot 4X = 12X^3$, then $\dfrac{dY}{dX} = 36X^2.$

Quotients of functions of X:

If $Y = \dfrac{f(X)}{g(X)}$,

$$\text{then } \frac{dY}{dX} = \frac{g(X)\dfrac{df(X)}{dX} - f(X)\dfrac{dg(X)}{dX}}{[g(X)]^2}.$$

If $Y = \dfrac{1}{X^2}$,

$$\text{then } \frac{dY}{dX} = \frac{X^2 \cdot 0 - 1 \cdot 2X}{X^4} = -\frac{2}{X^3}.$$

Composite functions of X:

Let $Z = F(Y)$ and $Y = f(X)$,
then

$$\frac{dZ}{dX} = \frac{dZ}{dY} \cdot \frac{dY}{dX}.$$

If $Z = 3Y^2$ and $Y = 2X$,

then $\dfrac{dZ}{dX} = 6Y \cdot 2 = 12Y$ or $24X$

because $Y = 2X$. Direct substitution gives

$Z = 3(2X)^2 = 12X^2$ or $\dfrac{dZ}{dX} = 24X$.

Partial Derivatives

In this section we shall state a simple rule, appropriate for a very large number of cases in economics. Assume that

$$Y = HX^K \tag{7}$$

where Y and X are free to vary, while H and K are not free to vary. We wish to take the derivative of Y with respect to X.

H and K may both be constants—numbers which obviously cannot be varied. In this case we get the usual direct derivative of Y with respect to X, dY/dX. But perhaps H and K are not strictly constants—are not numbers *incapable* of variation. H and K may be quantities that we are deliberately holding constant while we take our derivative. In a production function we may hold one physical variable constant while we take the derivative of output with respect to another physical variable. For instance, we may hold the quantity of land constant while we determine the derivative of output with respect to the quantity of labor. That is, we could vary land, but we do not want to; land is assumed to be the fixed input. In this case we get a *partial* derivative—partial because not all the quantities capable of variation are permitted to vary. The partial derivative is written $\partial Y/\partial X$.

Fortunately, the rule for taking the derivative is the same in both these cases. If the function is as shown in (7), then the derivative (simple or partial) of Y with respect to X is

$$\frac{dY}{dX} \quad \text{or} \quad \frac{\partial Y}{\partial X} = H \cdot K \cdot X^{K-1}$$

As long as H and K are not permitted to vary, there is no exception to this rule. For example, let

$$Y = 3X^2$$

Here $H = 3$ and $K = 2$; that is, 3 occupies the place that H occupies in (7), and 2 occupies the place that K occupies in (7). Since 3 and 2 are constants, the derivative of Y with respect to X will be a simple derivative. Applying the rule gives

$$\frac{dY}{dX} = 3 \cdot 2 \cdot X^1 = 6X$$

Next, suppose that

$$Y = X_1 X_2^3$$

where X_1 and X_2 are both quantities capable of being varied. We hold X_1 constant while we take the derivative of Y with respect to X_2. In terms of (7), $H = X_1$ and $K = 3$. Thus,

$$\frac{\partial Y}{\partial X_2} = 3X_1 X_2^2$$

This derivative is partial, since X_1 could be varied. If we hold X_2^3 constant, then $H = X_2^3$ and $K = 1$, so that we find

$$\frac{\partial Y}{\partial X_1} = X_2^3$$

The rule holds no matter how complex H and K may be. For instance, let

$$Y = (4X_1 + 3w + u)X_2^{(6X_1 + 4r)}$$

In terms of (7), $H = (4X_1 + 3w + u)$, and $K = (6X_1 + 4r)$. If X_1, w, u, and r are quantities capable of variation, then

$$\frac{\partial Y}{\partial X_2} = (4X_1 + 3w + u)\,(6X_1 + 4r)X_2^{(6X_1 + 4r - 1)}$$

You may find the function in the form

$$Y = X^K \cdot H \tag{8}$$

The order in multiplication does not matter, so you can rewrite (8) in the form of (7).

We end this section with a formal definition of partial derivatives.

When $Y = f(X_1, X_2)$ is a smooth, continuous surface (function), the partial derivatives can be defined as

$$\frac{\partial Y}{\partial X_1} = \lim_{\Delta X_1 \to 0} \frac{f(X_1 + \Delta X_1, X_2) - f(X_1, X_2)}{\Delta X_1}$$

$$\frac{\partial Y}{\partial X_2} = \lim_{\Delta X_2 \to 0} \frac{f(X_1, X_2 + \Delta X_2) - f(X_1, X_2)}{\Delta X_2}$$

where it must be remembered that X_2 is held constant when the derivative is taken with respect to X_1 and vice versa. Thus, these definitions are similar to the one given above for a function of one varible. Consider the graph in Figure I–11. The partial derivatives are to be evaluated at a point on the surface directly above OC of X_1 and OD of X_2 (the intersection denoted A). Then, the partial derivative of Y taken with respect to X_2 measures the slope of the surface above point A in the direction of the arrow CE. Likewise, the partial derivative of Y taken with respect to X_1 measures the slope of the surface above A in the direction of the arrow DF, holding X_2 constant at OD. Thus, partial derivatives are seen to be *marginal product equations* of *sub-production functions* that are parallel to the input axes. The existence of the partial derivatives implies there is a tangent plane that rests evenly on the surface at point A; that is, there are no cliffs, rocks, or rills on the surface. The values of the partial derivatives at the point of tangency determine the slope of the tangent plane—in the directions parallel to the axes.

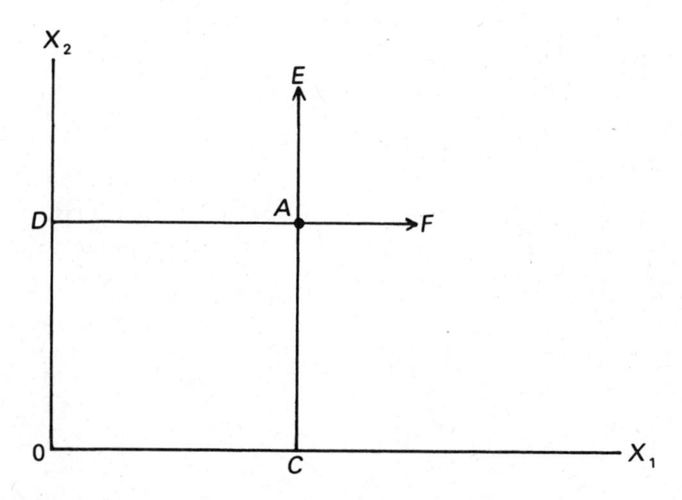

Figure I–11

EXPONENTS

Use of the Cobb-Douglas function requires some manipulation of exponents. In this section we apply the rules of differentiation to exponential functions and review some rules of exponents. If

$$Y = -3X^{1/3},$$

$$\frac{dY}{dX} = -3 \cdot \frac{1}{3} X^{1/3-1} = -1 \cdot X^{-2/3} = -X^{-2/3}$$

It may be useful now to recall the meanings of negative exponents and fractional exponents. (In the expression HK^K, K is the exponent of X—the power to which X is raised.)

$X^{1/a}$ means the ath root of X. Thus, $X^{1/2}$ is the square root of X, and $X^{1/3}$ the cube root. $X^{a/b}$ may be considered the bth root of the quantity X raised to the ath power, or the ath power of the bth root of X. Thus,

$$X^{a/b} = [X^a]^{1/b} = [X^{1/b}]^a \tag{9}$$

We can see that the three expressions in (9) are equivalent if we recall that

$$[X^c]^d = X^{cd}$$

If this sounds only vaguely familiar, test it. For instance,

$$[2^3]^2 = 8^2 = 64 = 2^{3 \cdot 2} = 2^6$$

Of course, it is also true that

$$[X^c]^d = [X^d]^c = X^{cd} \tag{10}$$

Notice that (9) is like (10) except that a fraction is involved. A negative exponent refers to the inverse operation, or moving a factor across the dividing line of a fraction, thus,

$$X^{-a} = \frac{1}{X^a}$$

and

$$\frac{1}{X^{-a}} = X^a$$

Seeing the nature of a negative exponent enables us to expand the range of problems that we can handle with the rule for differentiating exponents. For instance, let

$$Y = \frac{4}{X^3} \tag{11}$$

At first glance it may appear that the rule cannot be used. We must have the factor in X in the numerator. But (11) can be written as

$$Y = 4X^{-3}$$

so that

$$\frac{dY}{dX} = -12X^{-4}$$

FINDING EXTREME POINTS

Functions of One Variable

Derivatives are often used to find the points at which functions take maximum or minimum values, called extreme points. When a function of one independent variable takes a maximum or a minimum at a point, its tangent at that point will have a slope of zero. Thus, when searching for extremes, we can equate the derivative of a function to zero and solve, finding one or more particular values of X. The function will have tangents with zero slopes at these points; thus, they should lead to the extremes, if any exist. Points where the derivative is zero are called critical points. For example, consider the function $Y = X^2$, which has the derivative $2X$. But, $2X = 0$ will be true only for $X = 0$, and when $X = 0$ then $Y = 0$. The point (0, 0) is an extreme point. Now, $Y = X^2$ will increase without limit as X either increases or decreases; so the critical point is a minimum. Thus, in each case, the value of X for which a derivative is zero must be examined with care. While it is true that the derivative will always be zero at maximum (or minimum), it is not always true that a maximum (or minimum) will occur at an X value for which the derivative is zero. All extreme points are critical points, but all critical points are not extreme points.

The possible cases are shown in Figure I–12. The curve in the graph has three tangents with zero slope in the interval (a, b). At X_1 a minimum occurs: X_3 is a maximum. But at X_2 the function takes neither a maximum nor a minimum; X_2 is a critical point at which the tangent crosses the function and is called an inflection point.

Graphing the curve around a critical point will usually determine which of the three cases has occurred. We could also examine the slope

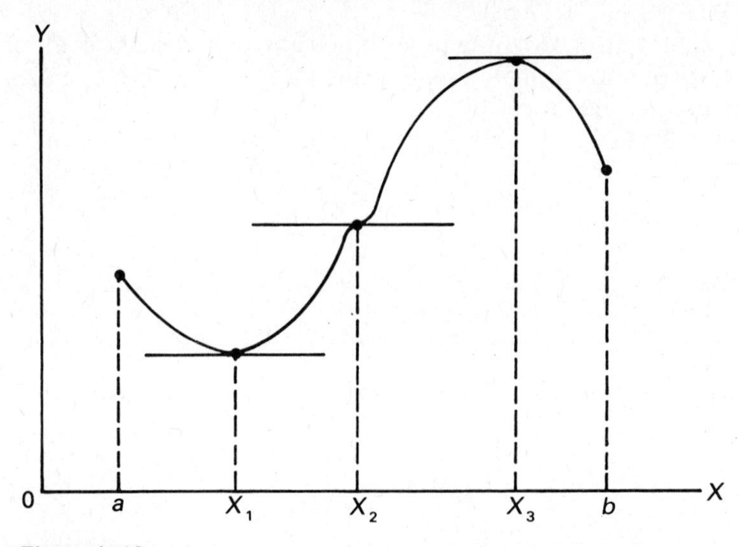

Figure I–12

of the curve on each side of the critical points using the derivative equation. Study of Figure I–12 would suggest the following table:

Point	*Extreme*	*Slope on Left*	*Slope on Right*
X_1	minimum	−	+
X_3	maximum	+	−
X_2	none	+	+

But the signs derived in the table suggest that if a minimum is to occur, the slope (derivative) of the function must be increasing through the critical point. The slope of a function is negative at the left of a minimum value, zero at the minimum value, and positive on the right. For a maximum, the reverse is true. If the derivative increases through a critical point, a minimum exists; if it decreases through a critical point, a maximum exists. Recall that the *MPP* is decreasing and equal to zero when *TPP* is a maximum.

All of this suggests that we examine the slope of the derivative at the critical points, not just the *value* of the derivative but also its *slope*. To do this, we take the derivative of the derivative; such a derivative is called the *second* derivative. What we have been calling *the* derivative we will now call the *first* derivative. The second derivative is obtained from the equation of the first derivative using all the same concepts described above.

To summarize our rules for finding and evaluating critical points: If the value of the first derivative is zero and the value of the second derivative is negative at a critical point, the critical point is a maximum. If the value of the first derivative is zero and the value of the second derivative is positive, the critical point is a minimum. If the value of the second derivative is zero (as at point X_2 in Figure I–12), further examination will be required.

For example, consider the function $Y = X^2$; then

$$\frac{dY}{dX} = 2X \quad \text{and} \quad \frac{d}{dX}\left(\frac{dY}{dX}\right) = \frac{d^2Y}{dX^2} = 2$$

where d^2Y/dX^2 is used to denote the second derivative of $Y = X^2$ with respect to X.

But

$$\frac{dY}{dX} = 2X = 0 \quad \text{for} \quad X = 0 \quad \text{and} \quad \frac{d^2Y}{dX^2} = 2 > 0 \text{ for all } X$$

implies that at the critical point, $X = 0$, the function has a minimum.

As a second example, consider the function

$$Y = 8X - X^2$$

Then

$$\frac{dY}{dX} = 8 - 2X \quad \text{and} \quad \frac{d^2Y}{dX^2} = -2$$

The first derivative is zero at $X = 4$ and the second derivative is always negative. The critical point that occurs at $X = 4$ is a maximum.

As a final example, consider the function

$$Y = 3X + 2X^2 - 0.1X^3$$

$$\frac{dY}{dX} = 3 + 4X - 0.3X^2$$

$$\frac{d^2Y}{dX^2} = 4 - 0.6X$$

The first derivative takes the value of zero at two points, $X = -0.7$ and $X = 14$. These points are found by using the quadratic equation as follows:

$$X = \frac{-4 \pm (16 + 3.6)^{1/2}}{-0.6} = 6.67 \pm 1.67(4.43)$$

Evaluating the second derivative at these points gives

$$X = -0.7: \quad 4 + 0.6(.7) = 4 + 0.4 = 4.4 > 0, \text{ minimum;}$$
$$X = 14: \quad 4 - 0.6(14) = 4 - 8.4 = -4.4 < 0, \text{ maximum.}$$

When the first and second derivatives are both zero, the tests must be carried further. Higher derivatives should be evaluated at the critical point. If the first nonzero derivative is even,

$$\frac{d^4Y}{dX^4}, \frac{d^6Y}{dX^6}, \frac{d^8Y}{dX^8}, \text{ etc.,}$$

then the sign of that derivative can be used to test for a maximum or minimum, exactly as above. If the first nonzero derivative is odd, the tangent crosses the function at that critical point. For practice, consider $Y = X^5$, $Y = X^6$, $Y = 3X^2 - X^6$, etc.

We should note that an inflection point need not be a critical point. Critical points occur at values of X for which the first derivative is zero. Inflection points occur at values of X for which the second derivative is zero; inflection points locate possible extreme point of the first derivative. It is not generally true that a function and its first derivative will have critical points at the same X values. For the function presented immediately above,

$$Y = 3X + 2X^2 - 0.1X^3$$

which has the shape of the classical production function, the second derivative

$$\frac{d^2Y}{dX^2} = 4 - 0.6X$$

will be zero at $X = 6.67$. At this point, which is an inflection point of the production function, the first derivative or marginal physical product

$$\frac{dY}{dX} = 3 + 4X - 0.3X^2$$

attains a maximum value of 17.34. But the inflection point $X = 6.67$ is not one of the critical points, $X = -0.7$ or $X = 14$. On the other hand, the function $Y = X^3$ has both a critical point and an inflection point at $X = 0$.

As an aid to understanding these concepts, the reader should study with care a cubic function such as

$$Y = 20 + 10X - X^2 + 0.03X^3$$

by plotting the function, its first and second derivatives, and locating all critical points and inflection points.

Functions of Two Variables

The critical points for functions of two variables may also be located using derivatives. When a surface takes a maximum or minimum value, then both first partial derivatives will have values of zero at that point. Regarding the derivatives as slopes along the sub-production functions that parallel the two axes, the maximum on the surface will occur only at a point for which both sub-production functions attain their maxima simultaneously. Rather than a tangent line, the surface will have a tangent plane that touches the surface at the critical point and has slopes in each direction determined by the values of the derivatives. At each critical point the tangent plane will be flat—horizontal in both directions.

Therefore, a surface such as $Y = f(X_1, X_2)$ will have critical points for values of X_1 and X_2 where

$$\frac{\partial Y}{\partial X_1} = 0 \quad \text{and} \quad \frac{\partial Y}{\partial X_2} = 0$$

Examples are presented in Appendix II under the heading "Unconstrained Maximization" and will not be repeated here.

As before, the critical points may be maxima, minima, or a third alternative, called now a saddlepoint. There is a second derivative test that is easily used but not readily justified intuitively. We will state the conditions without extensive discussion.

When the first partial derivatives are zero at a point and the second partial derivatives, when evaluated, take the following signs

$$\text{(a)}\ \frac{\partial^2 Y}{\partial X_1^2} < 0,\ \frac{\partial^2 Y}{\partial X_2^2} < 0 \quad \text{and} \quad \left(\frac{\partial^2 Y}{\partial X_1^2}\right)\left(\frac{\partial^2 Y}{\partial X_2^2}\right) - \left(\frac{\partial^2 Y}{\partial X_1 \partial X_2}\right)^2 > 0$$

then the critical point is a maximum. When

$$\text{(b)}\ \frac{\partial^2 Y}{\partial X_1^2} > 0,\ \frac{\partial^2 Y}{\partial X_2^2} > 0 \quad \text{and} \quad \left(\frac{\partial^2 Y}{\partial X_1^2}\right)\left(\frac{\partial^2 Y}{\partial X_2^2}\right) - \left(\frac{\partial^2 Y}{\partial X_1 \partial X_2}\right)^2 > 0$$

then the critical point is a minimum. The notation can best be explained with an example. Consider the equation:

$$Y = 6X_1 - X_1^2 + 8X_2 - 1/2X_2^2 + X_1X_2$$

Then

$$\frac{\partial Y}{\partial X_1} = 6 - 2X_1 + X_2 \quad \text{and} \quad \frac{\partial Y}{\partial X_2} = 8 - X_2 + X_1$$

The second partial derivatives are the derivatives of the first partials taken with respect to the same variables. Thus,

$$\frac{\partial^2 Y}{\partial X_1^2} = -2 \quad \text{and} \quad \frac{\partial^2 Y}{\partial X_2^2} = -1$$

The third type of second partial derivative is called a crosspartial and is

$$\frac{\partial}{\partial X_1}\left(\frac{\partial Y}{\partial X_2}\right) \quad \text{or} \quad \frac{\partial}{\partial X_2}\left(\frac{\partial Y}{\partial X_1}\right)$$

which is rewritten

$$\frac{\partial^2 Y}{\partial X_1 \partial X_2} \quad \text{or} \quad \frac{\partial^2 Y}{\partial X_2 \partial X_1}$$

Therefore,

$$\frac{\partial^2 Y}{\partial X_1 \partial X_2} = 1 \quad \text{and} \quad \frac{\partial^2 Y}{\partial X_2 \partial X_1} = 1$$

To find the crosspartial, the derivative is taken first with respect to one independent variable and then the second derivative is taken with respect to the other independent variable. For continuous, differentiable functions, the order of differentiation is not important because the result will be the same.

We now apply the tests to our example. The first partial derivatives are set equal to zero and solved simultaneously. The equations to be solved are

$$-2X_1 + X_2 = -6$$
$$X_1 - X_2 = -8$$

and only one solution, $X_1 = 14$ and $X_2 = 22$, exists. A critical point will exist at these input values. The second derivative test gives

$$\frac{\partial^2 Y}{\partial X_1^2} = -2 < 0 \quad \text{and} \quad \frac{\partial^2 Y}{\partial X_2^2} = -1 < 0$$

and

$$\left(\frac{\partial^2 Y}{\partial X_1^2}\right)\left(\frac{\partial^2 Y}{\partial X_2^2}\right) - \left(\frac{\partial^2 Y}{\partial X_1 \partial X_2}\right)^2 = (-2)(-1) - (1)^2 = 1 > 0$$

The critical point is a maximum.

When a critical point of a function of two variables is neither a maximum nor a minimum, it is called a saddlepoint. If one second derivative is positive and the other negative, then the surface attains a minimum on a line parallel to one axis and a maximum on a line parallel to the other axis. The shape of this surface will appear much like a saddle, with the critical point located in the saddle seat. The minimum will occur on a line paralleling the length of the "horse"; the maximum will occur on a line perpendicular to the first line and extending to the right and left of the "horse." This type of saddlepoint is easily detected because the second partial derivatives are of opposite sign. A more subtle saddlepoint may also occur. Using a production function as an example, suppose both inputs display diminishing returns when one is increased while the other is held constant, but that there are increasing returns along an expansion path. Such a function will have a critical point that is a saddlepoint. Using the quadratic equation presented above, we can change the response by increasing the coefficient on the interaction term, X_1X_2, to (say) 2. Then we have

$$Y = 6X_1 - X_1^2 + 8X_2 - 1/2X_2^2 + 2X_1X_2$$

This function displays diminishing returns to each input and will have convex isoquants. But the test using the second partial derivatives is

$$\left(\frac{\partial^2 Y}{\partial X_1^2}\right)\left(\frac{\partial^2 Y}{\partial X_1^2}\right) - \left(\frac{\partial^2 Y}{\partial X_1 \partial X_2}\right)^2 = (-2)(-1) - (2)^2 = -2 < 0$$

As input use increases, the positive interaction between the variables counteracts the diminishing returns to each input. For example, on an expansion path where $X_1 = X_2$,

$$Y = 6X_1 - X_1^2 + 8X_1 - 1/2X_1^2 + 2X_1X_1 = 14X_1 + 1/2X_1^2$$

and, as might be expected, the function has no critical points in the positive orthant. The coefficient on the interaction term can be adjusted to yield constant returns along an expansion path. Let the coefficient on X_1X_2 be equal to $\sqrt{2}$. Then the partial derivative test presented in the preceding paragraph will be zero and constant returns will exist along the expansion path $X_2 = \sqrt{2X_1}$.

Suggested Readings

Allen, R. G. D. *Mathematical Analysis for Economists,* New York, St. Martin's Press, 1956. Chapters 5, 6, 7 and 8.

Chiang, A. C. *Fundamental Methods of Mathematical Economics,* Second Edition New York, McGraw-Hill Book Company, 1974, Chapters 6 and 7.

INTRODUCTION TO NONLINEAR PROGRAMMING

APPENDIX II

Linear programming is a special case of a more general set of programming techniques. The unique aspect of linear programming is, of course, obvious. Both the objective and the constraint functions are linear. In general, this does not need to be true; the world is not necessarily linear.

This section presents an introduction to nonlinear programming. A very general introduction to nonlinear techniques, such as that contained in Hadley, requires advanced mathematics. However, an introduction can be presented using only the techniques of elementary calculus, provided the reader will accept intuitive explanations of some rather difficult mathematical concepts.

The outline for the remainder of this appendix will be as follows: First, the concepts of unconstrained maximization will be reviewed. Second, maximization subject to a constraint will be demonstrated again, but it will be used to develop the Lagrangean multiplier technique. The dual will be presented in a more general form and applied to illustrate the dual solution in linear programming. Third, inequality constraints will be introduced using a series of examples. These lead to what is known as the Kuhn-Tucker theorems, although those theorems will not be presented in their general form. Finally, an example of a nonlinear programming problem will be illustrated using calculus.

UNCONSTRAINED MAXIMIZATION

The production function presented in Chapter 4 was the quadratic (parabolic) function

$$Y = 18X_1 - X_1^2 + 14X_2 - X_2^2$$

And, although not discussed at the time, it is clear that the non-negativity restrictions $Y \geqq 0$, $X_1 \geqq 0$ and $X_2 \geqq 0$ must be applied to this equation if it is truly to represent a production function. Thus, even though we will call what follows *unconstrained* maximization, economic theory will always dictate some type of constraint or restriction to ensure that mathematical equations will conform to economic reality. With that caveat, the unconstrained maximum can be found by setting the first partial derivatives of the production function, taken with respect to the two inputs, equal to zero and solving the resulting two equations simultaneously. Thus the input amounts that result in the maximum yield are found by solving

$$\frac{\partial Y}{\partial X_1} = 18 - 2X_1 = 0, \quad \text{or} \quad X_1 = 9$$

$$\frac{\partial Y}{\partial X_2} = 14 - 2X_2 = 0, \quad \text{or} \quad X_2 = 7$$

and the resulting maximum yield, found by substituting these input values back into the production function, is 130. In this case, because X_1 and X_2 do not occur in both equations, simultaneous solution of the equations is not necessary. As can be seen from the graphs in Chapter 4, this solution is a maximum.[1] Inputs and outputs are positive.

Farmers may be interested in maximizing profit rather than output. The profit equation is expressed as Profit $= TR - TC$ or, for our example,

$$\text{Profit} = P_Y Y - (P_{X_1} X_1 + P_{X_2} X_2) - TFC$$

where $Y = H(X_1, X_2)$ and TFC, fixed cost, is a constant. Maximizing this with respect to the inputs results in

$$\frac{\partial \text{ Profit}}{\partial X_1} = P_Y \cdot \frac{\partial H}{\partial X_1} - P_{X_1} = 0 \quad \text{or} \quad VMP_{X_1} = P_{X_1}$$

$$\frac{\partial \text{ Profit}}{\partial X_2} = P_Y \cdot \frac{\partial H}{\partial X_2} - P_{X_2} = 0 \quad \text{or} \quad VMP_{X_2} = P_{X_2}$$

Referring back to the example in Chapter 5, the price of output was assumed to be \$0.65 per unit, the price of X_1 was \$9, and the price of X_2 was \$7. Inserting these prices and the derivatives above in the maximizing equations gives

$$\frac{\partial \text{ Profit}}{\partial X_1} = \$0.65(18 - 2X_1) = \$9, \quad \text{or} \quad X_1 = 2.1$$

$$\frac{\partial \text{ Profit}}{\partial X_2} = \$0.65(14 - 2X_2) = \$7, \quad \text{or} \quad X_2 = 1.6$$

The resulting yield would be 53.17, quite different from the unconstrained maximum output. The resulting maximum profit is

$$(\$0.65)(53.17) - \$9(2.1) - \$7(1.6) - FC = \$4.46 - TFC$$

[1] We will not consider the sufficient conditions for a maximum or a minimum in this section. The sufficient conditions require more mathematical background than we will utilize here. Examples chosen will always meet the sufficient conditions.

MAXIMIZING SUBJECT TO A CONSTRAINT

To begin this section, we will work with a familiar example. Consider the problem in Chapter 4, using the same production function as above. When $P_{X_1} = \$2$ and $P_{X_2} = \$3$, a study of Figure 4–10 shows that the least-cost combination of X_1 and X_2 needed to produce a yield of 105 is 6.2 and 2.8, respectively. We will work this as a maximizing problem, assuming that $\$20.80 = \$2 \cdot 6.2 + \$3 \cdot 2.8$ of capital is available.

The constrained maximization problem for this example is to maximize revenue subject to the budget constraint, given the production function, prices, and the total available capital. The total revenue function, $P_Y Y$, is the objective function, and the total outlay function, $\$2X_1 + \$3X_2 = \$20.80$, is the constraint. Geometrically, the problem is to find the point on the isocost line that produces the maximum revenue. Whereas the unconstrained maximum can be *any* input combination, the constrained maximum *must* be one of the points on the isocost line. Thus, the potential set of solutions is constrained at the outset of the problem.

The most direct solution to this problem is to solve the constraint equation for one variable as a function of the other. For example, the budget constraint can be expressed as

$$X_1 = 10.4 - 1.5X_2$$

which is the equation of the isocost line in Figure 4–10. This form of the constraint can then be substituted directly into the total revenue function. When the unconstrained total revenue, TR, is given by

$$TR = P_Y \cdot Y = \$0.65(18X_1 - X_1^2 + 14X_2 - X_2^2)$$

then constrained total revenue, TR_C, will be given by

$$TR_C = \$0.65[18(10.4 - 1.5X_2) - (10.4 - 1.5X_2)^2 + 14X_2 - X_2^2]$$
$$TR_C = \$0.65[79.04 + 18.2X_2 - 3.25X_2^2]$$

The equation in brackets represents the equation of the production function restricted to the set of points that lie on the isocost line for \$20.80.[2] Note that the price of output will not affect the quantities of X_1 and X_2 that will ultimately be the solution; it only affects the magnitude of the resulting total revenue. The portion of the production function represented parametrically within the brackets is the equation of the parabola stretching from points A and B in Figure 4–11.

[2]Although total revenue is a function of both inputs, this "reduced" function can be referred to as a parametric representation where X_2 is the parameter. The equation could also be written as a function of the parameter t where $X_1 = 10.4 - 1.5t$ and $X_2 = t$.

With this as background, the problem is to maximize revenue subject to the total outlay constraint. To do this, set the derivative of the constrained total revenue to zero as follows:

$$\frac{\partial TR_c}{\partial X_2} = \$0.65(18.20 - 6.50\,X_2) = 0$$

So

$$X_2 = 2.8$$

and

$$X_1 = 10.4 - 1.5 \cdot 2.8 = 6.2$$

The maximum revenue that can be attained when \$20.80 is available is \$0.65 · 105 = \$68.25, and profit, before fixed cost, is \$47.45.

The above technique can be represented symbolically. In general, we have the objective function $Y = f(X_1, X_2)$ and the constraint function $g(X_1, X_2) = b$. Using arrows to represent causality and the function name, g or f, to show the functional relationship, we would have in the unconstrained case

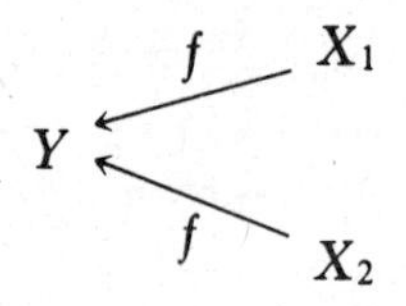

where X_1 and X_2 both influence Y but are independent of each other. X_1 and X_2 are free to take any values whatsoever. In the constrained case, we have

f X₁
Y ↑g
f X₂

where X_1 and X_2 are connected by the budget constraint. Increasing X_2 causes a reduction in X_1 because only \$20.80 is available. Thus, X_1 and X_2 are no longer independent variables in the production function. When the constraint function, g, can be solved for X_1 as a function of X_2, say, $X_1 = h(X_2)$, then the objective function can be expressed (in constrained form) as a function of X_2 only. Thus,

$$Y_C = f[h(X_2), X_2] = F(X_2)$$

and the procedure above was to maximize Y_C as a function of X_2. But when two so-called independent variables are related, the objective function can be differentiated with respect to the causal variable. Assuming the isocost function expresses X_1 as a function of X_2, then for $Y_C = f(X_1, X_2)$ we would have

$$\frac{dY_C}{dX_2} = \frac{\partial f}{\partial X_1} \cdot \frac{dX_1}{dX_2} + \frac{\partial f}{\partial X_2}$$

where, in diagrammatic form, the rates of change may be expressed as

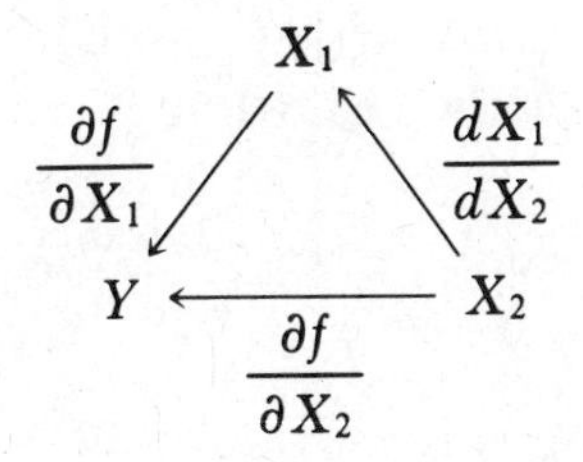

Thus, a change in X_2 stimulates a direct change in Y, expressed as

$$\frac{\partial f}{\partial X_2}$$

and an indirect change in Y, expressed as

$$\frac{\partial f}{\partial X_1} \cdot \frac{dX_1}{dX_2}$$

In the second case, a change in X_2 necessitates a change in X_1 which causes a change in Y. Returning to our example of this section, we substitute the derivatives directly into the above equation

$$\frac{dY_C}{dX_2} = \$0.65[(18 - 2X_1) \cdot (-1.5) + (14 - 2X_2)]$$

where

$$X_1 = 10.4 - 1.5X_2$$

so that

$$\frac{dY_C}{dX_2} = \$0.65(18.20 - 6.50X_2)$$

Setting this equation to zero and solving will again give the same answers.

Direct substitution of the constraint function into the objective function depends on being able to solve the constraint function explicitly for one input as a function of the other, that is, expressing $g(X_1, X_2) - b = 0$ as $X_1 = h(X_2)$. Sometimes this is not possible even though the function h theoretically exists. Sometimes it is possible, but the function h and its derivatives are complex, to say the least. In such cases a more general solution is available using implicit differentiation. Thus, we know that

$$g(X_1, X_2) = b$$

and we also know that dX_1/dX_2 exists, but we can't find it directly. But by differentiating g implicitly, regarding X_1 as a function of X_2, we get

$$\frac{\partial g}{\partial X_1} \cdot \frac{dX_1}{dX_2} + \frac{\partial g}{\partial X_2} = 0$$

since $\partial b / \partial X_2 = 0$. Solving this we find that

$$\frac{dX_1}{dX_2} = - \frac{\frac{\partial g}{\partial X_2}}{\frac{\partial g}{\partial X_1}}$$

For our revenue-maximizing example, recall that

$$g(X_1, X_2) = \$2X_1 + \$3X_2 = \$20.80$$

but

$$\frac{\partial g}{\partial X_1} = 2, \quad \frac{\partial g}{\partial X_2} = 3$$

so that

$$\frac{dX_1}{dX_2} = - \frac{3}{2} = -1.5$$

which agrees with the solution we obtained by solving g for X_1 and differentiating directly. Finally, we have the derivative of the constrained objective equation

$$\frac{dY_C}{dX_2} = \frac{\partial f}{\partial X_1} \cdot \frac{dX_1}{dX_2} + \frac{\partial f}{\partial X_2} = 0$$

which we set equal to zero because we want to maximize the constrained value. But we can now express dX_1/dX_2 as the ratio of the partial derivatives of g, so we have

$$\frac{dY_C}{\partial X_2} = \frac{\partial f}{\partial X_1}\left(-\frac{\dfrac{\partial g}{\partial X_2}}{\dfrac{\partial g}{\partial X_1}}\right) + \frac{\partial f}{\partial X_2} = 0$$

The beauty of this is that all derivatives are obtained by partial differentiation of functions we know. We do not have to explicitly solve the constraint for one variable as a function of the others or attempt to differentiate it directly.

In summary, our solution to the maximization problem subject to the constraint is given by

$$\frac{\partial f}{\partial X_2} - \frac{\partial f}{\partial X_1}\left(\frac{\dfrac{\partial g}{\partial X_2}}{\dfrac{\partial g}{\partial X_1}}\right) = 0$$

and

$$g(X_1, X_2) = b$$

where the first equation gives the marginal conditions and the second determines the value of X_1, given X_2; that is, it ensures that the constraint will be met. But remember that we have not substituted directly to obtain the marginal conditions (the first equation). Hence, X_1 may occur in that equation. Simultaneous solution of the two equations will ensure the correct result. For our maximization example, these equations give

$$0.65(14 - 2X_2) - 0.65(18 - 2X_1)\left(\frac{3}{2}\right) = 0$$

$$\$2X_1 + \$3X_2 = \$20.80$$

which has the solution $X_1 = 6.2$ and $X_2 = 2.8$.

The equations we just derived are often rewritten

$$\frac{\dfrac{\partial f}{\partial X_2}}{\dfrac{\partial g}{\partial X_2}} = \frac{\dfrac{\partial f}{\partial X_1}}{\dfrac{\partial g}{\partial X_1}}$$

$$g(X_1, X_2) = b$$

The resulting values must then be substituted into the objective function to determine the value of the constrained maximum.

MINIMIZING SUBJECT TO A CONSTRAINT

We have just solved a constrained maximization problem several different ways. Actually, all the methods presented were the same in concept—only the method differed. The same techniques apply to minimizing subject to a constraint. We will now investigate a constrained minimization problem.

Consider the problem depicted in Figure 4–10. The economic problem is to determine the least-cost combination of inputs for an output of 105. The production function is the same one used throughout this section, and the prices of the inputs are $P_{X_1} = \$2$ and $P_{X_2} = \$3$. Mathematically, the problem is to minimize

$$TVC = \$2X_1 + \$3X_2$$

subject to

$$105 = 18X_1 - X_1^2 + 14X_2 - X_2^2$$

The production function, with Y set equal to 105, is now the constraint. We cannot select *any* combination of inputs; we can only select from those combinations that will produce 105 units of output. From these, we will select the combination that costs the least.

The direct method of solution would require that the constraint be solved explicitly to express, say, X_2 as a function of X_1 and that this function be substituted directly into the variable cost function. But the explicit solution of the production function in this case would be the isoquant equation, which is a rather complex expression with an algebraically messy derivative.

However, the derivatives of the production function are easily found and, in this case, are simple algebraic forms. Thus, we can determine the solution by recalling that the conditions for a minimum are

$$\frac{\dfrac{\partial f}{\partial X_2}}{\dfrac{\partial g}{\partial X_2}} = \frac{\dfrac{\partial f}{\partial X_1}}{\dfrac{\partial g}{\partial X_1}}$$

$$g(X_1, X_2) = b$$

which in this example will be

$$\frac{3}{14 - 2X_2} = \frac{2}{18 - 2X_1}$$
$$18X_1 - X_1^2 + 14X_2 - X_2^2 = 105$$

because f is now the variable cost equation and g is the production function with b equal to 105. (The first equation represents the marginal conditions presented in Chapter 4. Thus, we have again derived the marginal concepts—but in a different manner.) The first equation can be solved for X_1 to give

$$X_1 = \frac{13}{3} + 2/3X_2$$

This is, in fact, the isocline equation for $P_{X_1} = \$2$ and $P_{X_2} = \$3$. This equation can then be substituted into the constraint (production function with Y equal to 105), and the resulting expression can be solved using the general solution for the quadratic formula. The answers are $X_1 = 6.2$ and $X_2 = 2.8$, as in Figure 4–10.

The question the reader might ask is how do we know that the extreme point is a minimum rather than a maximum? This comes from the nature of the problem. In this case, the production function chosen has isoquants convex to the origin so that the extremal must be a minimum. More generally, the sufficient conditions must be examined to ensure the minimum.

Notice that we have now found the solution to this problem two different ways. First, we maximized revenue subject to a budget constraint of \$20.80. This led to a solution that was equivalent to maximizing yield subject to the constraint because P_Y did not affect the solution. Second, we found the same answer by minimizing variable cost subject to an output constraint of 105.

THE LAGRANGEAN FUNCTION

The problems presented previously are often set up in a form known as a *Lagrangean function.* The Lagrange technique is more general in many respects and yields additional information of value. In particular, it enables us to derive the dual solution and the shadow prices of linear programming.

We have the objective function

$$Y = f(X_1, X_2)$$

subject to the constraint

$$g(X_1, X_2) = b$$

which has the maximizing (minimizing) solution(s) at values of X_1 and X_2 where

$$\frac{\partial f}{\partial X_2} - \frac{\partial f}{\partial X_1}\left(\frac{\dfrac{\partial g}{\partial X_2}}{\dfrac{\partial g}{\partial X_1}}\right) = 0$$

and

$$g(X_1, X_2) = b$$

But notice that the first condition can be written

$$\frac{\partial f}{\partial X_2} - \left(\frac{\dfrac{\partial f}{\partial X_1}}{\dfrac{\partial g}{\partial X_1}}\right)\frac{\partial g}{\partial X_2} = 0$$

and it is true by definition that

$$\frac{\partial f}{\partial X_1} - \left(\frac{\dfrac{\partial f}{\partial X_1}}{\dfrac{\partial g}{\partial X_1}}\right)\frac{\partial g}{\partial X_1} = 0$$

so we proceed to define a new variable λ as

$$\lambda = \frac{\dfrac{\partial f}{\partial X_1}}{\dfrac{\partial g}{\partial X_1}}$$

so that the necessary conditions for an extreme value can be written

$$\frac{\partial f}{\partial X_1} - \lambda\frac{\partial g}{\partial X_1} = 0$$
$$\frac{\partial f}{\partial X_2} - \lambda\frac{\partial g}{\partial X_2} = 0$$
$$g(X_1, X_2) - b = 0$$

We now have three derivatives expressed in three variables (X_1, X_2, λ). Can we find a function for which these equations are the first derivatives? The answer is yes. The function, known as the Lagrangean function, is

$$f(X_1, X_2, \lambda) = f(X_1, X_2) - \lambda[g(X_1, X_2) - b]$$

To maximize (or minimize) the function f subject to g, set up the Lagrangean function, equate its partial derivatives with respect to X_1, X_2, and λ to zero, and solve them simultaneously. The partial derivatives are, of course, the three equations presented above in this paragraph and do therefore determine the solution. The Lagrangean multiplier is known as λ.

The Interpretation Of λ

The Lagrangean multiplier, λ, that seemed to be conjured out of thin air, in fact has a legitimate and very useful interpretation in economics. λ measures the amount the objective function would increase when the constraint value, b, is increased. In terms of derivatives,

$$\frac{df(X_1, X_2)}{db} = \lambda$$

This interpretation is arrived at intuitively as follows: The values of X_1, X_2, and Y at the constrained optimum are determined by the amount of the constraint. When maximizing revenue subject to the variable cost constraint, b was \$20.80. When minimizing variable cost subject to the yield constraint, b was 105. In the first instance, if we increased b to \$21.80, how much would total revenue increase? In the second, how much would variable cost increase if output were increased to 106? λ provides a numerical answer to these questions.

Conceptually, the solution is obtained by regarding b as a variable rather than a constant. Increasing b will permit increases in X_1 and X_2, which will in turn cause increases in total revenue or variable cost. Thus, diagramatically

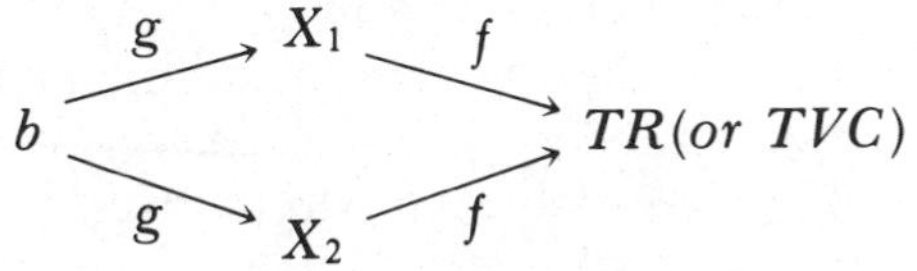

$$\begin{array}{ccccc} & \overset{g}{\nearrow} & X_1 & \overset{f}{\searrow} & \\ b & & & & TR(or\ TVC) \\ & \underset{g}{\searrow} & X_2 & \underset{f}{\nearrow} & \end{array}$$

Increasing b relaxes the constraint, g, causing X_1 and X_2, through f, to increase TR (or TVC). In the maximizing example, we would have

$$\lambda = \frac{dTR}{dTVC}$$

the addition to total revenue resulting from an addition to total variable cost; while in the minimizing example we would have

$$\lambda = \frac{dTVC}{dY}$$

the addition to total variable cost resulting from an addition to output. Therefore, on the expansion path, for the minimization example

$$\lambda = \frac{dTVC}{dY} = \text{marginal cost}$$

Finally, the fact that λ is the total derivative of f with respect to b can be derived. Consider the objective and constraint functions simultaneously:

$$Y = f(X_1, X_2)$$
$$b = g(X_1, X_2)$$

where X_1 and X_2 are regarded as functions of b. Then, by differentiating implicitly,

$$\frac{dY}{db} = \frac{\partial f}{\partial X_1}\frac{dX_1}{db} + \frac{\partial f}{\partial X_2}\cdot\frac{dX_2}{db}$$
$$1 = \frac{\partial g}{\partial X_1}\frac{dX_1}{db} + \frac{\partial g}{\partial X_2}\cdot\frac{dX_2}{db}$$

Multiplying the second equation by λ and subtracting it from the first equation will yield an equation in which most terms will vanish at the point of the constrained optimum, yielding the result.

The Dual Of The Lagrangean

We discussed the dual of the linear programming problem in detail. The dual solution exists for all problems that can be solved using the Lagrangean technique. We will state the concept of the Lagrangean dual without proof; the proof is complex and beyond the scope of this discussion.

Assume a particular point (X_1^0, X_2^0) yields the (constrained) maximum of the objective function $f(X_1, X_2)$ subject to the constraint $g(X_1, X_2) = b$. This is the primal. Then the dual problem can be stated as: The point $(X_1^0, X_2^0, \lambda^0)$ gives the solution that minimizes $F(X_1, X_2, \lambda)$

subject to

$$\frac{\partial f}{\partial X_1} - \lambda^0 \frac{\partial g}{\partial X_1} = 0$$

$$\frac{\partial f}{\partial X_2} - \lambda^0 \frac{\partial g}{\partial X_2} = 0$$

where all derivatives are evaluated at (X_1^0, X_2^0) and

$$\lambda^0 = \frac{\dfrac{\partial f(X_1^0, X_2^0)}{\partial X_1}}{\dfrac{\partial g(X_1^0, X_2^0)}{\partial X_2}}$$

Many problems in economics have duals. For example, variable cost can be minimized subject to a particular output (total revenue) level; or output (total revenue) can be maximized subject to a specific level of variable cost. The answer obtained will be identical if the constraint levels are consistent. In the examples above, it was shown that the answer would be the same if, when maximizing, variable cost was limited to $20.80 or, when minimizing, output was limited to 105. This example of the dual was worked above and will not be repeated here.

It is useful to illustrate the dual by using a linear programming problem from Chapter 9. One problem was to maximize $3Y_1 + 2Y_2$ subject to

$$Y_1 + 0.5\,Y_2 \leqq 4$$

$$Y_1 + Y_2 \leqq 5$$

When set up as a Lagrangean function, we will get

$$F(Y_1, Y_2, \lambda_1, \lambda_2) = 3Y_1 + 2Y_2 - \lambda_1(Y_1 + 0.5Y_2 - 4) - \lambda_2(Y_1 + Y_2 - 5)$$

which has the first order partial derivatives

$$\frac{\partial F}{\partial Y_1} = 3 - \lambda_1 - \lambda_2 = 0$$

$$\frac{\partial F}{\partial Y_2} = 2 - 0.5\lambda_1 - \lambda_2 = 0$$

$$\frac{\partial F}{\partial \lambda_1} = Y_1 + 0.5Y_2 - 4 = 0$$

$$\frac{\partial F}{\partial \lambda_2} = Y_1 + Y_2 - 5 = 0$$

Solving the third and fourth equations simultaneously gives $Y_1 = 3$ and $Y_2 = 2$, which gives a total profit of \$13 (see Figure 9–5A). Solution of the first two equations gives the shadow prices of $\lambda_1 = \$2$ and $\lambda_2 = \$1$.

The dual of this problem would be to minimize $F(Y_1, Y_2, \lambda_1, \lambda_2)$ subject to $3 - \lambda_1 - \lambda_2 = 0$ and $2 - 0.5\lambda_1 - \lambda_2 = 0$. However, the function $F(Y_1, Y_2, \lambda_1, \lambda_2)$ will simplify to $4\lambda_1 + 5\lambda_2$ permitting the problem to be restated as

$$\begin{aligned} \text{minimize} \quad & 4\lambda_1 + 5\lambda_2 \\ \text{subject to} \quad & \lambda_1 + \lambda_2 = 3 \\ & 0.5\lambda_1 + \lambda_2 = 2 \end{aligned}$$

but this is the dual problem stated in Chapter 9.

This example is very special. In this application, we knew that the solution would be positive and the equalities would hold at the point of solution. In general, linear programming problems cannot be cast in the Lagrangean mold because of the possible existence of corner solutions and even negative solutions. That is, the Lagrangean technique does not in general permit corner or boundary solutions and does permit negative solutions. Throughout this section the examples have been carefully tailored to avoid violating these assumptions. But many examples could be chosen that do in fact violate these conditions—conditions that are necessary in economics. In the next section we present a glimpse of the more general techniques.

Inequality Constraints

We will now incorporate the non-negativity requirements and inequality constraints into our analysis. First, we will consider maximizing a function of one variable, $Y = f(X)$, subject to $X \geqq 0$. If the function f takes a maximum at a point, say, X_0, where $X_0 \geqq 0$, then it will be true (necessary) that

$$\frac{df(X_0)}{dX} \leqq 0 \quad \text{and} \quad \frac{df(X_0)}{dX} \cdot X_0 = 0$$

That is, either the first derivative is zero at X_0, or X_0 is zero, or both happen. This can be demonstrated using geometry. Refer to Figure II–1; in each case, we seek the maximum of the function for the set of points $X \geqq 0$. In Figure II–1A, the function is linear and decreases throughout the feasible set—therefore, $df(X_1)/dX < 0$ and $X_0 = 0$. In B, the function does take a maximum but outside the feasible set, so again at X_0, $df(X_0)/dX < 0$ and $X_0 = 0$. In Figure II–1C, the function's

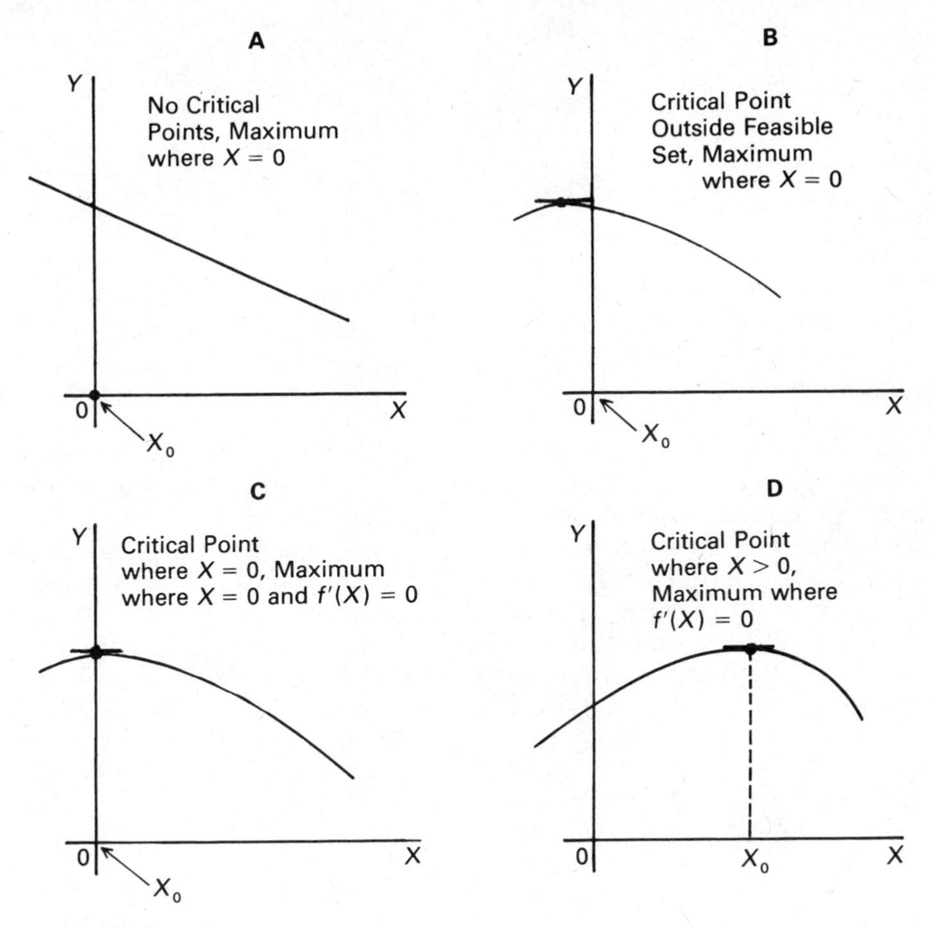

Figure II–1

critical point is at zero, so it is true that $df(X_0)/dX = 0$ and $X_0 = 0$. In case D, the critical point is within the feasible set, so that $df(X_0)/dX = 0$ even though $X_0 > 0$. In each of the four cases, the necessary conditions stated above will hold.

The one variable case can be immediately generalized to two or more variables. Assume now that the function $Y = f(X_1, X_2)$ has a maximum subject to $X_1 \geqq 0$ and $X_2 \geqq 0$. Then, again, it will be true that

$$\frac{\partial f(X_1^0, X_2^0)}{\partial X_1} \leqq 0$$

$$\frac{\partial f(X_1^0, X_2^0)}{\partial X_2} \leqq 0$$

and

$$\frac{\partial f(X_1^0, X_2^0)}{\partial X_1} \cdot X_1^0 = 0$$

$$\frac{\partial f(X_1^0, X_2^0)}{\partial X_2} \cdot X_2^0 = 0$$

Again, these conditions are based on the same reasoning presented for a function of one variable. If the function has a critical point (assumed here to be a maximum) where X_1 and X_2 are positive, the first derivatives will vanish. If the maximum is on a boundary (either axis) or in the second or fourth quadrants, the first derivative in question will be zero or negative, but in either case the value of the variable, X_1^0 or X_2^0, will be zero. If the maximum occurs at the origin (or in the third quadrant, which is outside the feasible set), then $X_1^0 = X_2^0 = 0$ and both first partial derivatives will be zero (or negative).

We now complete this example by adding the inequality constraint. The problem now is to maximize an objective function $Y = f(X_1, X_2)$ subject to $g(X_1, X_2) \leqq b$ and $X_1 \geqq 0$, $X_2 \geqq 0$. The functions f and g can be linear or nonlinear; there are no constraints on their form. They must have derivatives, however. This problem satisfies the needs of the linear programming formulation, but it is much more general. When the functions f and g are not linear, these problems are called, not surprisingly, nonlinear programming problems.

The theorem starts with the Lagrangean function

$$F(X_1, X_2, \lambda) = f(X_1, X_2) + \lambda(b - g(X_1, X_2))$$

and the function f is assumed to take a maximum subject to the constraint g at the point $X^0 = (X_1^0, X_2^0)$. The Lagrangean extremal would be at $P^0 = (X_1^0, X_2^0, \lambda^0)$. Then there will exist a λ^0 such that the following conditions will always hold

$$\text{Condition 1:} \quad \frac{\partial F(P^0)}{\partial X_1} \leqq 0 \quad \text{or} \quad \lambda_{X_1} = \frac{\dfrac{\partial f(X^0)}{\partial X_1}}{\dfrac{\partial g(X^0)}{\partial X_1}} \leqq \lambda^0$$

$$\frac{\partial F(P^0)}{\partial X_2} \leqq 0 \quad \text{or} \quad \lambda_{X_2} = \frac{\dfrac{\partial f(X^0)}{\partial X_2}}{\dfrac{\partial g(X^0)}{\partial X_2}} \leqq \lambda^0$$

$$\text{Condition 2:} \quad \frac{\partial F(P^0)}{\partial X_1} \cdot X_1^0 + \frac{\partial F(P^0)}{\partial X_2} \cdot X_2^0 = 0$$

$$\text{Condition 3:} \quad \frac{\partial F(P^0)}{\partial \lambda} = b - g(X^0) \geqq 0$$

$$\text{Condition 4:} \quad \frac{\partial F(P^0)}{\partial \lambda} \cdot \lambda^0 = [b - g(X^0)] \cdot \lambda^0 = 0$$

Conditions 1 and 2 are based on the non-negativity restrictions already developed. Only the addition of λ^0, the value of the shadow price at the optimum, is new. It is instructive to work out all possible cases to ensure that condition 2 is always zero. Condition 3 follows because the extreme value of f will either fall on the constraint ($g(X^0) = b$) or within it ($b > g(X^0)$). Condition 4 says that either the constraint is fulfilled ($g(X^0) = b$) or that λ^0 is zero. If the extreme occurs where the constraint is not fulfilled—at a point interior to the feasible set—the shadow price is zero. The objective function cannot be increased by increasing b because f is already a maximum. There is a surplus of whatever is represented by the constraint.

Finally, the optimal value of λ, called λ^0, must be chosen to fulfill conditions 3 and 4. λ^0 will always be chosen to be the maximum of three values:

$$\lambda^0 = \text{maximum} \left[\frac{\dfrac{\partial f(X^0)}{\partial X_1}}{\dfrac{\partial g(X^0)}{\partial X_1}}, \frac{\dfrac{\partial f(X^0)}{\partial X_2}}{\dfrac{\partial g(X^0)}{\partial X_2}}, 0 \right]$$

That is, when the solution is found, X_1 will have the largest shadow price, or X_2 will have the largest shadow price (not ruling out, of course, the case where they are equal), or the shadow price will be zero.

The best way to illustrate all this is to work an elementary example. Consider an objective function of the form

$$Y = 8X_1 - X_1^2 + 12X_2 - X_2^2 = f(X_1, X_2)$$

with the constraint

$$5X_1 + X_2 = 5 = g(X_1, X_2)$$

and X_1 and X_2 are non-negative. We can regard Y as total revenue and X_1 and X_2 as inputs. The objective function is quadratic and the constraint is linear; we have what is often called a quadratic programming problem. Nonlinear programming problems do not have simple but general algorithms for solution such as the simplex solution for linear programming. The solution is found by analysis of the functions involved, and each situation may be different. In this case, it makes sense to begin by analyzing the properties of the objective function, f.

The first partial derivatives of f are

$$\frac{\partial f}{\partial X_1} = 8 - 2X_1$$

$$\frac{\partial f}{\partial X_2} = 12 - 2X_2$$

Setting these equal to zero and solving gives $X_1 = 4$ and $X_2 = 6$. The function takes an unconstrained maximum of $Y = 52$ at this point. The derivative for X_1 is positive when $X_1 = 0$ so the function is increasing in the X_1 direction (that is, when the X_2 axis is crossed on any line parallel to the X_1 axis). The same is true for X_2. Ridge lines for this function occur when $X_1 = 4$ and $X_2 = 6$. Thus, the constrained

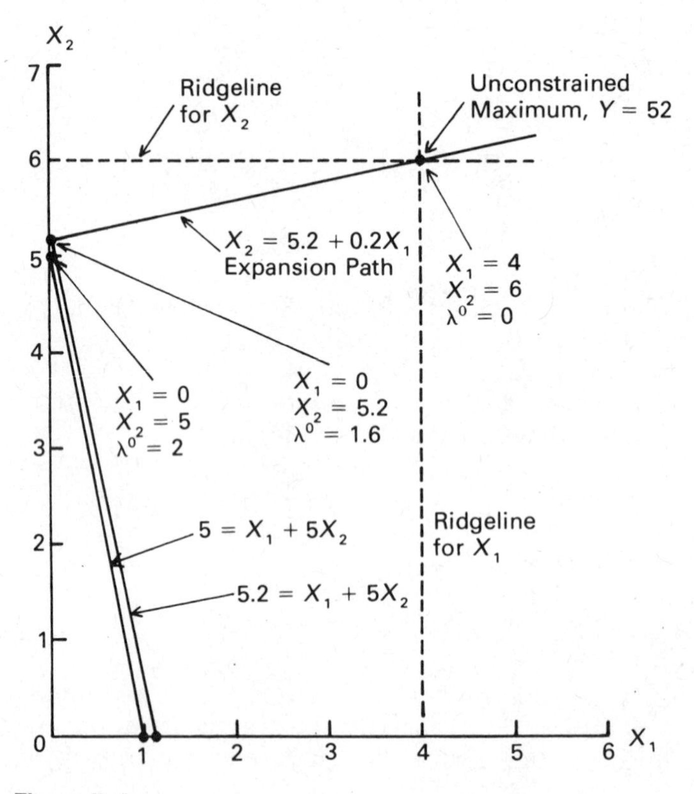

Figure II–2

maximum we seek will fall in the set of points enclosed by

$$0 \leqq X_1 \leqq 4$$
$$0 \leqq X_2 \leqq 6$$

This set of points is depicted in Figure II–2.

The constraint can be expressed as

$$X_2 = 5 - 5X_1$$

and when graphed appears as a straight line between the points $(X_1, X_2) = (0,5)$ and $(1,0)$. The unconstrained maximum of f falls outside the constraint. Therefore, the constraint will be satisfied at the constrained maximum and could be on the constraint line within the first quadrant or on the endpoints. To solve, we can substitute the constraint, g, directly into the function, f, as follows

$$Y_C = 8X_1 - X_1^2 + 12(5 - 5X_1) - (5 - 5X_1)^2$$
$$Y_C = 35 - 2X_1 - 26X_1^2$$

for $0 \leqq X_1 \leqq 1$. That is, this is the value of function above the constraint only. Visual examination of this function shows that it is maximized within the set when $X_1 = 0$. X_2 will be 5 and output will be 35. This is a corner solution. Examining the conditions which should hold when the isoquant for $Y = 35$ is tangent to the isocost line, we find

$$\frac{MPP_{X_1}}{MPP_{X_2}} = \frac{P_{X_1}}{P_{X_2}}$$
$$\frac{8 - 2X_1}{12 - 2X_2} = \frac{8}{2} \neq \frac{5}{1}$$

Thus, the tangency condition is not fulfilled. By examining the necessary conditions for a constrained maximum we can determine λ^0.

$$F(X_1, X_2, \lambda) = (8X_1 - X_1^2 + 12X_2 - X_2^2) + \lambda(5 - 5X_1 - X_2)$$

so that

$$\frac{\partial F}{\partial X_1} = 8 - 2X_1 - 5\lambda \leqq 0$$
$$\frac{\partial F}{\partial X_2} = 12 - 2X_2 - \lambda \leqq 0$$
$$\frac{\partial F}{\partial \lambda} = 5 - 5X_1 - X_2 \geqq 0$$

Substituting the optimal values of X_1 and X_2 in the first two equations yields (ignoring the inequalities)

$$\lambda_{X_1} = \frac{\dfrac{\partial f}{\partial X_1}}{\dfrac{\partial g}{\partial X_1}} = \frac{8 - 2X_1}{5} = \frac{8}{5}$$

or

$$\lambda_{X_2} = \frac{\dfrac{\partial f}{\partial X_2}}{\dfrac{\partial g}{\partial X_2}} = \frac{12 - 2X_2}{1} = \frac{2}{1} = 2$$

We chose $\lambda^0 = \max\{1.6, 2, 0\}$. But since the constraint is satisfied, λ^0 cannot be zero. If we chose λ^0 to be 1.6, the second partial derivative is not satisfied. Thus, $\lambda_0 = 2$. The largest shadow price is associated with X_2. Increasing X_2 will increase the objective function more than increasing X_1. We will not review them, but the student should check that $(X_1^0, X_2^0, \lambda^0) = (0, 5, 2)$ do satisfy all the necessary conditions for the constrained maximum.

From the standpoint of economic theory, the expansion path for this example starts at the origin and extends out the X_2 axis. After $X_2 = 5.2$, the expansion path becomes $X_2 = 5.2 + 0.2X_1$. Thus, 5.2 units of X_2 will be used before X_1 will be used.

Suppose the constraint is increased to 5.2 (b is increased to 5.2). The constrained maximum will be at the point $(X_1, X_2) = (0, 5.2)$. What will λ^0 be? Again, using the equations above,

$$\lambda_{X_1} = \frac{8 - 2X_1}{5} = \frac{8}{5} = 1.6$$

$$\lambda_{X_2} = \frac{12 - 2X_2}{1} = \frac{1.6}{1} = 1.6$$

and the constraint is again satisfied. λ^0 is chosen to be 1.6. This is again a corner solution, but we have moved out the expansion path to the point where X_1 now enters into the least-cost combination. Another way to explain it is that as the use of X_2 increased, its marginal productivity dropped until it became feasible to use the high-priced input, X_1. Substituting into the tangency conditions for a least-cost combination, we find

$$\frac{MPP_{X_1}}{MPP_{X_2}} = \frac{8 - 2X_1}{12 - 2X_2} = \frac{8}{1.6} = \frac{5}{1} = \frac{P_{X_1}}{P_{X_2}}$$

They are satisfied.

As the constraint is increased further, the constrained maximum will move along the expansion path, and it will always be true that $\lambda_{X_1} = \lambda_{X_2} = \lambda^0 > 0$. Suppose that b were increased to 26; then the constrained optimum would be $(X_1, X_2) = (4, 6)$ and would coincide with the unconstrained optimum. At that point

$$\lambda_{X_1} = \frac{8 - 2X_1}{5} = \frac{8 - 8}{5} = 0$$

$$\lambda_{X_2} = \frac{12 - 2X_2}{1} = \frac{0}{1} = 0$$

and the constraint is satisfied. λ^0 would be chosen to be zero.

Finally, if the constraint value is increased past 26, to, say, 30, the constrained optimum remains at $(X_1, X_2) = (4,6)$. The constraint is never satisfied, $b > g(X_1, X_2)$ so that $\lambda^0 = 0$. The shadow prices of the inputs are zero. However, all the necessary conditions for the constrained optimum are satisfied. For example,

Condition 1: $$\frac{\partial F}{\partial X_1} = 8 - 2X_1 - 5\lambda = 8 - 2 \cdot 4 - 5 \cdot 0 \leqq 0$$

$$\frac{\partial F}{\partial X_2} = 12 - 2X_2 - \lambda = 12 - 2 \cdot 6 - 0 \leqq 0$$

Condition 2: $$\frac{\partial F}{\partial X_1} \cdot X_1 + \frac{\partial F}{\partial X_2} \cdot X_2 = 0 \cdot 4 + 0 \cdot 6 = 0$$

Condition 3: $$\frac{\partial F}{\partial \lambda} = 30 - 5X_1 - X_2 = 30 - 5 \cdot 4 - 6 = 4 \geq 0$$

Condition 4: $$\frac{\partial F}{\partial \lambda} \cdot \lambda = [30 - 26] \cdot 0 = 0$$

We have just worked an example of a nonlinear programming problem in detail. Although it is a special case, it does serve to illustrate many of the general principles. In general, neither f or g have to be linear but can assume any form. The theory remains the same, but the search for the solution becomes much more difficult. Fortunately, computer software abounds with algorithms designed to solve special problems.

Suggested Readings

Chiang, A. C. *Fundamental Methods of Mathematical Economics,* Second Edition. New York: McGraw-Hill Book Company, 1974, Chapters 9, 11, 12, and 20.

Hadley, G. *Nonlinear and Dynamic Programming.* Reading, Mass.: Addison-Wesley Publishing Company, Inc., 1964, Chapters 3 and 6.

Takayama T. and Judge, G. G. *Spatial and Temporal Price and Allocation Models.* Amsterdam: North-Holland Publishing Company, 1971.

ANSWERS TO SELECTED PROBLEMS AND EXERCISES

APPENDIX III

CHAPTER 2

2–1 (a) $MPP = 1 + 8X - 0.6X^2$, $APP = 1 + 4X - 0.2X^2$
(b) $X = 6.67, 10, 13.4$
(c) $10 \leq X \leq 13.4$

2–2 (b) No
(c) $MPP = 0.125, 0.083, 0.063, 0.05$
$APP = 0.25, 0.167, 0.125, 0.10$

2–3 (a) 100
(b) 10
(c) 20
(d) 36
(e) 0.375

2–4 (a) $AVC = 2 - 2Y + Y^2$
(b) $Y = 1$
(c) $MC = 2 - 4Y + 3Y^2$
(d) $MC = AVC$ when $Y = 1$

2–6

X	*TFC*	*TPP*	*APP*	*MPP*	*TVC*	*AVC*	*MC*	*AFC*	*ATC*
0	$40.00	0	—	—	$ 0	$0	$0	$40.00	$40.00
1	40.00	4	4	4	5.00	1.25	1.25	10.00	11.25
2	40.00	10	5	6	10.00	1.00	0.83	4.00	5.00
3	40.00	15	5	5	15.00	1.00	1.00	2.67	3.67
4	40.00	18	4.5	3	20.00	1.11	1.67	2.22	3.33
5	40.00	20	4	2	25.00	1.25	2.50	2.00	3.25

2–8 $\epsilon_P = 0.75, 0.333, 0$

2–10 (a) $TVC = Y^3 - 2Y^2 + 2Y$, $MC = 3Y^2 - 4Y + 2$
(b) $Y = 1$
(c) $Y = 0.667$
(d) $Y \geq 1$

2–12 greater

2–13 (a) $\epsilon_P > 1$
(b) $\epsilon_P = 1$
(c) $\epsilon_P = 0$
(d) $\epsilon_P = 1$
(e) $\epsilon_P < 0$
(f) $\epsilon_P < 1$

CHAPTER 3

3–1 (a) $MPP = 0.5X^{-0.5}$ or $MPP = 0.5/\sqrt{X}$
(b) 0.25, 0.125, 0.10
(c) $X = 4$

3–3 4 units per acre to 10-acre field and 3 units per acre to 5-acre field

3–4 (a) $X = 50$
(b) $\epsilon_P = 0.89, 0.75, 0.57, 0.33, 0$
(c) $X = 25, 37.50, 43.75, 47.50$

3–5 $TC = (1/8)Y^3$, $MC = (3/8)Y^2$, $AC = (1/8)Y^2$

3–6 (a) $Y = 25$
(b) $MC = AVC$
(c) No, $107.5 < 108$

3–7 $X = 300, 200, 100, 0, 0$

3–8 5 units of X to each enterprise

3–9 (a)

Nitrogen Price per Pound of N ($)	Sorghum Price per Bushel ($)				
	1.50	*2.00*	*2.50*	*3.00*	*3.50*
0.00	200	200	200	200	200
0.20	167	175	180	183	185
0.40	133	150	160	167	171
0.60	100	125	140	150	157
0.80	67	100	120	133	143

(b) when P_X/P_Y ratio $\geq .8$

CHAPTER 4

4–1 (a) Combinations 6–7
(b) Combinations 3–4
(c) Combinations 2–3

4–2 (a) Increase the use of X_1
(b) Use more X_1
(c) Use more X_2 for technical unit Y_1

4–3 $X_1 = 16, X_2 = 16$

4–4 $X_1 = 12, X_2 = 12; X_1 = 6, X_2 = 96$

4–5 (a) $MPPX_1 = 0.2$
(b) $4X_1 = X_2$

4–6 $Y = 81$

4–7 (a) $X_1 = 1, X_2 = 1$
(b) $TVC = 9Y^{5/3}$, $AVC = 9Y^{2/3}$, $MC = 15Y^{2/3}$

4–8 Expansion effect

4–9 $X_1 = 2.4, X_2 = 5.3$

4–10 $X_1 = 250, X_2 = 500$

4–11 $X_1 = 1, X_2 = 1$

CHAPTER 5

5–2 Reduce X_2 and increase X_1.

5–3 (a) Any combination of wheat and milo
(b) All milo
(c) All milo

5–4 (a) 52 or 50 units of Y_1 and 17 or 23 units of Y_2
(b) 46 or 40 units of Y_1 and 28 or 32 units of Y_2
(c) 32 or 22 units of Y_1 and 35 or 37 units of Y_2

5–5 (a) Allocate 100 pounds of nitrogen to corn
(b) Allocate 133.6 pounds to corn and 66.4 to sorghum
(c) Allocate 158.6 pounds to corn and 141.4 to sorghum

CHAPTER 9

9–1 $Y_1 = 1.4, Y_2 = 1.5$

9–2 590 pounds of corn and 410 pounds of wheat

9–3 $Y_1 = 62, Y_2 = 47$

INDEX